The Genesis Strategy

Climate and Global Survival

"Some questions can be decided, even if not answered"

– Justice Louis D. Brandeis

The Genesis Strategy

Climate and Global Survival

Stephen H. Schneider

With Lynne E. Mesirow

PLENUM PRESS • NEW YORK AND LONDON

Library of Congress Cataloging in Publication Data

Schneider, Stephen Henry.
The Genesis strategy.

Includes index.
1. Climatic changes. 2. Human ecology. 3. Man–Influence on nature. I. Mesirow, Lynne E., joint author. II. Title.
QC981.8.C5S36 301.31 75-40321
ISBN 0-306-30904-1

We are happy to acknowledge permission to reprint some excerpts from: *The End of Affluence,* by Paul R. Ehrlich and Anne H. Ehrlich, © 1974 Paul R. Ehrlich and Anne H. Ehrlich, Ballantine Books (Div. of Random House), New York, 1974; *The Doomsday Syndrome,* John Maddox, © 1972 Maddox Editorial Ltd., McGraw-Hill, New York, 1973; *An Inquiry Into the Human Prospect,* by Robert L. Heilbroner, © 1974 W. W. Norton, New York, W. W. Norton, New York, 1974; *The New Industrial State,* by John Kenneth Galbraith, © 1967 John Kenneth Galbraith, Houghton Mifflin, New York, 1968; *Fragile Ecosystems,* E. G. Farnworth and F. B. Golley, eds., © 1974 Springer-Verlag, New York, Springer-Verlag, New York, 1974; *Famine 1975!,* by William and Paul Paddock, © 1967 William and Paul Paddock, Little, Brown, and Company, Boston, 1967; *We Don't Know How,* by William and Elizabeth Paddock, © 1973 William and Elizabeth Paddock, Iowa State University Press, Ames, 1973; *Mankind at the Turning Point,* by Mihajlo Mesarovic and Edward Pestel, © 1974 Mihajlo Mesarovic and Edward Pestel, Readers Digest Press/E.P. Dutton & Company, New York, 1974; *Times of Feast, Times of Famine,* by E. LeRoy Ladurie, © 1967 Flammarion, France, Doubleday, New York, 1971; *Snowpack, Cloud Seeding, and the Colorado River,* by Leo W. Weisbecker, © 1974 University of Oklahoma Press, Norman, University of Oklahoma Press, Norman, 1974; *Citadels of Mystery,* by L. S. and C. C. deCamp, © 1964 L. S. and C. C. deCamp, Ballantine Books, 1973, formerly published under the title *Ancient Ruins and Archaeology,* Doubleday, New York, 1964; *The Waning of the Middle Ages,* by J. Huizinga, © 1954 Edward Arnold, London, Doubleday, 1954; *By Bread Alone,* by Lester R. Brown with Erik P. Eckholm, © 1974 Overseas Development Corp., Prager, New York, 1974; *Weather and Climate Modification,* Wilmot N. Hess, ed., © 1974 Wiley, New York, Wiley, New York, 1974; *The Changing Climate,* by H. H. Lamb, © 1966 H. H. Lamb, Methuen, London, 1966.

A Division of Plenum Publishing Corporation
227 West 17th Street, New York, N.Y. 10011

United Kingdom edition published by Plenum Press, London
A Division of Plenum Publishing Company, Ltd.
Davis House (4th Floor), 8 Scrubs Lane, Harlesden, London, NW10 6SE, England

Printed in the United States of America

To

Doris and Samuel Schneider

and

Lilli and Raymond Mesirow

In the beginning—and ever since—they have been there

Acknowledgments

This book is a collaboration between a scientist and a writer, and the contributions of each reflect roughly the individual's specialties. Although Lynne Mesirow is at the root of a number of the ideas and opinions in this book, the bulk of the scientific content, personal experiences, and opinions relate to me; thus responsibility for these must necessarily fall on my shoulders. For that reason, the first person is often used, even though both of us may be talking on a number of occasions.

The world predicament—a problem of immense proportions—is a central topic of the book, and its multifaceted nature necessitated a wide circulation of the manuscript. Thus, we are deeply indebted to a number of experts from a spectrum of disciplines, whose suggestions and critical comments proved invaluable.

In particular we want to thank Paul Ehrlich, whose cryptic marginal comments on our original draft should (if not banned in Boston) someday be published as the "Annotated Genesis Strategy" in the *National Lampoon;* Lester Brown, who encouraged us to write this book; William Kellogg, who, more often than not, was the first person confronted with my ideas; and Harlan Cleveland, John Holdren, Halstead Harrison, Margaret Mead, Carl Sagan, and Louis M. Thompson, whose help has been absolutely vital. Walter Orr Roberts, in addition to his comments on two lengthy draft manuscripts, is largely responsible for this work in that he steered me in the direction of food–climate problems and the world predicament several years ago, and since then has provided intellectual and personal encouragement as well as important opportunities to interact with others of like interests.

My parents, Samuel and Doris Schneider, provided, in addition to their constant personal support and help with the manuscript, a continuous stream of *New York Times* clippings, many of which appear here; and Lynne Mesirow's parents, Raymond and Lilli Mesirow, have been comparably helpful, but their scissors were put to work mainly on the *Chicago Tribune*.

Other helpful suggestions have come from Reid Bryson, Robert Chervin, Ralph Cicerone, James Coakley, Stella Melugin Coakley, Paul Crutzen, Alex Dessler, Robert Dickinson, Anne Ehrlich, Michael Glantz, Judith Goode, Gilbert Hersh, Mike MacCracken, Clifford Mass, Jane Pisano, Lee Schipper, Roy Wessel, Gilbert White, Edwin Wolff, and Eileen Workman.

Publication of this book is, in large measure, due to the tireless efforts of our editor, Thomas Lanigan, who supported it steadfastly from its conception and through all intermediate stages.

Eileen Workman, my secretary, typed a large portion of the manuscript; her unfailing support, help, and cheerfulness under all circumstances have been greatly appreciated. Also, the very rapid and accurate typing help from Ruby Fulk, Sharon Carroll, and Nadine Perkey made meeting certain deadlines possible. And the help from the NCAR Photographics Group and from William Hemphill, Michael Shibao, and Justin Kitsutaka of the NCAR Graphics Group has been valuable. The suggestions of Susan Forman, and Peter Schneider of The Total Picture, New York, are appreciated.

National Center for Atmospheric Research (NCAR)
Boulder, Colorado
November 1975

STEPHEN H. SCHNEIDER

Preface

In a scientifically exacting world scientists must assume responsibility for the consequences of science and technology. . . . The individual member of the educational and scientific estate may wish to avoid responsibility; but he cannot justify it by the claim of higher commitment.

—John Kenneth Galbraith
The New Industrial State

Scientists can no longer afford to be naïve about the political effects of publicly stated scientific opinions. If the effect of their scientific views is politically potent, they have an obligation to declare their political or value assumptions, and to try to be honest with themselves, their colleagues and their audience about the degree to which these assumptions have affected their selection and interpretation of scientific evidence. Once scientific opinion enters into the public domain, the possibility of political neutrality disappears, but this does not mean that objectivity should be thrown to the winds.

Harvey Brooks, 1973
Dean, Engineering and Applied Physics
Harvard University

Imagine a huge ship—let's call it *Titantic II*—crossing the North Atlantic. While the passengers enjoy the comfort for which they are paying so dearly, the first mate at the helm believes he sees the dreaded form of an iceberg about three miles ahead through the thick fog. He hurries to tell the captain about it. As a veteran of countless uneventful voyages, the captain is skeptical of the danger. Nevertheless, he proceeds quickly to the helm, but fails to see any obstacle ahead in the fog. He chides the mate, "We've never encountered icebergs in these waters before, and after all, other mates have thought they were seeing icebergs that turned out to be little more than dense patches of fog."

The mate remains uneasy; he knows that the captain—anxious to collect his bonus for making good time on the voyage—has ordered the ship to proceed at full speed and has taken a short-cut through potentially icy northern waters. The mate knows, too, that it takes about two miles to stop the mighty ship at full speed and that the iceberg could be as little as two and a half miles away. He summons up his courage and asks the captain to throw the juggernaut into reverse—or at least slow down the ship. "But this is Saturday night," the captain snaps. "We would surely upset the tables, bowl over the dancers, likely injure some passengers, and certainly damage our reputation. Let's wait until we're sure there's really something out there." The captain and mate continue to argue, neither of them certain of what might be ahead. And, finally, when the distant form materializes unquestionably as an iceberg, they are only a mile from it, and probably too close to avoid it.

By early 1974 the media began to report the sighting of a new "iceberg" in the distance: climate change and related misfortunes. Could chemicals in seemingly harmless aerosol spray cans eventually destroy part of the earth's ozone shield and thereby threaten life on earth? Would a new cycle of droughts related to a global cooling trend portend chronic famine and world chaos? Was climate modification becoming a threat as a weapon of war? It was easy to dismiss such matters as science fiction, highly improbable events put in the spotlight by sensationalist journalists and mad scientists.

But for those who were willing to listen, there was a debate raging, worthy of notice. As in other modern debates about aspects of the "world predicament," such as population, environment, and resources, articulate "prophets of doom" were presenting strong evidence to support their theories of imminent catastrophe. Meanwhile, with equal passion, choruses of "Pollyannas" were countering that history is replete with dire predictions of doomsday, none of which has yet come to pass.

Inevitably, newspaper editors and other purveyors of public

information began to take sides. Some hailed the courage of the sentinels of doom, while others accused them of overemphasizing disaster, thereby sapping our spirit. Often the issues were oversimplified by arguments that we can continue to overcome our problems by keeping our spirits high, working hard, and using more technology.

Ironically, it was a question by a reporter that abruptly challenged my own oversimplification of this dilemma and forced me to reexamine the significance of the interrelatedness of the problems that threaten the world—a realm in which scientists too must live!

It was May, 1975. I had agreed to participate in an interview with a reporter for a Denver TV news program. Our topic was the potential threat to climate of aerosol spray cans.

The usual script should have read: How many cases of skin cancer will so much ozone destruction cause? Will an ice age begin if we continue to spray away? Should we ban the can?

But the reporter asked bluntly, "Do you think that too many scientists today, like yourself, are crying wolf when there may be no such animal threatening us?"

"Dirty pool," I thought to myself since I was unprepared for the question. "Should I counter with something like 'Nothing is certain but death and taxes'?" He was waiting for a definite answer, and all I could offer was a statement of probability—the odds for human catastrophes related to the world predicament.

My response, I knew, would satisfy neither of us. I could only warn him that the debate among seemingly irreconcilable "experts" about the seriousness, even the timing, of a host of prospective crises was a reflection of the uncertain state of scientific knowledge. Uncertainty, however, does not imply that there are no problems; nor does it deserve a "wait and see" attitude. Nor does political action require knowledge of the exact location of each tree behind which a wolf may be hiding. Rather, knowledge of the probability that wolves do lurk in the forest should be sufficient information for deciding whether to take preventive action.

Long after the reporter and technicians had gone, I continued my own inner dialogue. One thing was becoming clear: The reporter's question must be answered.

I was reminded of President Harry Truman's advice in 1951, when he was warned by a visitor to his office that the specter of hunger in faraway China, Formosa, and Korea was posing a serious threat to our way of life.

Typically, Truman cut through the verbiage.

"You are right," he replied. "The lesson is clear. We must apply it to the rest of the free world before it's too late. But the load is on the shoulders of the man in the street. Tell *him* what you've just told me." [1]

In the spirit of Truman's advice, I will discuss in these pages issues related to the world predicament because I believe that more people need to understand that there really are dangers ahead. Moreover, the degree of uncertainty that often exists in the scientific component of most survival issues should not delay consideration of actions to prevent, or at least forestall, plausible catastrophes.

Perhaps we can learn from a story in the Old Testament that seems especially applicable to such dilemmas. In the Book of Genesis, Joseph interpreted one of Pharaoh's dreams as a warning that seven years of famine were to follow the seven years of feast that Egypt would enjoy. Pharaoh, choosing not to take the chance, heeded Joseph's advice and had food stored up against the possibility of a food shortage. Events happened as predicted. In this story we see the vindication of Joseph's warning against the famine "wolf" (which, no doubt, is the best long-range weather forecast ever made) and of Pharaoh's action to protect his "flock" by storing food to hedge against the *possibility* of famine.

As the United States enters its third century as a nation, there are those who warn that the hungry wolf of biblical times is lurking in the woods again. They advise that we adopt the proven "Genesis Strategy" (as I dubbed Joseph's plan in an article in 1974 [2]) as the prudent response to the prospect of climate-induced famine.

However, the world has grown much more complex since Joseph's time. Agricultural technologies exist that were not available to the people of Genesis—tools such as modern irrigation systems, artificial fertilizers, genetic-hybrid seeds, insecticides, and more efficient and sophisticated means of transporting and storing food, to name a few. Today, building and maintaining food reserves are expensive procedures that dramatically stabilize market prices once the reserves have been established; accordingly, many producers and most food traders resist adoption of this strategy, calling it an obstacle to profit incentives or at least unnecessary in today's technological world.

The Genesis Strategy of maintaining large margins of safety to secure our means of survival can be applied to many of the urgent questions of human survival. One of these questions is climatic change, a phenomenon that deserves wide attention for several reasons: (1) It is often global in scope and concretely demonstrates potentially high risks to world stability; such risks can result from shortsighted political practices that stress short-term national solutions to problems that are often both long-term and worldwide. (2) Issues of climatic change have been underplayed or incompletely treated in most debates about the world's survival. (3) Nuclear technology was unleashed on the world before enough people had adequate appreciation of its dangers to demand strict control, but there is probably still time left to anticipate and even to prevent a number of climate-related crises, in addition to a climate-food disaster, that may lie ahead.

To be sure, there is plenty of controversy and conflicting evidence surrounding each potential crisis. It is important to present opposing views, even though this may necessitate some technical details. But I am convinced that ill-founded certainty is far worse than a realistic appreciation of the confusing issues of human survival. The public must be better informed if it is to be politically wise about these issues.

Any reduction in the number and degree of uncertainties surrounding the scientific testimonies of honest experts will go a long way toward helping modern technological societies make the wisest decisions. We must recognize all the while, however, that not every sighting of an iceberg will be followed by a collision. In addition, we must realize that the research needed to reduce the uncertainties we confront will be difficult, time-consuming, and expensive.

But my chief concern for the future is political rather than scientific; it is that some wolves will attack long before we are certain enough of their existence to feel compelled to effect difficult political actions. Such actions, however expensive or unpleasant, may be vital to hedge against dangerous plausibilities long before scientific certainty about their magnitude and timing is established. Since most predictions of threats to our survival are by no means guaranteed to come true, I feel compelled to go into considerable detail on the scientific bases of a number of dangers, in order to show that concern may nevertheless be justified. Although I will attempt to be as objective as possible in discussing scientific evidence, I have been unable to avoid injecting my own personal philosophies into some of the discussions, particularly those addressing the question of whether present scientific evidence justifies immediate action. Realizing that total objectivity is impossible, I have tried throughout this book to state my biases openly and to help the reader separate personal or political philosophy from scientific opinions.

I am deeply convinced of the real dangers that societies face in the years ahead and the need to clarify the importance of the scientific component of these dangers. To convey my sense of urgency, I have decided to eschew the traditional role of the scientist to advance knowledge quietly, and instead to write a book that mixes politics and science and often goes beyond the confines of my academic training. I agree with John Kenneth Galbraith's advice to scientists (and to those others who claim the title of "intellectual") to step past the traditional, and comfortable, boundaries of familiar scholarship and deal publicly in the real world of politics—especially when politics impinge on one's own scientific expertise.

The world predicament, crucial as it is to our future survivability, often appears to be a problem that we are powerless to influence as individuals. However, much of the dilemma can be traced to our rationalizations that we are not technically qualified or significant enough to have any impact on the solution to our growing problems. Thus, we abdicate our personal involvement and influence to leaders and specialists, who are supposed to understand these bewildering issues "better" than we do. But *there is a personal message here* directed to everyone, and I have tried to weave it into these pages: Most of the crucial issues of human survival that will confront humanity over the next few decades will call for ethical and political value judgments—decisions on *how to act in the face of uncertainties.* In few cases will these decisions be based on issues clear enough to be decided easily by an input of scientific truths comprehensible to only a handful of specialists. Human value judgments are too important to be left exclusively to the experts. Don't underestimate the worth of your contribution.

Finally, I return to the parable of *Titanic II* and today's world predicament. Although the comparison to the real world is inexact—for the world has no captain, but rather a vast array of quarreling mates and numerous throttles—the crucial question to my mind is not whether we can *prove conclusively* that disaster really lurks ahead, but rather whether we can afford to be unprepared for its not unlikely occurrence. The odds for a number of possible world disasters, though difficult to quantify, may often be greater than one in ten. Thus, those who expect that the future is necessarily secure even if present trends continue or prefer to "wait and see" before taking action are not so much high spirited optimists, as global-survival gamblers, willing to play Russian roulette with the future of the world.

Place your bets.

Contents

Part I

Climate, Technology, and Human Survival: An Overview

Part II

Climate-Related Crises

Part I

Climate, Technology, and Human Survival: An Overview

There is a question in the air, more sensed than seen, like the invisible approach of a distant storm, a question that I would hesitate to ask aloud did I not believe it existed unvoiced in the minds of many: "Is there hope for man?"

—ROBERT L. HEILBRONER
An Inquiry into the Human Prospect (1974)

The world is unhappy. It is unhappy because it does not know where it is going and because it senses that if it knew, it would discover that it was heading for disaster.

—VALERY GISCARD D'ESTAING
President of France (1974)

1

The Weather: No Longer a Subject for Small Talk

. . . the ways in which the atmosphere of the earth might be affected in the foreseeable future by human activities of various kinds are not at present likely sources of disaster. Indeed, the prospect is still remote.

—John Maddox
The Doomsday Syndrome (1972)

Climatic change poses a threat to the people of the world. The direction of the climatic change indicates major crop failures almost certainly within the decade. This, coinciding with a period of almost non-existent grain reserves, can be ignored only at the risk of great suffering and mass starvation.

—Consensus statement of the Workshop on "The Impact of Climatic Change on the Quality and Character of Human Life," International Federation of Institutes of Advanced Study (1974)

The Next Five Years: The Weather Reigns

Mark Twain is reputed to have said, "Everybody always talks about the weather, but nobody does anything about it." [1] When you think about it, the weather may be the most universal topic of conversation. No matter who you are or what you do, the weather is an uncontroversial, uncomplicated topic. Or is it?

In the next few decades, weather woes may well become one of the most controversial and complex of the world's political issues, perhaps replacing pollution as the most noted environmental problem. How could the universal topic of small talk suddenly lose its triviality and become such a serious concern?

Climate has not actually changed its ways overnight. Rather, we are just realizing that our vulnerability to climate changes is considerable. And we are only beginning to understand that recent climatic conditions have been very favorable for growing food, and that far worse conditions have been experienced in the past few hundred years. A steadily increasing and reliable supply of food is, of course, essential if we are to feed adequately a world that now uneasily supports more than 4 billion people, with some 80 million mouths to feed *each year* (an increase equivalent to almost 40 percent of the entire U.S. population). Despite spectacular improvements in food production resulting from modern agricultural technology, farmers have been hard pressed to keep abreast of population increases, most of which occur in the tropical and subtropical parts of the world where appropriate technologies are scarcely available and perhaps a billion people are already suffering from malnutrition.[2] Even more foreboding is the fact that world food stockpiles have decreased since the 1960s, despite the tremendous efforts of the proponents of the "Green Revolution," a matter I shall treat later. Thus, there is little food stored to cushion the shock of the kinds of weather problems that so suddenly and unexpectedly damaged crops in 1972, 1974, and 1975, and there is growing evidence that such damaging weather may occur more frequently in the next decade than in the last one.

Feeding large, growing populations has depended substantially on mechanized agriculture, which, in turn, demands chemical fertilizer for high productivity. Industrially produced fertilizer is a byproduct of petroleum, a resource whose price has been rising. Thus, some poorer countries that cannot afford enough fertilizer will have even more difficulty achieving self-sufficiency in future food production, even under optimum climatic conditions.

The possibility that favorable growing conditions will not persist—or even occur sufficiently often—in the 1970s and 1980s leads us to a major difficulty confronting the world in this decade and a principal theme of this book: The relative stability of the climate generally exhibited in the late 1950s and most of the 1960s cannot be expected to continue through the 1970s and 1980s.

Let us look briefly at the climatic trends since 1900 to illustrate why certain important granaries of the world experienced favorable climatic conditions during the middle of the twentieth century, a period that has coincided with significant strides in agricultural, medical, and other technologies and a dramatic increase in world population.

Since 1900, the average global temperature at the earth's surface has risen slightly more than 0.5° Celsius (about 1° Fahrenheit). Reid Bryson, director of the Institute of Environmental Studies at the University of Wisconsin, has argued that the period from 1930 to 1960, which (up to 1975) has been defined as the climatic "norm" or reference period by international agreement, was actually the most abnormally warm period in a thousand years.[3] Since 1940, and particularly during the 1960s, there has been a sharp reversal of this warming trend in the Northern Hemisphere. (There are relatively few long-term temperature records in the Southern Hemisphere, and thus we know little of its recent climatic history.) In fact, average temperatures in the Northern Hemisphere have recently appeared to be moving toward lower levels reminiscent of times immediately prior to the twentieth century. This temperature change has been documented mostly for the northern half of the Northern Hemisphere—that is, the colder portion nearer the polar icecap.[4] The consequences of this kind of regional temperature variation can, however, have global overtones, although a few tenths of a degree Celsius decrease in average hemispheric temperature may initially appear quite insignificant.

For example, Hubert Lamb, a noted British climatologist, has found evidence that the growing season for a number of recent years in some northern countries like Canada, Iceland, and even England has been shortened by a few weeks.[5] In addition, there is (admittedly meager) recorded evidence over the last few centuries tending to show that the variability of the weather from year to year has been higher in brief eras of cooler northern temperature than in warmer periods, such as the one that marked the middle decades of this century. In other words, now that we appear to have entered a cooler period in high latitudes, there have been suggestions that we will suffer more droughts, floods, temperature extremes, or other manifestations of climatic variability.[6] Unfortunately, that probability is still difficult to quantify, as we shall see later on, but the consequences for agriculture of an increase in variability are much more easily demonstrated.

With some exceptions, such as the recent half dozen years of drought in the Sahelian zone of Central Africa, most of the years between 1956 and 1972 were blessed with remarkably favorable climatic conditions for agricultural production, particularly in North America. For example, in the five midwestern wheat-belt states in the United States, data compiled by Donald Gilman of the U.S. National Weather Service, Extended Forecast Branch, show that the period from 1955 to 1973 was marked by summer rainfall generally greater than the long-term average and summer temperatures slightly lower than this average,[7] precisely the optimum conditions for high crop yields in this important agricultural region.

Several years following both 1910 and 1950 were marked by just the opposite conditions: high summer temperatures and low rainfall, which are the characteristics of crop-damaging drought. A longer period of nearly ten years in the 1930s was dominated by drought conditions so severe that the affected area, the southcentral United States, was labeled the "dust bowl." (History and literature are filled with accounts of the hardships and dislocations caused by the unfortunate coincidence of the dust bowl and the Great Depression.[8]) Similar climatic effects have been documented by agricultural climatologists for corn- and soybean-growing regions in the U.S.

The year 1972 was a bad one for crops in many places outside of North America, and some weather experts think it could well presage a return to times of higher climatic variability and the attendant likelihood of severe food disasters. In that year a coincidence of climatic disruptions occurred in many parts of the world, among them damaging floods in the American Midwest and elsewhere; the warming of the coastal waters of Peru, which resulted in a near collapse of the economically and nutritionally important anchovy catch; the continuation of drought in the Sahel, with the consequent migration and starvation of some of its population; a few-weeks delay in the onset of the life-giving monsoon rains in much of India; and a drought in the Soviet Union so serious that it led directly to the infamous Soviet wheat purchases from the United States and Canada.

In view of these adverse conditions, it is not surprising that in 1972 world food production was reduced by roughly 1 percent from the previous year, the first such reduction in more than a decade, a period during which total food production increased by nearly 3 percent per year. The variable weather of 1972 reduced grain reserves from enough to feed the world for sixty-nine days to about only a month's supply by mid-1974, according to food expert Lester Brown.[9] The price of wheat also rose dramatically because of its short supply, making additional purchases more difficult than ever in the poorer countries of the hunger belt, a term referring to the belt of highly populated relatively poor nations spanning the globe in most tropical and subtropical zones.[10]

Although 1973 set records for worldwide agricultural production, 1974, like 1972, was a year of productivity decline. The United States and Canada were hard hit by the weather for the first time in more than fifteen years with a combination of excessive spring rainfall, intense July drought, and early fall frosts. Corn yields per acre in the United States dropped by roughly 20 percent, despite continuing optimistic predictions of the U.S. Department of Agriculture for a record bumper crop right up to harvest time.[11] This setback in world food production occurred at a very critical time. Since fertilizer was relatively scarce and expensive as a consequence of petroleum price increases and the Arab oil embargo, food reserves were already dangerously low, and inflation was rampant in most parts of the world.

Regardless of any long-term steps taken to improve the food situation, we must face the uncomfortable reality that for the immediate future the likelihood of food shortages, food price rises, inflation, and famine will be governed by the vagaries of the weather at least as much as by the policies of agricultural bureaucracies. The issue can be simply stated: The margin between world food production and demand is so narrow, and the level of reserves so low, that even a 1 percent reduction of global food production can have serious ramifications. Higher prices would be inevitable, and shortages in well-fed countries could portend famines elsewhere. The seriousness of a small reduction in the world food supply is perhaps easiest to appreciate when one considers that the world population is increasing at the rate of about 2 percent annually. This means that a no-growth food-production rate from one year to the next is equivalent to a 2 percent per capita reduction in food consumption. For the malnourished billion, then, stable or decreasing production is disastrous.

It is not at all improbable that weather-induced fluctuations in food production could occur again, possibly with serious consequences. That is what happened in 1972 and 1974 when world food productivity dropped, reversing a decade-long upward trend. Food reserves dwindled, prices soared, and the world was left with little margin of safety to insure against another widespread crisis, a food crisis potentially far more serious in human terms than the energy "mini-crisis" [12] in 1974. To minimize the impact of short-term climate variability, an adequate food reserve could be achieved by 1980, but it will require a combination of favorable weather, international cooperation, and perhaps large-scale capital investments in agricultural production and land reforms to encourage production. However, such precautionary measures, *even if they are taken*, may not remove the weather from its place near the top of the list of world worries for the 1980s.

The following sections provide a brief explanation of why I make this assertion.

Nature Is Getting Competition: Inadvertent Climate Modification

Climate has always played an important role in human societies. In prehistoric times, humans tended to respond defensively to the harsher aspects of their environment by seeking shelter from the elements. And when populations grew or the immediate environment changed and could no longer support a particular group or tribe, they often migrated to a more suitable area, perhaps displacing the indigenous population, or paid the terrible price of starvation.

As their numbers and their technological prowess increased, people began to act increasingly *upon* the environment, rather than primarily reacting to it. They domesticated animals and cleared and cultivated the land. Settlements proliferated.

With the coming of the industrial age a sizable portion of the land in many parts of the world was cleared at an accelerated pace, thereby modifying the earth's natural capacity for storing and evaporating moisture and for absorbing solar energy. These human-induced modifications of the environment often slightly altered the local climate from what it had been when the land was in its natural state.

Industrialized peoples, with their higher standards of living, rapidly became dependent on the extensive use of energy to turn their wheels, heat their buildings, produce their electricity, and grow, store, and transport their food. Most of the energy liberated was (and still is) derived from fossil fuels—coal, gas, and oil—whose burning is accompanied by the disagreeable by-products that we now call pollution.

One form of such pollution that affects the entire atmosphere is the release of carbon dioxide (CO_2) gas. Even though it makes up a small fraction (less than one one-thousandth) of the gases that comprise the atmosphere, CO_2 is crucial in determining the earth's temperature because it traps some of the earth's heat (to produce the so-called greenhouse effect). Human activities have already raised the CO_2 content in the atmosphere by 10 percent and are estimated to raise it some 25 percent by the year 2000.[13] In later chapters, I will show how this increase could lead to a 1° Celsius (1.8° Fahrenheit) average warming of the earth's surface, which is roughly equivalent to twice the warming that occurred "naturally" in the first half of this century. This 1°C warming could reduce the masses of ice at the poles, and thereby raise the height of the world's oceans, or affect climate in the temperate zones or other latitudes.

Another form of atmospheric pollution results from the introduction of dust and smoke particles, which, when suspended in the air, are called atmospheric aerosols. The word "aerosol" is merely a term used to describe the suspension of any kind of particle in a gas. These particles can be solids like dust, sand, ice, and soot. Or, they can be droplets like the water particles in clouds and fogs or the liquid chemicals that are dispensed as droplets from aerosol spray cans. The air contains trillions upon trillions of aerosol particles, which, like CO_2, comprise only a minute fraction of the total atmospheric mass. Despite their relatively small volume, aerosols can affect our climate, primarily by absorbing and scattering back to space some of the sunlight that could have otherwise reached the surface.

Aerosols get into the air by many processes, some natural and some anthropogenic (i.e., human-induced). Even though naturally produced aerosols comprise the majority of all atmospheric aerosols, the anthropogenic ones may be critical since the human activities that create them are increasing each year. Aerosols can sometimes be recognized from the ground as an increase in the sky's haziness and a brownish hue, a condition well-known to those who live in or near cities or naturally dusty places. (It should be noted that most people think of air pollution as being primarily localized in urban regions and detrimental mainly to health and aesthetics. But, as we shall see, scientists are beginning to study the evidence that such pollution can spread extensively over the globe and perhaps affect the climate as well.)

Industry is not the only human activity that causes aerosols. They are also produced in great quantities by the widespread primitive agricultural practice of slash-and-burn, whereby large quantities of vegetation are purposely burned to clear land. I recently flew over Mexico when extensive burning was being conducted in several valleys and was amazed at how the smoke had spread as far to the horizon as I could see, obscuring everything between the mountain peaks below and thus altering the natural radiation heating patterns over a large area of land. Similar practices are common to many regions with generally unmechanized agriculture.

There is some evidence that human activities, such as slash-and-burn agriculture, that raise the concentration of atmospheric aerosols may have already affected the climate.

A consensus among scientists today would hold that a *global* increase in atmospheric aerosols would probably result in the cooling of the climate; however, a smaller but growing fraction of the current evidence suggests that it may have a warming effect. A few climatologists believe that the 0.5°C warming observed in the first half of this century may have resulted in part from the increase of CO_2 in the atmosphere and in part from the relative absence of aerosols due to little volcanic activity; they also maintain, in turn, that an increase in the quantity of aerosols from both human and natural (predominantly volcanic) sources has caused the subsequent cooling of the past few decades.

My own feeling is that "consensus" is an inadequate way to do science and that climatic theory is still too primitive to prove with much certainty whether the relatively small increases in CO_2 and aerosols up to 1975 were responsible for this climate change. I do believe, however, that if concentrations of CO_2, and perhaps of aerosols, continue to increase, demonstrable climatic changes could occur by the end of this century, if not sooner; recent calculations suggest that if present trends continue, a threshold may soon be reached after which the effects will be unambiguously detectable on a global basis. Problematically, by that point it may be too late to avoid the dangerous consequences of such an occurrence, for *certain proof* of present theories can come only *after* the atmosphere itself has "performed the experiment."

Ban the Can?: Should We Let the Atmosphere "Perform the Experiment"?

Seemingly innocuous compounds present in the atmosphere can often have dramatic, unforeseen side effects, particularly when they are present in large quantities. One bizarre example is the recently discovered effect that chlorofluorocarbons (CFCs) have on the atmosphere's ozone layer. Ozone is a gas that filters out much of the sun's ultraviolet radiation, which is harmful to life. CFCs have a variety of commercial names, such as Freon, which is used by Du Pont. These compounds are used as a propellant in some aerosol spray cans and as a refrigerant in cooling devices. CFCs migrate slowly upward through the troposphere (lower atmosphere) to the stratosphere, where they are destroyed by ultraviolet solar radiation. (The stratosphere is the lowest region of the upper atmosphere between about ten and fifty kilometers above sea level.) These CFCs were thought to be inert (i.e., will not react with other things) and thus an ideal choice for large-quantity commercial use. But it turns out that the ultraviolet radiation impinging on the stratosphere is sufficiently intense to break down these compounds. In the breakdown process, combinations of fluorine and chlorine atoms are released, and they, in turn, destroy ozone molecules.

Ozone is a gas continuously created by sunlight in the stratosphere. It not only absorbs most of the ultraviolet radiation of the sun, but also controls the temperature of the stratosphere—and thus indirectly affects the climate at the earth's surface. Removal or substantial reduction of the ozone layer would likely lead to enormous increases in the rate of skin cancer in humans and largely unknown but generally destructive effects in other animals and plants. A National Academy of Sciences publication [14] reported that studies of the effects of a simulated 50 percent decrease in ozone showed that for some species of plants growth decreased from 20 to 50 percent, chlorophyll content declined 10 to 30 percent, and degenerative changes in cell structures occurred. Harmful mutations increased seven- to twentyfold in some preliminary experiments. The studies also revealed that ultraviolet has a greater effect on young plants than on mature ones and is especially influential on single-celled algae. It should be noted, however, that these results were not completely conclusive and that a great deal of further research is badly needed. There is still much uncertainty and controversy over the actual seriousness and immediacy of the ozone/chlorofluorocarbon threat and whether regulation and control of the multibillion-dollar industry is warranted. Weighing the immediate and quantifiable economic hardship that harsh controls could bring the industry against the potential (but uncertain) long-range damage to people and the environment that supports them is a difficult political question. And it is an issue typical of the climate-related problems and choices we will examine throughout these pages.

Chlorine Water Purification and Nitrogen Fertilizer: Mixed Blessings?

But chlorofluorocarbons may not be the only culprits. Equally perplexing is the recent discovery that the atmosphere contains large amounts of ozone-destroying chlorine atoms in the form of carbon tetrachloride (CCl_4), a compound that has been used for years as a household cleaner. Thus, CCl_4 might turn out to be just as serious a threat to the ozone layer as the CFCs. The important question is, Where did so much CCl_4 come from and how will its future concentration change relative to that of the CFCs?

Another [15] speculation is that the chlorine compound chloroform may be associated with the purification of drinking water or sewage treatment; both processes often use chlorine. If chloroform could get to the stratosphere and destroy ozone, then the populations of all countries will be confronted with a serious long-term dilemma. If improvement of living and health standards for the growing billions, especially in less developed countries, is to be accomplished and maintained—and I have no doubt that this is an essential international goal—then a massive effort to upgrade water and sewage purification everywhere is imperative. However, since this could entail a huge increase in the use of chlorine (which in water may react chemically to produce chloroform), there arises the question of potential ozone depletion, which could endanger life on the entire planet.

Even more telling than the speculation on the possible water chlorination–chloroform case is the potential to destroy ozone by vastly expanded use of nitrogen fertilizer. Whereas the chlorination–chloroform example is, for the moment at least, merely wild speculation, there is some hard evidence connecting the increased usage of nitrogen fertilizer to an increase in the production of the gas nitrous oxide (N_2O), otherwise known as laughing gas, in the soil. It is known that N_2O can destroy ozone; therefore, when fertilizer-produced N_2O works its way up into the stratosphere, it, too, can reduce our protective shield against ultraviolet radiation. The most difficult question is, How great are these effects? Preliminary calculations suggest that they could be enormous, although there are still great uncertainties to be cleared up.[16] Moreover, industrially produced fertilizer is the key ingredient in modern scientific farming, and, indeed, its use is projected to increase by hundreds of percent in the next twenty-five years.[17] Thus, a discovery and warning that a great increase in fertilizer use could ultimately destroy part of the ozone shield would, if heeded, dash hopes that the developing countries could achieve self-sufficiency in food production soon. Such a discovery could be a staggering blow to those who are counting on expanded use of this particular technology to solve one important aspect of the world predicament.

These examples serve as useful illustrations of the drawbacks of a number of possible grandiose technological "solutions" to spur development and lessen poverty for the poorer billions—*approaches that may be highly effective on a limited scale, but must be carefully re-examined when they expand to the point that their environmental impact becomes global in scale.*

The medical field provides another example of how moderate

use of technological aid can be very valuable, whereas large-scale usage can be very harmful. For example, although powerful drugs in controlled dosage can sometimes cure or control disease, they can also cause serious side effects that require the patient to use still other forms of medicine. In addition, too much of a "helpful" drug can be lethal. That this wisdom applies to the earth was recognized a century ago by the great American Indian leader, Sitting Bull, who commented on the habits of the white settlers at the Powder River Council in 1877: "They claim this Mother Earth of ours for their own and fence their neighbors away from them. They degrade the landscape with their buildings and their waste. They compel the natural earth to produce excessively and when it fails, they force it to take medicine to produce more. This is evil."

Sitting Bull was a man of the land. Even without a working knowledge of the geochemistry of the nitrogen cycle, he anticipated that excessive forcing of the natural system could be an evil. The problem today is estimating the amount of forcing that is excessive since the well-being of present populations, whether we like it or not, already depends on forcing nature; and subsistence of projected numbers of people into the twenty-first century demands that even more medicine be force-fed to the earth; in the form of fertilizer or chlorine, for example.

The effects on ozone of the chloroform-water purification or N_2O-fertilizer connections remain to be determined, but a general principle emerges: The bigger the technological solution, the greater the chance of extensive, unforeseen side effects and, thus, the greater the number of lives ultimately at risk.[18] In a macabre sense the situation here has some similarities to the well-known remark of an American officer in Vietnam, who, pointing to a bombed-out, enemy-infiltrated village, explained to a war correspondent that the village had been destroyed in order to be saved.

Heat Pollution: Big Is Bad

Another form of industrial pollution is the heat released into the environment as a by-product of energy production and use. This form of pollution is expected to continue its rapid growth because an increase in per capita energy consumption generally accompanies improved *material* living standards. And such development is a goal of almost all countries. In countries such as the United States, however, where per capita energy consumption is already very high, it is not at all clear that increased energy use will cause a higher quality of life.[19]

Since solar energy heats the earth and is the driving force behind the weather, it is important to compare the magnitude of energy released into the environment by human activities to the solar energy absorbed by the earth. This is true because climatic changes would likely result from any widespread change in the surface energy budget (i.e., natural balance of energy flows at the earth's surface) by more than about 1 percent of the total sunlight energy absorbed at the surface. Even such a seemingly small alteration to the natural processes can be important because our climate system is in a very delicate balance among many opposing forces. Today the total production and use of energy by humans is only slightly more than one one-hundredth of 1 percent of the sun's heat absorbed at the earth's surface; thus, no significant *global* climatic warming has occurred yet as a direct result of our release of heat.

However, in regions of high population density and widespread affluence, such as most large metropolitan areas, the energy consumed is primarily produced locally (and certainly consumed locally). This concentration of energy production and use, which contributes to heat and air pollution, combined with the alteration of the natural environment with buildings, pavements, and treeless expanses of land, has already produced significant alterations in urban climatic environments. As a result of such local heating, which is called the urban heat island effect, temperatures in the centers of large cities are considerably warmer than the surrounding areas, especially at night or in winter, when the buildings and pavement retain the heat they absorbed during the day. As an urban megalopolis grows, its heat island becomes regional in scale, and the total effect on the climate magnifies accordingly. Already the region of twenty-one metropolitan areas stretching from Boston to New York to Washington consumes enough energy and thus releases enough heat to be seen at night as almost one continuous lighted region from a satellite in earth orbit (see Figure 26). Heat released to the environment in this megalopolis has grown to equal an average of some 5 percent of the net energy input by natural radiation processes [20] in that region, which is already enough to create substantial pockets of warming.

Looking ahead fifty years or more, some technological optimists [21] envision a world in which 20 billion people will live comfortably, consuming twenty kilowatts of energy per person (nearly twice the current U.S. consumption figure). Such a scenario involves more than a 400 percent increase over present world population and, at the same time, would represent a vastly increased level of energy consumption (about ten times greater than today) on the global level. A deeper look reveals that the heat released into the environment by such an energy-abundant civilization would be nearly one-half of 1 percent of the solar energy that is absorbed over the entire surface of the earth. This is a threshold at which detectable inadvertent alterations to natural global climatic regimes might be expected; and severe local and regional effects can be anticipated *well before* global impacts are felt. Furthermore, this heating effect in no way takes into account the accompanying vast increase in CO_2 concentrations that would occur if fossil fuels were a significant fraction of the energy source, as they are today.

Although technological solutions to the long-term population–resource problems may seem attractive at first, the advocates of such schemes must be careful to consider a number of potential troublesome side effects.

Of first importance is the requirement that any scheme designed to provide resources adequate for maintaining a high population level must not be so lavish as to ignore the consequences of certain fundamental laws of nature, such as the laws of thermodynamics. More specifically, heat pollution combined with the carbon dioxide released from the use of massive quantities of energy can compete with the natural processes that distribute energy and affect the climate.

The second crucial point is that the massive use of energy or other large technologies to support a vastly increased world population subjects that population to a dangerous level of energy dependence; for should that energy production ever fail—whether as a result of miscalculation, acts of God, strikes, war, political control, or even sabotage—those whose ways of life are energy dependent will suddenly find themselves thrust into the strange and harrowing position of mandatory self-reliance. Those who will be unfortunate enough to discover too late that they are almost entirely dependent on services provided by other people and on external technologies to thrive, or even survive, may find it enormously difficult to continue without the customary aid and amenities, energy produced and food grown elsewhere, in particular.

The Population Factor

A third major concern is the potential impact of technologies on the natural ecosystems (which produce food, control pests, and recycle wastes, for example) through various forms of pollution, alteration of the landscape, overexploitation of resources, and so on. The damage to the ecosystems, be they climatological or biological, is proportional to the amount of stress placed on them.[22] This stress, in turn, depends on the product of three factors: the nature of the technology, the per capita use of that technology, and the number of people using it. Thus, the stress on the environment will increase proportionately to the increase in any of these factors separately, but a simultaneous increase in more than one factor will dramatically increase that stress since the impact is augmented by the product of two increasing factors. Thus, if large-scale technologies are to be used to increase the per capita consumption levels of growing populations, the environmental impact will increase substantially since it is proportional to the product of the per capita consumption level and the population size. The relevance of inadvertent climate modification to this issue is that climatic constraints set an outer bound—time limits within which the world must approach stationary population and consumption levels or otherwise risk damage to a most basic natural system, the climate.

Finally, this mention of timing in the food/population/technology/climate problem leads to an immediate concern: that the development and maintenance of massive new facilities to provide for increased consumption for more people can be monumentally expensive, and thus capital may well be a limiting factor on development in the interim. Although it is *possible* that technology could provide for basic human needs for all humanity in fifty or one hundred years, the immediate fear is that serious threats to major portions of the human species will occur in the present decade or the next one; the threats will come if the production of essential goods and services continues to be so closely balanced with absolute need that only the slightest imbalance—whether produced inadvertently by the collapse of an overtaxed ecosystem, or deliberately by political removal of a technological prop—could mean death or extreme deprivation for millions.

The danger exists already. Although questions about the ultimate size and character of a "steady-state" world fifty years hence require some answers now, the more immediate question is, How smooth will the transition be from the present unstable world condition of catapulting population growth and intensifying resource competition to the mature phase of a stationary population with adequate resources to assure adequate standards of living for most, the scenario envisioned by the technological optimists? The next question we might ask is, What are the levels and distribution of population and resources appropriate to a steady-state world, and can such a state possibly be achieved without interim catastrophe if present trends persist? Such matters and their relation to climate change occupy a major fraction of these pages.

Climate Control: A Cure Worse Than the Disease?

But what about technology? What miracles, if any, might be conjured up to alleviate climatic threats to the delicate balance between food supply and demand? Earlier I noted that increasing either the quantity of atmospheric CO_2 or the amount of energy used warms the climate and that increasing dust could warm or cool it. If the precise effects of these processes could be demonstrated, it is quite likely that some people would propose human countermeasures to offset the expected consequences of any natural or inadvertently induced climate change. As another example, if chlorofluorocarbons, nitrogen fertilizer, or carbon tetrachloride threaten the ozone shield, could we not find another chemical to act as an antidote to restore or even increase the ozone?

Similarly, it might be argued (as I will do strenuously) that regardless of human influence, natural climatic fluctuations can pose a serious threat to world food security in the next few decades. Shouldn't we then be directing our vast resources and growing knowledge of the atmosphere's workings toward engineering projects aimed at improving or controlling the climate? "Of course!" would be the reply from many involved in weather modification projects throughout the world. As a case in point, I recently attended a conference that adopted this point of view in its very title, Weather Modification—A Useable Technology—Its Potential Impact on the World Food Crisis.

More to the point, if a few unfavorable climatic years were to result in food shortages, might not those adversely affected want to propose, and perhaps insist on, climate modification schemes to reverse their poor luck?

As a hypothetical case, consider the not terribly remote possibility of several successive or closely spaced years of drought in the Ukraine, such as that of 1972, occurring and hampering Soviet goals of self-sufficiency in grains. The Soviets might then be strongly tempted to attempt to melt the few-meters-thick (three to ten feet) floating sea ice that caps the vast moisture reserve on the North Polar Arctic Sea. The winter sea ice cover exists in a very delicate balance, melting to a thickness of about one meter each summer before the winter freeze causes it to thicken again. Deliberately melting the ice might well be feasible by using any of several techniques, each of which might produce severe side effects with negative as well as positive consequences; and these effects might even be induced accidentally. For example, if the Soviet or Canadian rivers that carry fresh water runoff north to the Arctic Sea were diverted southward and used for irrigation, the Arctic Sea would be deprived of an important source of unsalty surface water. Since fresh water is more prone to freeze than salt water, a reduction in the amount of sea ice might rapidly follow, and, in turn, affect the local climate or even the climate in other countries. The possibility of interstate conflicts would grow much more likely under these circumstances, regardless of the intent with which the rivers were diverted.

Another way to diminish the amount of sea ice, or even to eliminate it altogether, would be to sow it with a cover of black soot dropped from aircraft. If this were attempted in early spring, the darkened ice surface might absorb enough additional energy from the sun to reduce or even melt the ice completely. Halstead Harrison, an atmospheric scientist at the University of Washington in Seattle, calculated [23] that "it would take 50,000 Boeing 747 aircraft sorties (say, 500 aircraft for 50 days of two sorties each) to dump the 10 million metric tons of soot necessary to cover 10 million square kilometers of icecap" to a small depth of one ten-thousandth of a centimeter. Harrison pointed out that the "cost at $0.10 per ton mile and 1000 miles, plus $100 per ton for the soot, would be about 2 billion dollars," a very small figure compared to some current military budgets. Whether it would work, however, is still an open question.

Other schemes have been,[24] and no doubt will continue to be, suggested. In any case, large-scale climate modification could be feasible in the not-too-distant future with the financial and technological resources of a major power.

But before we take comfort in our growing ability to bend nature to our purposes, we must remember that the atmosphere, the oceans, the land surfaces, and the snow and ice fields—which are the major components of the climate system—all act in concert to determine the climate. The forces that generate winds and rain at one particular place on earth are coupled in varying degrees to those forces at places on the other side of the earth, a relationship meteorologists call teleconnections. Although atmospheric science has not yet progressed to the point that cause-and-effect processes in the climate system can be reliably traced, enough has been learned about individual forces to conclude confidently that every place on earth is connected to some extent by the climate system to every other place. Although one cannot be sure, for example, about the *specific* consequences that melting the Arctic Sea ice would have on the climate thousands of miles away, we know enough to recognize that there is a good chance that it could cause significant departures from the climates we observe today in various parts of the earth.

The effects of such an ice-melting project might well prove beneficial to some in the North (possibly by increasing precipitation in areas that border the Arctic Sea), but it could also alter temperature and rainfall patterns at latitudes as far south as the United States and even the tropics. And the project could actually prove detrimental in the zone where its maximum benefits were to be expected.

Because of climatologists' inability to say much with certainty about the climate system, it is very difficult to predict what the ramifications of a particular climate modification exercise might be for different places on earth. Therefore, *any unilateral attempt to alter the climate, even within the region of one's national sovereignty, has the potential for creating national and international conflicts*. The difficulties that even generally friendly neighbors such as Mexico and United States have encountered over water rights to the Colorado River, which they share, provide a useful analogy.

Another useful example is provided by Walter Orr Roberts, atmospheric physicist, astronomer, and author. He pointed out that while Hurricane Inez threatened the U.S. Gulf Coast with grave damage in 1968, the "rains brought by the same hurricane filled the reservoirs of large portions of the Mexican altiplano with sufficient water to assure favorable crops for at least a season." [25] The storm never even hit the United States, as it turned out; although a U.S. team of meteorologists who were then studying the feasibility of hurricane modification did not tamper with that particular storm, it is not unlikely that Mexico would have registered strong objections had they tried.

As our understanding of the processes that determine the climate evolves, it may be learned that certain activities in one region inadvertently affect the climate elsewhere. If the effect were adverse, which it would undoubtedly be somewhere, then again the knowledge of or misinformation about climatic cause-and-effect linkages could well promote interregional disputes. It seems clear that the study of potential conflicts over possible climate modification schemes is timely, since there are no current international agreements that have proven capable of coping adequately with problems deriving from the need for international sharing of our atmospheric and oceanic resources.

I strongly urge that an international body be formed to consider possible areas of conflict, and to suggest the structure of future international authorities or agreements to deal with climate-related conflict situations—before they arise and before their potentially irreversible consequences are felt. The recent stirrings in the United Nations to ban weather warfare are an encouraging beginning, but a much more comprehensive mechanism is needed. A major threat to international peace and security could conceivably come even before the end of this century, not only from the much feared specter of nuclear war, but also from the less known, slower acting, and possibly more ominous potential of an altered climate, whether modified by accident or design, for "peaceful" purposes, or for war.

As with nuclear technology, the developing capability we possess to understand and modify the climate has almost unimaginable potential for human good; but it also possesses the corollary prospect of horrible misuse or mistake. Yet, unlike the public in the first days of nuclear technology, we still have the opportunity to develop climate consciousness and anticipate both the good and evil potentialities of understanding the climatic system. U.S. Secretary of State Henry Kissinger has repeatedly issued the urgent warning that if U.S.–Soviet agreements to limit the cancerous growth of offensive nuclear missiles are not reached by about 1977, we will be virtually helpless to stop the expensive spiral of deadly development that will proceed unchecked. Although climate technology is far less developed than the techniques for building and using nuclear arms (and thus we are not facing an imminent deadline), the analogy does suggest that it is not too early now to begin working toward agreements to safeguard the shared atmospheric resource and to limit the development of geophysical weapons.

It seems to me that humanity also has the obligation to steer rationally and altruistically the course of climatological technology, or any other technology that has potential climatological consequences, while we are still fortunate enough to be at the controls. To do otherwise might eventually put us in a situation where Mark Twain would be forced to revise his reputed complaint to a sardonic "Nowadays, everybody is doing something about the weather, but nobody is talking about it!"

2

The Great Debate over the World Predicament

The world is not on the brink of chaos and famine . . .

—Earl Butz
U.S. Secretary of Agriculture (1974)

There is very important climatic change going on right now, and it's not merely something of academic interest. It is something that, if it continues, will affect the whole human occupation of the earth—like a half billion people starving.

—Reid Bryson
Director, Institute of Environmental Studies, University of Wisconsin (1974)

How long halt ye between two opinions.

—1 Kings XVIII, 21

The Technological Fix: Panacea for the Future?

Life has become very comfortable for many people because of technology, the application of science to human endeavors. Medicine, a prime example, has in recent years enormously reduced pain and suffering, increased human longevity, and substantially eased fears of premature death or disability. A return to the "good old days" may for some seem a glamorous pilgrimage to an earlier uncomplicated life, but to suffer again the widespread incidence of death in childbirth, the smallpox and polio, and the host of other crippling or fatal diseases we have conquered would be unthinkable for most of us.

Technology has also created mechanized transportation systems and labor-saving devices in all sectors of modern life: food and energy production, office- and housework, entertainment and recreation, and so on. Machines also allow us considerable leisure time to relax and develop hobbies. Many of the drudgeries, miseries, and dangers found in pre-industrial societies have, for hundreds of millions, been banished by technology. Some call this technological repair of unpleasantness in the human condition a technological fix.

Of course, the majority of the world's inhabitants do not fully share the benefits of technology enjoyed by those in the richer developed countries (DCs). The millions of poor people living in DCs and the billions in the less developed countries (LDCs) still suffer from curable diseases, are blighted by seriously deficient diets, and generally have access to only a small fraction of the world's resources. Not surprisingly, this inequitable sharing of material wealth is creating growing tensions between rich and poor. Ironically, it is through technological means that the masses in LDCs often learn how the so-called other half lives. Television by satellite, transistor radios, movies, magazines,[1] and countless additional devices and structures developed in the DCs have helped stir discontent in the LDCs. For those whose major pursuit is devoted to survival, the visible paraphernalia of the rich often paint a picture of an enviable way of life in industrialized countries.

As the world's population passes the 4 billion mark, adding about 80 million new people each year, the competition for the planet's finite resources—fresh water, oil, food, recreational land, and so on—is becoming more intense. And as unchecked inflation, famine, resource embargos, and terrorism become more widespread and the threat of nuclear blackmail looms, the urgency and magnitude of the world's problems grow.

Addeke Boerma, director general of the United Nation's Food and Agriculture Organization, recently stated that the international community must soon come to terms with "the stark realities facing the people of this planet." He said, "Remember that, for one thing, prolonged deprivation leads people to desperation. Desperation often leads them to violence. And violence, as we all know, thrives on enlarged prospects of breaking down restraints including those of national frontiers." [2] Obviously, we cannot nourish a "generation of peace" on empty stomachs.

Despite the increasing influence that climate could exercise on our food production for the next few years, the crucial question over the longer term is, What can be done to improve the prospects for eliminating hunger and starvation beyond 1980? Nearly everyone who has looked seriously at food crises agrees that (1) food production can and must be immediately and sharply increased, and (2) population growth must *ultimately* be controlled. Beyond these points, however, there is no consensus on how rapidly we should proceed, how much it will cost, who should pay, or who should lead the effort. In fact, there is nothing today that even resembles agreement on what the ultimate number of people, standard of living, or distribution of populations and wealth should be when the world's population eventually stabilizes. Nor is there a consensus on how many people the earth is *capable* of supporting in a steady-state condition. One unfortunate result of the complex "great debate" over the prospects of human survival is a loss of a sense of urgency. Without urgency, there seems little hope for political action to implement the necessary steps toward hedging against the potential dangers that lie ahead.

Exponential Growth: The Pessimists' Fear

Discussions about the world predicament are not new. A loud, bitter, and growing debate has been raging for centuries over the fundamental questions of survival and the strategies to cure the world's ills. Perhaps thirty-two-year-old Thomas Robert Malthus touched off the modern debate with his famous *An Essay on the Principle of Population* in 1798. In that work he inquired about the fate of future generations, thereby setting the tone for scores of other debaters to follow—some employing the tools of logic, others the tools of probability or statistics, and more recently those who use computers. Malthus explained, "It has been said that the great question is now at issue, whether man shall henceforth start forwards with accelerated velocity towards illimitable, and hitherto unconceived improvement; or be condemned to a perpetual oscillation between happiness and misery, and after every effort remain still at an immeasurable distance from the wished-for goal."[3] Although the world population has more than tripled since Malthus's time, we are hardly better able to answer his "great question" today.

Malthus's famous argument is the mathematical one that population grows geometrically, while food supply grows only arithmetically. That is, food supply is supposed to increase by a fixed *amount* in a given period, whereas population *doubles* its size in a fixed time, thereby eventually outstripping the food supply. (Current population growth is about 2 percent per year, corresponding to a doubling time of about thirty-five years.)

An interesting way to envision such a geometric progression (also called exponential growth) is through a parable about a lily pond: A retired couple, who have always wanted their pond stocked with rare fish, invest much of their savings in the most expensive and unusual varieties. They place the fish and several lilies in the pond. However, they are warned that each lily reproduces itself every three weeks and that the fish cannot live if the lilies cover more than about three-quarters of the pond. The couple plans to watch the lilies and to begin cutting them when they cover half of the pond. Several seasons pass, and the lilies have gradually grown to occupy one-fourth of the pond. They are beautiful in themselves and provide a lovely setting for the fish. Just three weeks later the lilies cover one-half of the pond and the couple realizes they will soon need to begin keeping the lilies in check. "We'll cut the lilies in a month when we return from our vacation," they decide, reasoning that since it has taken dozens of weeks for the plants to cover the first half of the pond they will have plenty of time before the second half becomes covered. But when they return from their trip, they find to their horror that the pond is choked with lilies and the fish are dead. The couple didn't realize, of course, that the lilies covering half of the pond when they left on vacation would take only three weeks to reproduce themselves and double their number, and thereby cover the entire pond. In one three-week period, the lilies added as many of their number as they had in all of the previous three-week periods combined.

Another way to illustrate the impact of exponential growth is to present the parable differently. Suppose that it takes one hundred weeks to cover the pond with a kind of lily that doubles every week. Then ask the question, "What fraction of the pond is covered at the ninety-ninth week?" Many people guess a number like 90 percent. The answer, of course, is 50 percent. Regardless of the fact that it took the ninety-nine previous weeks to cover the first half of the pond, it will still take only one week for the lilies to double their numbers and cover the entire pond.

To be sure, once the pond is covered with lilies, there will be no more increase in their numbers, for the pads will have used up all the available room for growth, and new shoots will either die or crowd out old ones. Exponential growth has ended in this case by the natural checking mechanism of reaching the carrying capacity of their environment (i.e., overcrowding the pond), and only the fittest lilies—and none of the fish—will survive.

Similarly, Malthus argued that in the absence of population control the exponentially growing population of the planet would one day outpace the arithmetically growing food supply, and the inevitable course of natural population control—famine—would take over. In today's language, Malthus could be referred to as a founding "prophet of doom," who has been followed by many others of like mind. Some modern doomsayers even employ intricate computer models that show how so-called Malthusian checks could operate soon on our overpopulated and overpolluted world, resulting in great multiple catastrophies of famine, pestilence, and war. The modern debate has been stoked by such a computer study, headed by Donella and Dennis Meadows and published as the book *Limits to Growth.*[4]

More Technology: The Optimists' Hope

But every debate must have two sides. In this case, the group opposite the doomsayers could be termed technological optimists. These cornucopians or Pollyannas, as they have been called on occasion, point to the miracles of technology to support their arguments. They maintain that by rapid expansion and greater applications of technology we can attain a catastrophe-free transition to a world with high living standards for most, even at population levels several times the present ones. Some technological optimists assure us that the bountiful earth will provide enough arable land, water, steel, copper, energy, and even room to support a population five to ten times greater than today's—and with per capita energy consumption some ten times the present world average (about twice as much as present U.S. consumption).

The optimists have been accused of masking the urgency for curbing population growth by driving the ambitions of all nations toward unobtainable goals of greatly increased prosperity, economic growth, and expanded use of natural resources.

In return, some optimists delight in pointing out that although Malthus was correct in saying that population would grow exponentially, he was entirely wrong in thinking that food supply would be limited simply to arithmetic growth. They say that in the 1960s food production grew exponentially at a rate slightly ahead of population growth, and that there remains in the 1970s a great deal of potential to increase food production still further.

The optimists often argue that pessimists have erred before in predicting imminent disaster, that for years people have been parading with sandwich boards warning that the end is near. And despite all the fanfare, the end holds out on us, and of course never really arrives. The purpose of invoking the argument is merely to discredit the pessimists with the guilt-by-association tactic, to imply that today's forecasters of crisis or collapses will ultimately be proved as wrong as the fanatics who sport sandwich boards.

It is not fair to imply that the optimists are in favor of limitless population growth, for most would readily admit that their long-term objective is a world with a steady-state population; but it is reasonable to point out that most optimists do not attribute much of the world's present crises to population growth. Rather, they pin much of the guilt on those who have neglected or opposed the extensive use of technology to ease human problems, and who, by their negligence or negative actions, have allowed population increases to erase technological gains. The optimists' cure: more technology applied at a faster rate. Population growth can then be slowed sometime in the future under conditions of rising affluence and spreading literacy, the so-called demographic transition. After all, the cornucopians would argue, most industrialized countries reduced their population growth rates only after they attained affluence, so the present underdeveloped countries are entitled to do the same, especially since there is sufficient time remaining and resources to be tapped—with the aid of rapidly expanding technologies, of course.

Food production is a favorite example cited by the optimists. John Maddox, a British physicist and former editor of the scientific journal *Nature,* is representative of the optimist position. He argues: "There are several ways in which food production may continue to increase so as to better the pace of growth of population. Making fuller use of the land is an obvious place to start. In 1967, a committee of the President's Science Advisory Committee in the United States estimated only 44 percent of potentially arable land is at present cultivated. In many parts of the world, but especially in the tropics, vast areas of land are given over to unproductive tropical forests. The Amazon and Congo rivers, for example, include enough land at present not used for agriculture to provide 1,000 million acres of cultivable land, enough to feed 1,000 million people or more (given that, with present techniques one acre of cultivated land is needed to keep each person alive)." [5]

Each Side Blames the Other

When played off against the more virulent statements of doom, which are often quoted incompletely, the optimists' arguments are tempting. For instance, Paul Ehrlich, author of the best-selling book *Population Bomb,*[6] has flatly stated a view that many don't want to hear. "The battle to feed all of humanity is over . . . At this late date nothing can prevent a substantial increase in the death rate," he said, referring to the unnecessary starvation-related deaths of millions of malnourished children.[7] A Stanford University biologist, Ehrlich is a founder of the Zero Population Growth (ZPG) movement.

Often the debate grows bitter. Maddox vents a series of vituperative broadsides at the "latter-day doomsday men." He says, "Too often, reality is oversimplified or even ignored, so that there is a danger that much of this gloomy forboding about the immediate future will accomplish the opposite of what its authors intend. Instead of alerting people to important problems, it may seriously undermine the capacity of the human race to look out for its survival. The doomsday syndrome may itself be as much a hazard as any of the conundrums which society has created for itself." [8]

It is fascinating to hear the pessimists' point of view. Paul and Anne Ehrlich are perhaps the most well-known, outspoken, and witty. In *The End of Affluence* they write: "In the face of disasters, our political and social leaders frequently castigate the prophets of doom who fail to accentuate the positive and who want the lifeboats manned merely because the ship is sinking. Our entire society seems to suffer from a sort of mental block and may refuse to take action to correct its fatal course until it has passed the point of no return." [9]

Unfortunately, people tend to focus on the most sensational, and often superficial, aspects of the discussions among those holding conflicting views of the future. Thus, attention is often called away from the most important parts of their respective arguments: the suggested solutions to the world *problématique*. For example, the Ehrlichs have devoted hundreds of pages in numerous books [10] to analyses of various problems and suggestions to help avert potential catastrophies. Yet, when I hear them mentioned, it is usually in the same breath as phophecies of inevitable doom, despite the fact that most of their writing is concerned with potential problems and possible ways to avert or solve them.

Even though the sensational or accusatory aspects of the debate focus attention on the world predicament, they all too easily overshadow the important messages that lie beneath the glaring headlines. *This* is the worst danger of the often bitter debate.

To Feed the World: More Technology or Fewer People?

Let us return to the case of food, one of the optimists' chief bulwarks of optimism—and one of the pessimists' biggest fears of disaster. Technology, and scientific farming techniques in particular, have unquestionably increased world food production, most notably since World War II, although the food-producing system has been hard pressed to keep up with population increases in the less developed countries. Much of this improvement can be traced to the Green Revolution, the development and breeding of genetic hybrid plants that can produce dramatically higher yields per acre than more conventional varieties. Such "miracle" strains have been adopted extensively in heavily populated regions in the tropics and subtropics, such as China, Indonesia, the Philippines, India, and Mexico.

Unfortunately, as always, there is a catch. Let me mention the climatic one first. Deviation from optimal climatic conditions and the lack of large doses of fertilizer can, in some cases, actually reduce yields to below those of conventional species (although this may be avoided in many instances by careful selection of seed stock). Furthermore, suboptimal conditions almost always cause a larger decrease in the *amount* of expected yields with the miracle crops than with other varieties, even though the absolute or total yield may still be higher than for the conventional strains.[11] Thus, because they are more sensitive to environmental stress or disease than the conventional strains, the miracle seeds need to be backed up with substantial food reserves. This does not imply that miracle crops are a bad choice for farmers, but rather that their use can mean greater variability in expected yields and a need for optimal water and fertilizer conditions to achieve maximum yields. The solution to problems implicit in such variability is, I believe, to have adequate reserves of food and seed to provide a margin of safety in case of suboptimal environmental conditions. The problem, however, is to determine what constitutes an adequate reserve, a difficult determination that must balance the long-term benefits of a safety factor versus the immediate costs of building and maintaining reserves. The issue of reserves also has moral, economic, and political implications: To store food while people are hungry, are willing and able to pay for it, or are threatening to seize it if it is not handed over, adds still another dimension to this many-sided issue.

Despite the growing complexity of the climate/food matter, some people simplify certain of its aspects. Technology has reduced the dependence of crop yields on weather, I was told by a senior official of the U.S. Department of Agriculture (USDA) in June, 1974, and there is no real threat of famine (and, by implication, little need for building food reserves). We have an agricultural policy, he said, a policy of plenty. He was attempting to rebut arguments concerning the ominous prospects of increasing famine that Reid Bryson, a climatologist from the University of Wisconsin, and I had just presented to a gathering of several dozen White House policy-makers. We had argued that the risk of such complacency was food-price inflation in well-fed countries and widespread starvation in poor lands.

Although Bryson and I warned this group of the likelihood of climatic difficulties in 1974 before they actually happened, we were far from being the first to anticipate food shortages in the 1970s. Even when the Green Revolution was well under way in the mid-1960s, some were predicting that it would be a technological fix incapable of preventing famine in the 1970s—unless it were accompanied by controls on population growth and by land reform designed to maximize the use of scientific farming methods.

In 1967 William and Paul Paddock wrote a book—much ridiculed then—called *Famine 1975!*. They argued that technological panaceas such as synthetic foods, new seed strains, artificial fertilizers, modern irrigation projects, utilizing "vast untapped" jungles, or any other resource or product could not be developed in time. They said: "All of these combined can be the salvation of the twenty-first century. Any one of these could achieve a sudden leap forward in progress before the end of this century. Yet none, alone or in combination with each other, will have a major effect on the food crises of the 1970s.[12]

The Paddocks presented a graph incorporating a long-term projection of the U.S. Department of Agriculture that U.S. grain surpluses could keep up with both domestic and foreign demand well into the 1980s. On the same figure they plotted their projection that food supply and demand would cross in 1974, a year that would mark the onset of the "Times of Famine." (It appears that the USDA's optimism about the outlook for food, which Bryson and I encountered in 1974, is characteristic.)

In a more recent book, *We Don't Know How*, William and Elizabeth Paddock still argue that the Green Revolution will not work in time to prevent famines. This, they say, is because in underdeveloped areas, where the farmers' primary goal is subsistence, agricultural improvements take hold very slowly compared to the dramatic rate at which population increases. The Paddocks blame this on the judgment of leaders in poor countries, who are determined to have industrial development even at the expense of agriculture. "When such a thing as Green Revolution occurs," the Paddocks warn, "its name will be Disaster if it arrives ahead of a Population Control Revolution."[13]

They cite the comments of a Philippine farmer who had improved his crop production remarkably by using a miracle rice, and was speaking at an international conference in 1968 concerned with the "War on Hunger": "The farmer in his closing statements referred to his ten children and said that because of the high-yielding variety he and his neighbors would now have enough food for all, and they could enjoy seeing their women in the condition in which they were most beautiful—pregnant."[14] (In essence, the Paddocks are calling for massive changes in *culture*, and these are usually even slower in coming than the spread of scientific farming.)

Very few of those who developed the Green Revolution crops would attack the Paddocks' desire for decreased birth rates or increased attention to agricultural priorities in poor, developing countries. But few, if any, would agree that the entire Green Revolution is merely a technological fix, destined to cause worse difficulty in the long run by allowing more people to survive and propagate until overpopulation, crop failure, or both result in the starvation of many more than would have perished in the first place. Certainly, a solution to this dilemma cannot come exclusively from technological or social prophets. An evaluation of whether it is "better" to let people starve now, even though food may be available, to prevent possible greater suffering later presents a moral problem of immense dimensions. I return to this issue shortly.

It is important to note that although *Famine 1975!* was remarkably accurate in predicting some seven years in advance that food demand would catch up with the supply by 1975, the Paddocks probably were too pessimistic about the rapidity with which miracle crops would be able to improve agriculture in China, Mexico, India, or Pakistan, for example. After reading *Famine 1975!* more carefully, it seems clear to me that their prediction turned out to be extraordinarily accurate because of the essentially unforeseen effects of serious global weather problems in 1972 and 1974. Together with the rising food demand from growing populations and increasingly affluent nations, this spell of unfavorable weather quite overshadowed the gains made by the Green Revolution between 1967 and 1974.

Although the Paddocks warned of the vagaries of the weather, they did not accentuate this aspect. Rather, they anticipated slow progress in the implementation of Green Revolution methods in projecting 1975 as the beginning of the "Times of Famine." But only more recently has it been appreciated that the weather in the late 1960s and very early 1970s was unusually favorable for growing crops in major world granaries and that taking benevolent weather for granted is a dangerous habit. We will come back to the question of climatic influences on food production in later chapters, but before we leave the optimism versus pessimism debate, it might prove worthwhile to examine a major part of the pessimists' reply to the contention that plenty of arable land remains to be plowed.

Without giving further attention at this moment to the potential ecological consequences of such rapid and dramatic changes in the character of the earth or to the productiveness of such land, let's consider the capital investment that would be required to sustain a population increase of some 80 million people yearly by clearing 80 million acres of land annually. If it cost as little as $100 per acre to clear tropical forests, buy farm machinery, build agricultural settlements, and set up transportation systems, the capital investment would be a hefty $8 billion per year. If the cost of development were $1,000 per acre (a more realistic figure [15]), or even more, the yearly investment of at least $80 billion would be a significant fraction of the staggering $250 billion the world's nations spend on armaments. From whom would such capital be forthcoming for such nondefense purposes as growing food? The answer is "no one," I suspect, at least if present world alignments persist.

An important factor, often overlooked (and in specific reference to the Amazon basin clearance project mentioned earlier), is the evidence that without lush indigenous vegetation, the heavy rains that constantly fall in the tropics would wash away in only a few years most of the nutrients in the soil needed for high food productivity. Also, human populations in warmer regions are often plagued by a debilitating parasitic disease called schistosomiasis (also known as bilharzia). Greatly increasing human concentrations in the Amazon, for example, might well bring on severe increases in the incidence of this disease.[16] Schistosomiasis usually occurs in regions where untreated excrement is used for fertilizer.

Tiny worms called schistosomes get into the irrigation water supplies, and can be picked up by humans, after which they move through the bloodstream into the liver where they multiply and prosper. Schistosomiasis may well be the world's leading infectious disease,[17] plaguing hundreds of millions of people inhabiting tropical and subtropical regions of the world where agriculture is extensive and sanitation facilities are poor. Although it is rarely fatal, there is no simple cure for this disease, which debilitates its victims with recurring spells of diarrhea and fever.

But regardless of the costs or ecological consequences of clearing land, to obtain continuously high productivity from any land requires fertilizer. Norman Borlaug, Nobel Prize-winning developer of some of the Green Revolution strains, told *The New York Times* (June 21, 1974) [18] that it would take an annual investment of seven to eight billion dollars (in rapidly inflating 1974 dollars) just to meet increased demands for fertilizer. Even this, he contended, would merely "buy time" in the face of the current "monstrous population growth." And now, as noted earlier, there is growing evidence that massive applications of nitrogen fertilizer, the mainstay of the Green Revolution, may be damaging the ozone layer, which shields biological life from the sun's harmful ultraviolet rays.

Energy for Survival: More or Less of It?

I mentioned earlier that many technological optimists tout the massive use of energy, particularly nuclear power, as the primary means for improving the world condition. But other scientists, such as John Gofman and Arthur Tamplin, who were formerly with the Atomic Energy Commission—and indeed a great many other scientists of distinction [19]—have strongly challenged the need for and safety of nuclear power. On the other hand, scientists of comparable distinction [20] support the widespread use of nuclear power, arguing that reactors are safe enough for widespread use in order to produce much demanded power. However, the crux of this confusing argument between opponents and supporters of nuclear power is not so much each group's view of the probabilities of a nuclear accident; rather, it is whether taking that small risk is justifiable, given that other alternatives—including conservation—are available.

The Ehrlichs summarize the position of the nuclear power pessimists in their characteristic style: "Some technologists feel that foolproof systems for nuclear plants can be contrived, although it is a well-known law of technology (often credited to Edward Teller, 'father of the H-bomb') that a fool of sufficient magnitude can always be found to circumvent any such system. Both the technologists who favor nuclear power plants and those who oppose them agree there is no certain way to safeguard fissile materials (those capable of sustaining chain reactions required for either A-bomb or nuclear reactors) from being directed to clandestine uses. Nor is it possible to safeguard reactors completely against sabotage." [21] This latter point has been supported by a recent study of the General Accounting Office of the U.S. government,[22] although other studies reach opposite conclusions.[23] The greatest potential danger of this confusing debate is that it may prompt governments to direct the lion's share of funds earmarked for energy research toward the goal of providing a more definitive estimate of the probability of nuclear accident. Such a channeling of funds would likely be at the expense of development of alternative energy sources, most of which do not force the choice between the benefits of energy production and the risks of nuclear accident or sabotage.

But, regardless of the methods of energy production, one thing is clear: All production of unnatural energy or diversion of natural energy flows to produce power for human consumption can interfere with the processes that determine the climate.[24] This possibility raises the questions of climatic limits to growth, which will be explored in detail later.

There are indeed hosts of examples of seemingly unresolvable debates on the advisability and potential consequences of technological fixes. These range from increased reliance on efficient "supertankers" for delivering petroleum, to huge water diversion projects, and finally to climate control, the ultimate technological fix. The great debate will no doubt rage on; there has been little indication of an emerging consensus, or at least a cooling of temper, since Malthus fueled the debate. At the end of the eighteenth century, he decried the bickering among intellectuals over whether the world would survive: "Yet, anxiously as every friend of mankind must look forward to the termination of this painful suspense; and eagerly as the inquiring mind would hail every ray of light that might assist its view into futurity, it is much to be lamented, that the writers on each side of this momentous question still keep far aloof from each other. Their mutual arguments do not meet with a candid examination. The question is not brought to rest on fewer points; and even in theory scarcely seems to be approaching a decision." [25]

It is remarkable how little the flavor of the debate has changed in two centuries. But one certainty is illustrated by reversion to the parable of the lily pads. That is, the world, symbolized by the pond, is certainly much closer to its natural carrying capacity of humans in 1975 than it was in Malthus's time, and a plausible case can be made that we have passed the point at which such a large population can be sustained in the long run without intervening catastrophes. Quoting the Ehrlichs again: "To put it simply, mankind has blown its chance for a smooth transition to an equilibrium society. The general economic trend is going to be downhill from now on. There may be temporary reversals—renewed flows of oil, bumper harvests, partial recoveries—but as the end of the century approaches each decade will be worse than the preceding one for the average American, to say nothing of the average human being."[26]

Although the Ehrlichs' vision is undoubtedly a plausible scenario, they themselves point out that it is not *certain* that things must worsen as much as they fear. There are possibilities that new technologies may arrive, that attitudes will turn around and society's leaders will look seriously toward possible *solutions* to the world predicament posed by optimists and pessimists alike. For example, both sides agree that there are real dangers ahead and equally decry the slowness with which society is acting to attempt solutions in time. Their chief quarrel is over the technologies that should be brought to bear to solve the predicament. Unfortunately, this important debate is often obscured by the high visibility of their bickering. Neither the possible damage to our spirit inflicted by the pessimists nor the complacency brought on by the optimists are in themselves the real threats to timely, corrective actions.

The greatest harm may derive from the understandable confusion created in the minds of the public and the governmental leaders, who generally don't listen to much more than the irreconcilable banter of opposing "experts" and then conclude in exasperation that the debate is a stalemate. No wonder societies move slowly on issues of survival; many individuals who could become interested in global matters never develop a sense of urgency about them because the issues are clouded in discussions that are often too specialized and complex to grasp. In addition to that, the proposed solutions are usually tentative and expensive, if feasible at all, as the opposition is always quick to point out.

However, policies of inaction for whatever reason almost guarantee that both the haunting expectations of the pessimists and the worst fears of the optimists will be realized. On the other hand, it may not be too late to apply some of the suggestions of the technologists in an attempt first to postpone, and then to prevent, catastrophe; but these technologies must be chosen so that their impact on the physical and biological environment does not merely bring on larger-scale versions of the looming disasters the pessimists foresee. At a minimum, we are certainly able, and morally obliged, to act decisively to reduce the magnitude of possible disasters, even if it may be too late to fully avert them.

The Genesis Strategy—An Approach to Action

The overview of climatic change and its relation to other aspects of the human survival question presented earlier may appear to imply that there are no clear-cut answers to problems possessing both scientific and moral dimensions. One might infer that the only basis for consensus among the "world watchers" is that real danger lies ahead if we continue our present course, but that the potential seriousness of these dangers—their distance and the actions needed to avert them—are items of hot debate. Is there anything that could be done now to hedge against the worst plausibilities, such as weather-induced famine, ecological undermining of our food-producing systems, nuclear accident, or a new ice age? And, given the uncertain estimates of future risks, how much attention should be paid the wildly divergent proposals being offered as solutions to the world's predicament? How much might these alternatives cost, which are most likely to work, and who should pay?

These pages are not intended to reveal some new knowledge to help the world choose the true answers from the various offerings of the optimists and the pessimists (except perhaps in areas related to climate change). Rather, what is offered to the debaters on each side and to their listeners is a principle of action as old as recorded history, to be taken perhaps as a reminder that old truths may still be applicable to new problems. As you will recall from the preface, I call it the Genesis Strategy.[27] Humanity's dependence on the capriciousness of the weather has been recorded throughout history and literature, and as I mentioned, one of the best-known stories illustrating this point comes from the Book of Genesis. The Bible tells of Joseph's warning to Pharaoh:

> Behold, there come seven years of great plenty in the land of Egypt. And then there shall arise after them seven years of famine; and all the plenty shall be forgotten in the land of Egypt . . . .
>
> Now therefore let Pharaoh look out to a man discreet and wise, and set him over the land of Egypt. Let Pharaoh do this and let him appoint overseers over the land, and take up the fifth part of the land of Egypt in the seven years of plenty . . . .
>
> And when all the land of Egypt was famished, the people cried to Pharaoh for bread . . . . and Joseph opened all the storehouses, and sold unto the Egyptians; and the famine was sore in the land of Egypt. And all countries came into Egypt to Joseph to buy corn, because the famine was sore in all the earth.

The principle itself is neither new nor remarkable. It is fundamentally a call for prudence, negotiation, and margins of safety in all our future planning so that we are adequately prepared for such probable misfortunes as extended periods of bad weather and their harmful effects on crops—crises that weather history warns us will recur.

It is unfortunate that our faith in the ability of technology to insure against even age-old risks such as drought permitted the world to be caught dangerously short of food reserves going into 1975, thousands of years after Joseph suggested his practical plan. A "margin of safety," "a stitch in time saves nine," "an ounce of prevention is worth a pound of cure," "the Genesis Strategy"—each is a formulation of the same principle: that future survival should depend on systems sufficiently flexible and elastic to sustain moderate failures in one or several parts without causing catastrophe.

The principle is unassailable, yet the practice is controversial. How much margin of safety do we need, how much will safety cost, who will pay now, and who will benefit later? The principle applies not only to food/climate/reserves risks, but also to an entire spectrum of technological and survival controversies—water-diversion projects, supertankers, nuclear safety, pesticides, population growth, industrial development, deforestation, genetic engineering, and climate modification, to name a few. But economic viability is perhaps the central issue of the debate because it is the standard by which most governments, capitalist or socialist, are judged by their peers and their constituents. (One should recognize, however, that the argument is often misleadingly couched in terms of economic growth, which is not a necessary condition for economic viability.)

Economic growth, which implies the universally sought-after goals of increased employment and income, is often built on the idea of economies of scale, a principle according to which efficiency is maximized through deployment of large centralized technologies. But the Genesis Strategy embraces the safety factor afforded by the stability inherent in the principle of diversity. Herein lies the political conflict: When the system is such that diversity reduces efficiency, to what extent should efficiency be sacrificed for diversity in order to gain stability? This is nothing more than an updated version of a classical political dilemma: To what extent should today's governmental leaders pursue policies that enhance the stability of the future (e.g., building a food reserve), recognizing that this may be accomplished only by an immediate economic sacrifice that is politically risky (e.g., a loss of income from storing instead of selling that food).

Another aspect of the Genesis Strategy principle with economic overtones is the preservation of nonrenewable resources for future generations; that is, recycling our wastes. Unfortunately, real progress in this direction may be difficult to attain without some shifts in our economic structure, changes that will be prompted primarily by a change in political consciousness. For example, a team of engineers headed by David Bendersky of the Midwestern Research Institute in Kansas City, Missouri, did a study on the feasibility of recovering wastes, that is, recycling garbage. They presented their results in 1972 in a paper to the First International Meeting of the Society of Engineering Science. Despite the fact that present technologies are able to recover marketable products, these engineers were not very optimistic about the prospects for widespread use of these methods in programs to recycle garbage. "The primary obstacle to the economic success of any resource recovery system lies in the marketability of its recovered products," they reported. And, they concluded, "Unless adequate markets are known to exist or can be firmly established for the recovered materials, capital committed to any resource recovery system must be considered a high risk investment." [28]

Here is an example where the Genesis Strategy suggests a principle of action: Government could encourage the competitive position of resource recovery systems through tax advantages. Or, better still, research expenditures could be made to help develop improved recycling technologies that might even become economically competitive without special incentives. Few technological optimists or pessimists would quarrel with the desirability of expanding the role of recycling technologies. However, passing legislation that provides for incentives for recycling technologies would require heavy public pressure on the political system. It would be unfortunate if public pressure (that both sides would endorse) were diffused by the ongoing debate, thereby causing the issue to be ignored.

Decision-Making with Uncertain Inputs

Before discussing further specifics, let's examine briefly the ways most of us approach decisions on complex issues, where the points of the argument often lie beyond our individual knowledge and understanding. To begin with let us ask, How is a decision made? Do we rely mostly on facts, figures, common sense, or merely faith in authorities? We all undoubtedly use some personal blend of each of these aids. We gather information, analyze a situation, and then try to assess the relative merits of each approach. Perhaps we ask advice of our friends or colleagues, listen to debates among "experts," or consider the suggestions of political or religious leaders.

From a personal perspective, more often than not I make decisions, such as voting on obscurely worded amendments to my state constitution, without feeling that I understand the issue very well. Nonetheless, pulling down the lever is the moment of truth, and, once accomplished, one hopes that the right decision had been made.

Of course, it is possible to study many of these difficult questions more than is usually done, but that takes precious time. So we often follow the advice of those whose opinions we respect in the hopeful, but perhaps naïve, expectation that they have had more time to study these perplexing issues. The severest problem occurs when one of our respected sources is dead set against another, and again we are faced with a decision: Whom should we trust?

Many people probably don't know whose opinion to trust on the bewildering issues of climate change, technology, and human survival. Unfortunately, the remaining choice is to trust no one, thereby avoiding the issues, a course of action that inevitably translates into maintenance of the status quo. Since measures designed to enhance the likelihood of catastrophe-free, long-term survival can be as expensive as they are uncertain of efficacy, it is tempting to rationalize postponement of action on these perplexing matters until the scientists, economists, moralists, politicians, or other leaders of the debate are unanimous, or at least seem more certain, about what they are saying. At the same time, we often become preoccupied with more immediate crises—inflation, recession, taxes—and lose sight of the more distant calamities ahead. Unhappily, most democracies, like the United States, are in general responsive to relatively short-term concerns; and, as I will argue later, mechanisms must be devised immediately to encourage elected representatives (or even hard-fisted leaders in more totalitarian states) to place first priority on longer-term issues of human survival (perhaps by adopting some form of disaster insurance at an international level). A change in political consciousness, and ultimately political process, is essential.

Two-, Four-, and Six-Year Cycles

But such forward-looking consciousness is rare for elected public officials or their constituents. Returning to the case of climatic variability and a safe level of food reserves, I recall an incident in June, 1974, on the same occasion mentioned earlier, when Reid Bryson and I were arguing for just such a margin of safety with an Agriculture Department official before an audience of White House policy-makers. We pointed out that there has been over the past hundred years or so a drought of some significance in the Great Plains of the United States roughly every twenty to twenty-two years. Some researchers claim that these droughts are related to sunspot cycles, and others dispute this thinking as unfounded. We stressed that, regardless of possible causes for the droughts, the important consideration is that the next such drought is "due" in the mid-1970s, and thus prudence suggests that we be prepared with adequate food reserves. But food reserves are expensive and depress future prices for farmers, we were told by the official at the briefing. We agreed. However, he went on to repeat Agriculture Secretary Butz's belief that the government should stay out of the farming business regardless of the climatic risk. We protested that no one besides the federal government could voluntarily build up enough supplies to hedge against weather variability such as the twenty-two-year drought cycle. Then a voice from the back of the room taught us a lesson. "Around here," he said sarcastically—referring to the terms of office of the U.S. congressional representatives, president, and senators—"the only cycles that count are the two-, four-, and six-year cycles."

I recall responding that if the present political leaders concentrate only on the next election, and if we do not give more attention to longer-term questions now, the next crop of policy-makers filling chairs in that room ten years later will hear even less comforting information than we could give that day. The vociferous cynic in the crowd then asked, as I remember it, "But how will events ten years from now affect the results of the next election?"

I left the briefing far more pessimistic than I had been upon my arrival, when the very invitation to speak had served as a hopeful sign. But the debate following the presentation raised some disturbing questions with important implications: Could it be that bureaucracies like the U.S. Department of Agriculture, which are expected to provide the executive branch with the *technical information* needed to develop policy, are not necessarily neutral or value-free? It has even been suggested that some bureaucracies become advocates for special interests. In the USDA case, the allegation [29] is that the department favors the large farmer and agribusiness giants who oppose food reserves because reserves stabilize prices and limit profit incentives in times of scarcity. The critics contend that the Agriculture Department favors the largest food producers and the middlemen and may not always represent the most general interests of the United States regarding food policy. But, even if this is so, I still would have little objection to the USDA's special advocacy, *provided* there were other food "experts" in the government representing other U.S. food policy interests. For example, if the testimonies of food experts working for the State Department, or food relief agencies concerned with the prevention of famine abroad, or consumer interests concerned with domestic food prices could be allowed to challenge the technical advice from the food trading interests, this possible conflict of interest would be reduced.

At the conclusion of our formal remarks at the June, 1974, White House session, a debate broke out among various people in the audience, some from the State Department, others from the Agriculture Department, and even a few from the security agencies. The tenor of the argument made me wonder, Were the State Department, security agencies, and Council on International Economic Policy (which had arranged the briefing) using their climatologist consultants to rebut *on technical grounds* the Agriculture Department's apparent rebuttal of reports in the media warning that our food supplies were threatened by climatic variability? The occurrence of just such interdepartmental struggles over food policy in the federal government has been subsequently reported by several investigative news stories.[30]

The State Department, these relate, perceives part of the national interest as giving food away to prevent the collapse of governments saddled with famine problems. In the State Department's view, a policy of benign neglect allowing the demise of such countries could lead to widespread social hysteria and desperation. Such conditions could culminate in either internal or external conflict and perhaps in the collapse of "friendly" governments. The Department of Agriculture, on the other hand, is reported to view food primarily as a means to earn foreign exchange for the United States (to help pay the burgeoning energy-import bill, for example) and to increase profit incentives for farmers, processors, traders, and distributors. If these allegations of special interests are true, the organizational problem that arises is *not* the conflicting views of different departments; rather, it is that the responsibility for conveying factual information about U.S. agriculture to government policy-makers and to the public rests primarily with the USDA. If Bryson and I were called to the White House to rebut the technical arguments of the agency charged with providing agricultural facts to policy-makers, then this instance could be symptomatic of a serious organizational defect; a problem that suggests improvement in the governmental structure to facilitate transfer of technical information to decision-makers—and the public.

A weakness in the mechanism that provides state-of-the-art technical information to policy-makers would constitute a serious danger, one that could aptly be called an *information crisis*. It deserves the label *crisis* because a balanced presentation of conflicting information is essential both to prudent policy-making and public awareness. Of course, a knowledge of the complexity of issues increases the danger that people will be confused and adopt a wait-and-see attitude; and we must be careful that we do not allow complexity to lead to a preservation of the status quo.

Another area in which an information crisis may exist is the case of economic growth. Nearly all governmental economic advisers that one hears in public espouse a philosophy advocating significant increases in economic growth. Does this mean that the risks of growth—especially those leading to long-term environmental degradation and eventual resource shortages—are not presented to policymakers? It is rare that one hears from economists of the slow-growth philosophy; thus, we appear to be running the risk that potentially serious ecological or climatic disruption could follow in the long term, as an inadvertent consequence of the expanded use of technologies—primarily energy—at the high levels envisioned by many fast-growth oriented economic and demographic planners. On the other hand, it has been argued that a freeze in economic growth or in worldwide per capita energy consumption levels would be tantamount to freezing inequity between the rich, developed nations and the poor, developing ones. Furthermore, many argue that increasing the per capita standard of living, of which energy consumption is one important component, is one effective way to reduce long-term population growth rates and redress inequities in living standards, inequities that foment hostility and can lead to conflict.

Putting aside for the moment the question of whether the demographic transition is attainable with massive increases in per capita energy consumption, it still must be pointed out that this kind of massive increase in the use of energy has several dangers, and there seem to be far too few people pointing them out. First, there is risk of damage to ecosystems from heat released to the environment; second, there is the climatic risk discussed previously; third, there is the risk that a dangerous level of dependence on large-scale, centralized energy technology will be thrust upon billions, whose livelihoods would become tied to the stability of that energy supply. Unless the energy supply system possesses sufficient diversity and reserve capacity, any failure, miscalculation, or sabotage of parts of the system could cause a long, hard fall in accustomed quality of life for whatever large number of people may have grown dependent on that supply.

Admittedly, the collapse of any system cannot be forecast with certainty, and insuring against a hypothetical future collapse can seem impractical and overly conservative. After all, building up reserve capacity is expensive, and it diverts precious resources from the immediate needs of the people. And, in the same vein, diversity is often incompatible with efficiency (i.e., so-called economies of scale). Efficient systems, *while they last,* permit more resources to be made available and thereby help to ease some of our problems. Yet, despite these objections, the Genesis Strategy approach, which requires built-in fail-safe mechanisms, is an effective way to weave into our system a time cushion needed to assess stress situations as they arise, without forcing immediate crisis-response actions or suffering immediate catastrophe. This would be a desirable provision for any system that seems heavily oriented toward the short-term philosophy inherent in two-, four-, and six-year cycles. Thus, if present political practices and the short-term view persist, it appears unlikely that a sufficient number of advocates of the long view will be heard in government offices in time to prevent the mortgage of our future to technological fixes or unlimited growth strategies whose inherent risks have not been adequately assessed or hedged against. In short, we need to incorporate the action principle of the Genesis Strategy: A large margin of safety must be built into any plans to counter the social or economic risks of underdevelopment with grandiose, and perhaps risky, technological schemes.

Let me preview briefly a few examples of ways to adopt the Genesis Strategy, which are offered in the final chapter. One would be to expand the proven concept of checks and balances introduced two hundred years ago by the writers of the U.S. Constitution to insure that neither the legislative, executive, nor judicial branch of government would dominate the other two. The framers of the Constitution were acutely aware of the relative advantages and disadvantages of large, centralized governmental bureaucracies;[31] but they could not envision nuclear weapons, electronic spying devices, ecological collapse, or fertilizer/ozone problems. Nor could they foresee clearly that governmental bureaucracies would have, in addition to their specialized short-term missions, the awesome responsibility of providing unbiased and often complex technical information about issues critical to the nation's survival. Perhaps this charge is unduly burdensome for our governmental departments.

Thus, in what is intended to be in the true spirit of the U.S. bicentennial, I propose that the U.S. government update its system of checks and balances to account for the element that now promises to shape much of our future: large-scale technology. Increasingly, we need a long-range plan to avoid a dangerous dependence on some of the solutions it offers—solutions that seem attractive in the short run but could be very risky later. To reduce this danger, I propose consideration of the creation of a fourth branch of government, the Truth and Consequences Branch, whose mission would be neither administrative, legislative, nor judicial, but rather purely informational, serving in much the same way the present fourth estate (i.e., media) now operates. To obtain its information, the Truth and Consequences Branch would need no subterranean connections like those of certain columnists and muckrakers. It would be a democratically chosen arm of government possessing access to and concerned with the dissemination of *any* information (save top-security data) needed to sharpen or awaken public debate, particularly on critical issues of future survival.

Suggestions to create a branch of government dedicated to long-range planning are not new. For example, in their book *Ark II,* Dennis Pirages and Paul Ehrlich propose the addition of a "Planning Branch" to the U.S. government. Drawing from Rexford Tugwell's proposed new U.S. Constitution,[32] Pirages and Ehrlich suggest that the Planning Branch be charged with the tasks of "formulation and regular revision of five-, ten-, and fifty-year plans for America's future. These plans would provide the basis for a continuing policy dialogue about short-, middle-, and long-term political, environmental, economic, and social goals."[33] Although the objective of their Planning Branch and the Truth and Consequences Branch proposed here are indeed similar, I envision, as detailed in chapter 9, that the latter might also operate primarily as a source for public information on complex issues with a long-term technological component. In short, it could be an Affirmative Action [34] Agency, charged with the responsibility of promoting public education on issues of long-term survival.

Lifeboat Ethics and Triage: The Science of Inhuman Ecology

Still another element in a total Genesis Strategy would be a negotiated plan among nations to reduce the inequities in living standards among peoples in developed and underdeveloped areas. The goal of this plan is to diminish the likelihood of famine *and* to insure long-term food self-sufficiency for the less developed countries. The proposal advanced later on calls for immediate and massive corrective actions. Before treating it in great detail, however, it seems important to mention another serious argument against taking immediate action, one not yet given adequate attention in this discussion. That is, some argue that even if it were known how to prevent an upcoming famine disaster in some food-insufficient nation, it should not be attempted.

Perhaps the concept could be best summarized by the recently popular coinage of Garrett Hardin, biologist from the University of California at Santa Barbara—"lifeboat ethics." [35] The very term suggests the Darwinian principle of survival of the fittest. For the world food crisis, lifeboat ethics counsels deliberate avoidance of technological fixes or emergency aid coming from richer nations to forestall imminent catastrophe in poorer ones. This strategy would be appropriate in the absence of a comprehensive long-range plan encompassing such elements as birth control and increases in food production, the argument suggests, since disaster would likely occur anyway at a later date. Therefore, nothing should be done to save the situation in its earlier phases. Hardin suggests that if aid were given without fundamental corrections to the causes of food insufficiency, not only would the catastrophe occur inevitably, but more people would suffer in the end because those who were "saved" initially would have been reproducing in the meantime.

Kenneth Boulding, an eminent economist from the University of Colorado, has also written on this danger. He rephrased Malthus's warning that if the *only* checks to population growth were starvation and misery, these mechanisms would *eventually* keep the population in check, despite the application of technology. Boulding simply calls this Malthus's "Dismal Theorem." He has another proposition, however, that he calls the "Utterly Dismal Theorem." That is, "If the only check on growth of population is starvation and misery, then any technological improvements will have the ultimate effect of increasing the sum of human misery as it permits a larger proportion to live in precisely the same state of misery and starvation as before the change." [36] One can go back further than Boulding or Hardin to find concern over the same problem. Tolstoi wrote on the same subject: "The more is given the less the people will work for themselves, and the less they work the more their poverty will increase."[37] Finally, a classical reference of a similar nature can be found in Seneca: "If you stop supporting that crowd, it will support itself."[38] Similar arguments have been advanced by those opposed to social-welfare programs.

Hardin's bleak outlook on the world food–population problem may also be noted in his comments on the advisability of a world food reserve controlled by a world food bank. The food bank concept in its simplest form encompasses the idea of channeling food reserves through a central clearing house. Those nations faced with sudden food shortages from an unforeseen reduction in crop yields (resulting perhaps from poor weather or an attack of pests) could then withdraw from the food bank adequate supplies to help tide them over the crisis; and they could repay the bank later as their agricultural situations improved. However, as Hardin has written, the reality of present world food production and consumption patterns implies that the "depositors" of food are seldom the "withdrawers." Since those who tend to be withdrawers are often beset by *chronic* shortages in good times, they would hardly ever be in a position to make deposits. This idea, says Hardin, "is stretching the metaphor of bank beyond its elastic limits."[39] Why, he asks, should rich nations bail out poorer ones that are unable to commit themselves to controlling the factors that keep them chronically short of food—namely population growth and inadequate food production at home?

Hardin therefore constructs a metaphor that likens the world to a sinking ship in which the richer nations have gotten safely into lifeboats, filling nearly all the seats with their own kind, while the poorer nations' people are splashing about in the sea, trying desperately to climb aboard. In his metaphor there are more people in the sea than empty seats in the boats, and any attempt to allow all of them on board will only overload the boats and endanger everyone. "We may be tempted to try to live by the Christian ideal of being 'our brother's keeper,' or by the Marxian ideal of 'from each according to his abilities, to each according to his needs.' Since the needs are the same, we take all the needy into our boat, making a total of 150 in a boat with a capacity of 60. The boat is swamped and everyone drowns. Complete justice, complete catastrophe."[40] Hardin implies that an attempt to create a technological fix (such as giving food or bits of technology to nations chronically unable to feed themselves) that is not also a long-term solution is more dangerous in the long view than letting the hungry populations be controlled today by Malthusian checks, ugly as these may be. To do otherwise will only postpone the day of catastrophe to a time when even the rich nations will no longer be able to provide enough relief; and such a delay will result in an even more densely populated world in which there is more suffering (i.e., Hardin arrives at Boulding's Utterly Dismal Theorem). Furthermore, the even more cynical lifeboat advocate might point out that any delay in the arrival of the day of collapse will permit non-self-supporting nations to become more able to drag others down with them (by attaining nuclear capacity, for example).

Lifeboat ethics raises moral as well as practical political dilemmas of mind-boggling proportions. The problem now, as I see it, is that much of the world is so close to famine that the richer nations may soon be forced either to give immediate massive relief, or watch the bloated bodies and spindly legs of dying children in living color nightly as they eat dinner with their families, although, of course, television watchers always have the option of changing stations.

One moral evaluation of lifeboat ethics can be taken from a speech by Willy Brandt to the United Nations General Assembly. "Morally," said the Nobel Prize winner, then German chancellor, "it makes no difference whether a man is killed in war or is condemned to starve to death by the indifference of others." [41] However, acts of commission are more heavily weighed by present moral values than acts of omission (for example, the penalties for first-degree or premeditated murder far exceed those for manslaughter or second-degree murder). But even prevention of starvation need not always be done for purely moral or humanitarian reasons. There are also practical, selfish reasons for saving the lives of others. After World War I, in a disrupted Europe, John Maynard Keynes wrote: "Men will not always die quietly. For starvation, which brings to some lethargy and a helpless despair, drives other temperaments to the nervous instability of hysteria. . . . These in their distress may overturn the remnants of organization, and submerge civilization itself in their attempts to satisfy desperately the overwhelming needs of the individual. This is the danger against which all our resources and courage and idealism must now cooperate." [42]

Rich nations are probably taking a significant risk if they imagine that all poor countries faced with starvation will die with only a whimper. Not all will go down, in Keynes's terms, "with lethargy and helpless despair." This seems especially so now that many of them have discovered useful weapons from the arsenal of the rich: nuclear materials and weapons, resource embargos, terrorism, and alignments of nations to multiply their influences (for example, the recent cooperation among the Arab nations, who set aside traditional prejudices about each other to form an effective oil embargo against the developed, oil-dependent countries). And there is no consensus that population control is desirable. At the World Population Conference in Bucharest in the summer of 1974, the delegates from the People's Republic of China expressed fierce opposition to suggested population controls (despite their apparently successful programs at home), ostensibly because more poor people are needed 'for the fight against imperialism.'"[43]

Where does all this leave us now? Are we damned if we do offer only short-term help to starving countries to avert their immediate downfall and their possible frustration and violence against us, only to find that they are still overpopulating relative to their resource production and are needful of an ever-increasing amount of aid in resources? Or, are we damned if we don't help our fellow humans and risk the immediate and real possibility that their despair will lead to conflagration? Although practical considerations seem to thrust such a dilemma upon us, I believe there are yet other more human options still available and time in which to implement them. The lifeboat metaphor is, precisely, no more than that: a metaphor. It would be a logical error to recommend the strategy inherent in lifeboat ethics based on a metaphorical situation, *if* that situation were an inexact analogy to the present world food–population situation. Before expanding on this point, let us proceed to another often invoked metaphor on this predicament.

Perhaps almost as morally horrifying as a decision to practice the abandonment strategy implicit in lifeboat ethics would be adoption of the World War I battlefield policy of *triage*. When in 1967 William and Paul Paddock predicted the "Times of Famine" in *Famine 1975!*, they recommended for the United States the solution of *triage*. They proposed, essentially, that the United States decide which countries to supply with food and which to ignore. They explained that *triage,* a French word used in military medicine, refers to a priority system for treating the wounded in a battlefield hospital where there is an overflow of casualties and where facilities are limited. Each of the wounded is classified as either unlikely to survive regardless of treatment, able to survive without treatment regardless of pain, or likely to survive with immediate treatment but to die without it. In *triage,* only the last category of patients is treated. "It is a terrible chore for the doctors to classify the helpless wounded in this fashion" they write, "but it is the only way to save the maximum number of lives. Call *triage* cold-blooded," the Paddocks say, "but it is derived from the hard experience of medical humaneness during a crisis." [44]

The Paddocks translated it into practical food-aid terms and suggest that countries like India with cumbersome bureaucracies and high population growth rates cannot be saved, and therefore food shouldn't be wasted on them. An oil-rich, yet food-poor, country like Libya probably has enough resources to get by on its own even in emergencies, so food needn't be sent there. A country like Pakistan, however, shows hope for self-sufficiency and should be given food in crises.

The Global Survival Compromise

Triage, while morally horrifying like lifeboat ethics, does suggest that limited aid is worthwhile, despite some uncertainty that the recipient will become self-sufficient. Thus, *triage* could be considered a moral alternative, since it is designed to maximize the number of lives saved *if* the situation truly resembles the battlefield hospital circumstances appropriate to the metaphor. As with lifeboat ethics, the important question here is whether the present world food–population situation is so hopelessly desperate that the *triage,* or even the lifeboat, metaphors apply.

Fortunately, strong arguments can be made that these metaphorical situations do not apply—yet. Thus, the metaphors may well be inexact analogies to the real world; but there is yet another alternative, both practically and morally superior to the horrifying policies of *triage* or lifeboat ethics. I will call this alternative the Global Survival Compromise: [45] namely, short-term emergency aid and longer-term technological assistance and capital investment from the rich. These would be *coupled to and contingent on mutually agreeable plans* between the donor and recipient nations, to bring about self-sufficiency for the recipients as rapidly as possible but within a fixed time, after which aid would be unnecessary and not forthcoming. The sacrifices needed to achieve success with such self-sufficiency plans will undoubtedly be rejected at first, then highly resented, and will always be very difficult to negotiate. Unfortunately, in thc absence of spectacular and unforeseen technological or demographic breakthroughs that would place food supplies comfortably ahead of food needs, the alternatives *eventually* seem to boil down to Boulding's utterly dismal collapse, to *triage*, or to Hardin's lifeboat ethics followed by "mere" dismal collapse. Despite the inescapable logic of the metaphors, the policy actions that these situations demand do not necessarily apply to the world situation in the 1970s. Perhaps the food–population state of the earth will call for *triage* in 1985 or lifeboat ethics in 1995, but today it cannot be proved exactly analogous to either metaphor.

The implicit and crucial assumption invoked by these metaphors is that the earth has already reached or surpassed its carrying capacity; that is, it is overpopulated to the point where further population growth would require that one person be sacrificed to allow another to exist. But no one knows precisely how large the carrying capacity of the earth really is today or could be in the next century. *If* sufficient seed stocks were available to be planted on well-managed lands and adequate fertilizer could be provided—even after fossil fuels are used up within perhaps a century or two—a reliable supply of irrigation water were on hand, and the climate remained reasonably stable, *then,* some optimists have argued, the earth may be able to support perhaps as much as ten times[46] the present 4 billion people. On the other hand, the large-scale clearing of lands, massive use of chemical fertilizers and pesticides, heavy pressures on oceanic fisheries, gigantic river-diversion projects, and sheer magnitude of heat and CO_2 released to the environment from the vast quantities of energy needed to support so large a population will likely, some pessimists say,[47] irreversibly destroy many of the fragile ecosystems upon which populations greater than today's 4 billion would depend.

These additional people (i.e., those beyond today's 4 billion) would suddenly become food unsupportable were such an ecological catastrophe to occur; in addition, inevitable damage from technological forcing of a large fraction of the world's food-producing systems could conceivably reduce the subsequent carrying capacity to a level *well below where it is today,* perhaps to 2 billion.

But, it seems unlikely that prevailing social and political structures will change sufficiently to free the vast quantities of resources needed to achieve a carrying capacity of 40 billion, even if this could be proved ecologically and climatically safe. Similarly, it would be a disastrous and improbable coincidence for most food-producing systems to collapse *simultaneously* under the stress of ecologically unsound technological pressures, assuming the resources to operate those technologies were available.

At what point, then, between these extremes is the earth's actual carrying capacity likely to rest? Unfortunately, knowledge of future technological devices, biological or climatological systems, or social and political ones is far too sketchy to permit a credible forecast of their behavior sufficient to pin down a likely value for the future carrying capacity. Even a determination of whether this value is closer to the optimistic projection of 40 billion or the pessimistic one of less than 4 billion is impossible to defend definitively today; and any attempt to make this determination would undoubtedly touch off a heated and confusing debate.

Furthermore, it should be apparent from this discussion that the carrying capacity is not a static quantity, but depends on a complex combination of factors such as the stability of the climate, the fraction of land under plow, kinds and amounts of fertilizers and pesticides being used, the mix of crop strains planted, the food storage and transport systems in service and many others. But these factors are also changing with time. Thus, the question of timing is all important, for even if the carrying capacity could, with specified technology and resources, be proved large enough to support safely a steady-state population of say, 12 billion, this knowledge would provide no guarantee that people will work to achieve that capacity in time to prevent the population from growing at a faster rate than the carrying capacity during the transition period, thereby creating interim catastrophes.

To return to the lifeboat metaphor, it is not yet proven that many of the current generation must be "set adrift" [48] in order to build long-term stable solutions to the world predicament. Rather than risk global convulsion from immediate starvation, to say nothing of the moral compunction, it is probably still possible to prevent immediate starvation *and* to help build self-sufficiency into developing countries in time to prevent utterly dismal collapse. (This view is held by most optimists and pessimists, save those who embrace lifeboat ethics.) But I readily admit that this is unlikely to happen without an immediate and massive transfer of technological, food, and economic aid from the developed countries to the less developed nations; nor can this happen without some additional environmental risk. Such transfers imply a substantial slowdown in the growth of prosperity for the rich, and perhaps even a period of politically unpopular negative economic growth. At the same time, poorer countries must agree to reciprocate with comparably dramatic moves to adjust embedded practices that inhibit the stabilization of population growth by creating a social and political climate compatible with eventual self-sufficiency. In the past, when deals that tied food or technological aid to population control were suggested, most poor nations reacted with outrage and indignation. But the Global Survival Compromise would, by its very name, imply a *negotiated* arrangement, acceptable to both sides, and not imposed by the donors. This concept of a negotiated solution has recently been promoted by Harlan Cleveland under the term "Planetary Bargain."[49] Cleveland is a political scientist and "public executive," formerly U.S. Ambassador to NATO and Assistant Secretary of State for U.N. Affairs.

In the summer of 1975, I attended a workshop on Planetary Bargaining organized by Cleveland at the Aspen Institute for Humanistic Studies. The workshop participants came from both developed and developing nations and represented political, journalistic, academic, and industrial institutions. They brought with them widely diverging views of what constitutes an acceptable quality of life and whose obligation it is to maintain basic human needs for all the world's peoples. Despite some often heated and divergent discussions on these questions, one common point of agreement emerged: Only by face to face bargaining, of which the workshop was a microcosmic example, does there seem to be much hope that either side will develop, first, an appreciation of the other's viewpoint, and, then, sufficient personal rapport to put together a compromise agreement that could work to the long-term benefit of both sides.

No doubt, this negotiated approach to planetary bargaining will produce few immediate agreements, unless both sides show more willingness to compromise. Unfortunately, time is probably on the side of catastrophe, for any delay in working toward a solution to the world predicament is likely to allow the world to move steadily closer to *triage* or lifeboat conditions, after which compromise to avert catastrophe becomes moot.

This, then, is why a Global Survival Compromise, painful as it may at first appear to both sides, is imperative.

For the poor it could provide, at long last, a chance for self-sufficiency and escape from the cycle of chronic deprivation and catastrophe—and this would not require that developing nations adopt the development models of the rich.[50] For the rich, the rewards would be both humanitarian and practical; that is, a better and safer world for themselves and their children—who, as the Paddocks put it, "will suffer if we fail to obtain a fair return on this forfeiture" of our resources [51]—and also for the entire human family. This would be a true Global Survival Compromise.

It must be emphasized that any strategy to create a solution to the world predicament must begin immediately and be rapidly implemented (that is, in a generation or two). As stressed repeatedly here, large-scale technological solutions to problems of underdevelopment and overpopulation are inherently risky, and they grow in risk as they increase in size. Nevertheless, a few such risks are *now* worth considering since the social, economic, and, ultimately, military dangers of a widening gulf in the material living standards between rich and poor could well prove greater than the present or near-future environmental risks from carefully chosen technological solutions, provided a stable world equilibrium is reached rapidly. Should such an equilibrium not be achieved before world population grows hopelessly past the potential carrying capacity—which admittedly it *may* already have—then our big technological "solution" will only subject vast populations to even more hazardous dependencies and bring us ever nearer Boulding's utterly dismal collapse.

Herein lie my fundamental objections to lifeboat ethics. Hardin presents the lifeboat metaphor and then follows it immediately with a discussion of the dangers of a food bank in today's world; he leads his readers through a coldly logical metaphor and then immediately takes up the food situation on earth. It is easy to feel trapped by the circumstances of the metaphor and think that its inescapable message should lead to an immediate hard line on food aid. But the world food situation today has not been proven to be analogous to a lifeboat situation. It may take decades, even with existing technology, to reach carrying capacity, not accounting for future improvements that the optimists see emerging already. But even if we are near or beyond the carrying capacity, the lifeboat strategy, which counsels those in the boat to abandon the swimmers outside, may still fail for the simple reason that the *swimmers may sink the boat*, particularly if they band together and attach themselves to it in unison.

Even though I cannot agree with the lifeboat strategy of total abandonment, or the *triage* strategy of selective abandonment for today's world, we should not dismiss the horror inherent in either solution just because they may now be inapplicable to the real world; rather, we should redouble our efforts, perhaps through a Global Survival Compromise, to take advantage of the precious little time that may remain and use it to improve that real world situation.

Our hope must be that in a generation or two we will know better what a stable carrying capacity can be, and if that turns out to be lower than the population at that time, the world will have to voluntarily work to reduce its numbers. This could be considerably easier in the future if the world begins now to raise considerably its levels of literacy and standards of living. Despite the *possible* danger of utterly dismal collapse that could follow the Global Survival Compromise, it seems morally and practically superior to the *guaranteed* danger of practicing lifeboat ethics today, with no certain proof that the carrying capacity has been exceeded.

If the proponents of *triage* or lifeboat ethics really believe that the theories are apt analogies to the world today, then it must be pointed out that no one knows when the carrying capacity will be reached. If they offer their metaphors merely to shock the world into immediate actions, then they should state their intentions plainly. But to accept despair when hope is possible might create a self-fulfilling prophecy, since inherent in continuous pronouncements of inevitable doom is a serious danger of inaction. If one were booked on the *Titanic* and knew its fate, then why not go first-class? The lesson from these metaphors of inhuman ecology is to act with urgency to prevent them from becoming a certainty.

To summarize, the Global Survival Compromise calls for three basic conditions, the terms of which should be worked out in a negotiated plan:

1. The rich countries must provide short-term emergency relief and longer-term technological and economic assistance to poorer countries that agree, in return, to work toward achieving self-sufficiency in food production and population stability; the poorer countries would work toward achieving these goals within a predetermined period of time, after which aid would cease, but would bring about change toward population and food stability within the framework of their own particular cultures.

2. All nations must work urgently to create a stable-equilibrium world with adequate food supply and manageable demand, thereby reducing significantly the probability of large-scale catastrophes. This plan is morally and practically preferable to Malthusian checks on present food-insufficient populations.

3. Reserve capacity must be built into the major technological systems necessary to survival in an effort to prevent sudden catastrophic collapse should one or more of the technological props of such survival suddenly be withdrawn. (This is not an abandonment of technology, but rather a call to more careful planning of its use.)

The proposed Global Survival Compromise, carefully implemented, averts lifeboat ethics and *triage* on practical grounds in favor of immediate, massive, and controlled development in the less developed countries. But the immense moral anguish that lifeboat ethics and *triage* would engender is, to me, an equally powerful deterrent.

The shocking content of arguments in favor of these solutions to world problems brings to mind an intense but instructive dialogue between two of Dostoevsky's brothers Karamazov about the horrors of life and the possibility that a God who could permit such suffering exists. Ivan remarks to his tormented brother Alyosha: "Let me tell you, novice, that the absurd is only too necessary on earth. The world stands on absurdities, and perhaps nothing would have come to pass in it without them." Ivan recounts horror stories that typify the absurdity of this world: "A Bulgarian I met lately in Moscow told me about the crimes committed by Turks and Circassians in all parts of Bulgaria through fear of a general rising of the Slavs. They burn villages, murder, outrage women and children, they nail their prisoners by the ears to the fences, leave them so till morning, and in the morning they hang them—all sorts of things you can't imagine. . . . These Turks took a pleasure in torturing children, too; cutting the unborn children from the mother's womb, and tossing babies up in the air and catching them on the points of their bayonets before their mother's eyes. Doing it before the mother's eyes was what gave zest to the amusement." [52]

After further brooding, the Karamazov brothers construct a hypothetical solution to the world's suffering that personalizes the difficult moral choices inherent in the practice of lifeboat ethics or *triage*. "Tell me yourself, I challenge your answer," Ivan says earnestly. "Imagine that you are creating a fabric of human destiny with the object of making men happy in the end, giving them peace and rest at last, but that it was essential and inevitable to torture to death only one tiny creature—that baby beating its breast with its fist, for instance—and to found that edifice on its unavenged tears, would you consent to be the architect on those conditions? Tell me, and tell the truth." [53]

Dostoevsky's story presents a choice that is, of course, merely another metaphor, and this choice is not one we confront in reality; but it provides an emotionally tangible, moral likeness to lifeboat ethics or *triage* that permits another occasion to stress my main argument: Because the world situation is not necessarily analogous to the morally difficult choices inherent in the abandonment metaphors, we should respond to them by doing all in our power to prevent their becoming reality, especially since there is still a very good chance that enough time remains and resources could be made available to create a stable-equilibrium world without these horrors. If the people of the world act quickly, decisively, and in concert, they may yet avoid mutual collapse. If they indulge in continuing present trends, their fate is almost inevitable.

I hold few illusions that the nations of the world will easily band together politically through some type of Global Survival Compromise and act in time to prevent a few terrible human disasters. But even if they don't, there are nonetheless ample reasons to expect that much can be done to reduce the impact and number of potential future tragedies. Nearly all writers on the world predicament argue that the situation can be much improved even if they can't agree on whether some amount of catastrophe is inevitable. Individual efforts to reduce the danger can indeed make a difference and, thus, we should not be discouraged individually from trying simply because others do not promptly join us. In the end, though, only our collective labors will yield any real chance of total success.

To return to the central theme, there are no certainties now with which to resolve the experts' conflicting testimonies about the chances of our survival or even to answer simpler questions of anticipated levels of our future carrying capacity—or prosperity. Therefore, it shall be argued repeatedly, with an emphasis on the climatic aspects of the problem, that the most sensible policies to insure our future well-being are those that hedge against the very real possibilities of catastrophe. And those policies need not be without hope for improving the quality of life for the earth's inhabitants.

Part II

Climate-Related Crises

> *A major climatic change would force economic and social adjustments on a worldwide scale, because the global patterns of food production and population that have evolved are implicitly dependent on the climate of the present century . . . Our vulnerability to climatic change is seen to be all the more serious when we recognize that our present climate is in fact highly* abnormal, *and that we may already be producing climatic changes as a result of our own activities. This dependence of the nation's welfare, as well as that of the international community as a whole, should serve as a warning signal that we simply cannot afford to be unprepared for either a natural or manmade climatic catastrophe.*
>
> —Report of the Panel on Climatic Variation,
> U.S. National Academy of Sciences (1975)

The next five chapters are about the climate, and provide the background material necessary to build a case for immediate massive action on a worldwide scale. We can indeed reduce the chances for a number of possible climate-related catastrophes, not all of which can be counted on to disappear upon further study, or simply fail to materialize as we continue our habit of maintaining "business as usual." The crises involving climate change, added to the existing body of difficulties that constitute the world predicament, may very well demand of human societies levels of global cooperation hitherto unknown. No project confronting humanity has ever been more urgent.

3

Climate History – What if It Repeats?

The earth's climates have always been changing, and the magnitude of these changes has varied from place to place and from time to time. In some places the yearly changes are so small as to be of minor interest, while in others the changes can be catastrophic, as when the monsoon fails or unseasonable rain delays the planting and harvesting of basic crops.

—Report of the Panel on Climatic Variation, National Academy of Sciences (1975)

I don't believe the world's present population is sustainable if there were more than three years like 1972 in a row.

—F. KENNETH HARE
University of Toronto and Director General of Research on the Environment, Canada (1974)

The United States has had so little variability in weather and grain production in the past two decades (until 1974) that an attitude of complacency has developed.

—LOUIS M. THOMPSON
Associate Dean of Agriculture, Iowa State University (1975)

Ice Ages and In-Between

The skyline of Manhattan Island, crammed with skyscrapers more than a thousand feet high, is a spectacular sight. Yet, few realize that a mere seventeen thousand years ago—one grain of sand in the hourglass of geological time—when human civilization was first stirring and the island might have been bought for far less than twenty-four dollars in trinkets, Manhattan was the final stopping point in the southward march of a huge continental ice sheet. That massive glacier, more than a mile thick in places, stretched several thousand miles over the northern United States and across most of Canada to the Arctic. Temperatures in parts of the "tropics" may then have averaged as much as 5°C (9°F) less than today, a decrease sufficient to alter substantially the biological productivity of the region. The now heavy summer monsoon rains in India were then little more than a sprinkle. In Europe, ice sheets covered Scandinavia, the British Isles, and much of northern Europe and Asia (Figure 1).

Those huge blocks of ice contained vast amounts of water that had once been part of the oceans. Thus, the oceans were tens of feet lower than today's levels, and the edges of the land masses extended much farther into the sea than they do today. When the ice sheets on the continents eventually melted, the water flowed into the oceans, raising the water levels and flooding the exposed shores of the continents almost to their present margins. The ocean floor near today's shoreline, which we call the continental shelf, was actually the exposed edge of the land mass fifteen to twenty thousand years ago.

Contemporary variations in the size of the polar glaciers would also be accompanied by changes in sea level, an ominous prospect for most seaport cities or low-lying countries, were the ice to melt. We readily gain perspective on how rapidly large-scale variations in sea level might develop during major climatic shifts when we consider that over the past several hundred thousand years the rates of sea-level change have often been higher than ten millimeters (four-tenths of an inch) per year, although now they are probably less than one millimeter per year.[1] If the climate were to return suddenly to ice age conditions, the oceans might again change heights at about ten millimeters per year, and we would confront an immensely serious shift in the rate of change in sea level. The same could happen in the other direction if there were a sudden warming. Ten years of sea-level change at that rate would result in a total change of ten centimeters or about four inches. If either of those processes continued for a geologically brief thirty years, a wink in geological terms, the change in sea level would approach thirty centimeters (about one foot), and after a hundred years, would be one meter (nearly three and one-third feet). Even that seemingly negligible change could, if the change were an increase, spell disaster for the city of Venice. During the transition from the last Ice Age to our present interglacial period, the total sea-level change was about one hundred meters, but that transition took ten to fifteen thousand years to complete fully. How often have such things happened?

The earth's climate and geography have changed radically over the ages. But, from the standpoint of human experience, even the greatest of those changes scarcely seem significant. The comings and goings of the earth's major glaciations have always been separated by tens to hundreds of thousands of years. In a typical temperature record for the past hundred thousand years (Figure 2d), warm periods (interglacials) often alternate with cold periods (Ice Ages or glacials), and it is clear that we are now enjoying a warm period.

But the climate also varies over much shorter periods, and there are many small-scale, short-term fluctuations in climate occurring within the longer-term trend (Figure 2a-d). This is a good illustration of the basic lesson of climatic history: The climate varies on many time scales, and although the longest variations are often the greatest, the most significant and disruptive for people can occur over very short times—as we shall soon illustrate.

Unlocking Climatic Secrets

How can we tell what happened to the climate centuries before weather records were kept? Because there were no weather instruments or data files ten thousand years ago, climatologists must turn to Nature's own records and learn how to decipher them. A history of the Ice Age can be obtained from the ice itself, particularly the ice that remains on Greenland and the Antarctic Continent (the South Pole): Ice sheets thousands of feet thick still remain there as intact remnants—climatological Rosetta stones, really—of a million years of the variable history of climate. Climatologists are drilling holes thousands of feet deep and withdrawing cylindrical ice samples called cores, which are later analyzed in the laboratory for certain chemical elements that indicate what the temperature may have been at the time each layer of ice was formed.[2]

Another method of unlocking the natural record of climate history entails sinking tubes deep into the ocean floor and withdrawing cores of ocean sediment. The sediment includes the remains of hundreds of species of tiny plants and animals, many of which lived near the surface of the ocean thousands of years ago, much as they do today. Since different micro-organisms inhabit water of differing depth, temperature, and salinity, painstaking analyses of the relative concentration of certain species at various sediment depths can lead to a reconstruction of ocean-temperature patterns over thousands of years. The depth at which the fossil organisms are found in the sediment helps to indicate the approximate time they were deposited.

At the peak of the Ice Age, frigid water-surface temperatures, now confined to the iceberg-infested waters of the high northern latitudes near Greenland, extended as far south as Spain (Figure 3).[3] Today there are few people much concerned by the approach of the next ice age. And since ice ages take thousands of years to develop, why should we worry?

There are several reasons to worry. Even though the ice sheets may require centuries to build up—or disappear—the periods of transition from one climatic era to another can be accompanied by very rapid shifts in climate. About ninety thousand years ago, a shift to near glacial conditions developed in less than a century. The theory goes that a "snow blitz" occurred; that is, the heavy snows of winter did not melt over the summer, thus creating a mini-ice age. Biological species in the oceans, even at relatively warm locations such as the Gulf of Mexico, changed in only a few hundred years from warm-loving to cold-loving varieties.[4] This possibility of a dramatically fast shift to near ice age conditions was popularized by a BBC television production aired in the United States in the winter of 1975 on Public Television, "The Weather Machine." [5] However, the otherwise excellent documentary failed to emphasize that ninety thousand years ago the earth was already in a partial ice age, and thus the conditions under which such a snow blitz occurred at that time were different from the climatic conditions now (Figure 2d). None of this proves that another snow blitz is impossible today, but it does point up the kinds of difficulties one can encounter trying to learn the lessons of climatic history.

There can be other extremely rapid shifts in climate or sea level related to glaciers. One theory suggests that part of the massive Antarctic ice sheet disintegrated and slipped into the sea, causing a dramatic rise in sea level and sending ice far out into the ocean. Recent evidence presented by University of Miami paleoclimatologist Cesare Emiliani suggests that such a glacial "surge" occurred at the end of the last Ice Age about 11,600 years ago (the date set by Plato for the great prehistoric flood that destroyed Atlantis). Whether this finding explains Noah's Ark is of more than intellectual interest, for some scientists today think the Antarctic ice sheet may again be disintegrating.[6]

Climatic Connections

To take another example, several thousand years ago a minor cool period in the Arctic caused the tree line in Canada to shift southward. The tree line is the maximum northward extent to which trees can penetrate and survive the cold, dry polar region. Climatologists have been able to trace the movement of the tree line by analyzing soils to find out where the trees once grew. Using similar techniques, Reid Bryson and his colleagues at the University of Wisconsin have discovered a striking coincidence between the expansion of a pool of cold air in the northern latitudes, as evidenced by southward movement of the tree line, and the occurrence of a long, dry period in northwest India. Bryson has found that the Indus River basin, which is now a productive wheat belt across Pakistan and Northwest India that contributes significantly to the struggle to keep these countries fed, was desert during those periods of a cooler arctic zone. Bryson also points out that the transitions between wet and dry periods in northwest India have occurred very rapidly relative to the duration of ice ages.

Today, extending north of the Canadian forests into the barren tundra is an area containing fossil forest soils, which very strongly indicates that those forests once extended much farther north than they do today. "Where the fossil soil has been buried by windblown sand, a layer of charcoal at the top of the soil profile has been preserved," Bryson reports. "Dating of this charcoal shows that the forest extended its maximum distance poleward in 1900 B.C., but its northern edge had retreated at least 200 km southward by 1800 B.C."[7] Climate changes can alter the landscape drastically within the period of a century, as they did in the Canadian Arctic and the Indus River area.

If such a drastic shift occurred today in an area that finds hundreds of millions of people dependent on the productivity of this subcontinent region, the human consequences could be staggering.

It is tempting to conclude from such evidence that small changes in the size of the polar air masses are coupled to climatic shifts ten thousand miles away, such as desert growth in India. But the causal links have not yet, in my view, been conclusively established, let alone fully understood. The probability that these events occurred simultaneously only by chance is difficult to determine, but probably is not negligible. The important point in the context of human dependence on climate change is not merely to establish these interconnections with more certainty, but also to recognize that rapid climatic shifts can occur over large regions, regardless of the statistical validity of hypothesized connections. Climatic history is replete with evidence of significant climatic variations; and despite our relative ignorance of the causes of or interrelations among these variations, it is not improbable, as Bryson points out, that a change in one place may well be connected to a shift in another.

Lessons from Climatic History

It is not my intention in the following pages to offer a comprehensive survey of climatic history,[8] but rather to demonstrate by a progression of examples that the survival of human beings is still strongly dependent on the climate. Despite our civilization's powerful technology, which has greatly reduced many peoples' vulnerability to climatic variability, we remain at risk. However, the character of that risk has changed.

Since the effects of climatic variability on human affairs have been documented in many sources, it seems appropriate at this point to provide several examples to illustrate the effect of climatic variability on human societies. The richest historical record is principally that of the past thousand years in Europe, and the narrative will therefore concentrate on that period.

But first, a few stories about prehistoric peoples' bouts with climatic change.

In the thousand years preceding 1900 B.C., a great agricultural empire developed in what is now northwest India and Pakistan, with large cities lying along the Indus River and smaller settlements scattered through what is now the Rajputana Desert. After 1900 B.C., as Bryson points out,[9] the cities and farms were abandoned and parts of the area were buried under sand dunes. Seven hundred years later, Aryan nomads settled the area on different sites and sometimes even atop the dunes that had buried the earlier towns. (Recall that Bryson showed that this period of desertification in the Indus region coincided with a southward expansion of cold polar air into northern Canada.)

Some archaeologists and historians are now busily revising old theories about the fall of certain empires to place at least a share of the blame on climate, rather than faulting only the hordes of barbarian invaders. In addition to the Indus civilization, the Hittite, Mycenaean, and Mali civilizations of Africa are also being studied to determine the influence of climate-related changes on their fortunes.

Climatic conditions have surely shaped the evolution of many cultures. British climatologist and historian Hubert Lamb points out that when the Roman era began about two thousand years ago the climate was gradually growing warmer and drier in western Europe. With those favorable conditions a southern exposure was preferred for houses in Britain. At the same time, but far away, climatic conditions in Asia were reaching drought levels and are thought to have set off barbarian invasions of Europe.[10]

400–1400 A.D. The period 400–1200 A.D. was warm and dry, and, in northern waters, relatively storm-free. Celtic missionaries from Ireland were able to travel as far as Africa and Iceland. The Vikings settled Iceland and the fringes of Greenland and probably journeyed to America for timber-gathering voyages between Greenland and Labrador.

According to Emmanuel Le Roy Ladurie, a noted French climate historian, Norman burial sites that were excavated in 1921 were located in ground that was frozen all year. The excellent preservation of the material and wooden objects indicated a long history of year-round frost. According to the style of the clothing, the burials date back to about 1450. At that time, the ground may not have been permanently frozen because the roots of plants had bored through the biers, skeletons, and clothing. "Did the Norman colony dying out in the fifteenth century do so in a phase that was slightly less severe climatically than the 'modern' era (seventeenth century and after)? I merely raise the question,"[11] Ladurie says.

The mild period before 1200 even allowed the English to produce wine that posed a substantial, if temporary, threat to the trade of French vintners. Meanwhile, there was a record of unusually severe frosts in the Mediterranean; the Tiber River in Rome and even the Nile in Cairo froze over once or twice.[12] The climate was also wetter than it is now; for example, a bridge built in Palermo, Sicily, in 1113 was designed to span a river much wider than the one that flows today.

1200–1400 A.D. The period from 1200 to 1400 A.D. was marked by unusual climatic fluctuations in western Europe, with floods and droughts and very mild winters followed by severe ones. There was noticeably more polar ice in northern waters, and much storminess was reported.

According to Ladurie: [13]

> The winter of 1315–1316 was so wet the bishop of Winchester was obliged to have his oxen reshoed more than once after they had lost their shoes in the mud. And the summer of 1316 was so damp there was not enough good weather to shear the sheep . . . . On the death of each of his villeins,* men, women, and sometimes even children, the bishop of Winchester collected a tax called the *heriot*, and the annual number of these gives an idea of the number of deaths among the peasant population. This number was never so high during the century from 1245 to 1347 as during the two terrible famine years of 1315 and 1316 . . . . The terrified people organized processions to pray for deliverance from rain and the return of abundance. . . .

* A member of a class of half-free persons under the feudal system who were serfs with respect to their lord but had the rights and privileges of freemen with respect to others.

From 1348 to 1380, plagues dominated the mortality in western Europe. Perhaps 25 percent of the English population died. Although there have been suggestions that poor weather conditions contributed initially to the spreading of the plague, few have argued that it was a major factor once the epidemic had begun.

1400–1550 A.D. From 1400 to 1550, the climate became more favorable in Europe, although a few harsh winters after 1500 may have contributed to the abandonment of upland villages in Germany and England. On the average, milder weather allowed the introduction of some southern fruits such as apricots, figs, peaches, and quinces into English gardens. The Romans may have preferred southern exposure, but by 1500, "houses in England tended to be built on the northern slopes and face north or east, shunning the unhealthy south wind and the violent west wind and to admit only the gentle east and north wind which drives away pestiforous vapours." [14]

The voyages of discovery to the New World may have been encouraged by more years of reduced storminess over most of the oceans and the mild period in northern waters. Yet, there was a sharp reversal of favorable weather thereafter.

The Little Ice Age: 1550–1850. During the period from 1550 to about 1850, now called the Little Ice Age, the European and North American glaciers advanced farther than they had at any time since the Ice Age ended many thousands of years earlier. In England, houses were erected with an evident desire for shelter from all directions, Lamb notes.

For example, from the 900s through the 1400s, the Thames River froze over on the average of 1.16 times per century; but in the 1550s, 1600s, and 1700s, that average jumped to six times per century. The weather was apparently cold enough to allow for frequent ice-skating, and iron skates were introduced from Holland in 1662. "Henry VIII may have set the fashion for making sport of the freeze by driving down the Thames with his queen in 1536 to Greenwich. Queen Elizabeth was daily on the river ice in 1564–65 and sport began with the Court, which was then at Westminster," Lamb reports.[15]

The records of the Thames in London after the 1700s are not comparable because the London Bridge, built in 1831, allowed the water to move more freely; also, the use of locks and dams reduced the possibility of the river freezing over in London by preventing the passage of ice from the upper river. This is a good example of the danger of trying to infer too much climatic information from nonclimatic records. There was only one freeze reported in the nineteenth century (in 1874) and none has happened since, probably a result of both warmer times and free-flowing water.

During the Little Ice Age:

> There were years of distress in all northern countries; farms or farmland had to be abandoned to the ice in Iceland, Norway and the Alps. Growing of cereals completely ended in Iceland, only to be resumed in the 1920s and after. There were years of dearth in Scotland and Scandinavia, especially in the 1590s, 1690s, and 1780s, occasioned by poor summers as well as harsh winters. The woods

> were dying near the north-west coast of Scotland (Earl of Cromertie 1710), perhaps owing to increased windiness and salt spray, and these shores remain treeless today.[16]

While Europe suffered because of the unfavorable climate, so did places as far away as China (Figure 4).[17] In the cold, wet years of 1527, 1528, and 1529, extremely poor wheat harvests led to famine and riots in France. In Norwich, England (where, incidentally, Professor Lamb now studies climate history), the mayor's register from 1527 reads, "There was so great scarsenes of corne that aboute Christmas the commons of the cyttye were ready to rise upon the ryche men." [18]

Seventy years later, a great famine hit Europe for several years in a row (1594–1597). W. G. Hoskins (1964) wrote, "In Hungary it was said that the Tartar women ate their own children. In Italy and Germany poor people ate whatever was edible, fungi, cats, dogs, and even snakes. In England there were food riots in many counties . . . ." [19]

Meanwhile, in the New World a severe drought from roughly 1570 to 1590 temporarily turned the already dry southwestern United States into a desert. Ladurie notes:

> The volume of rain fell to 20 percent of its twentieth-century level, and the drought strangled the primitive maize-based agriculture and perhaps led to the abandonment of certain farms and villages. But climate does not have this sharp and simple effect on the European societies of the sixteenth century, and there its influence is mainly short term, affecting the level of harvests and agricultural production.[20]

Coming back again to Old World climatic history, the cold, wet years between 1690 and 1700 led to one of the worst famines in Europe during the seventeenth century. Conditions were so bad in 1693 that one-third of Finland's population died.[21] And, although poor weather was a principal cause of disasters throughout Europe, it was not the only one. Ladurie wisely warns against overinterpretation of climatic influences on history and points out that there were also non-climatic factors that led to the famine. Since the late 1680s all the European kings had been seriously at war with one another. The soldiers were given grain at the expense of the poor civilians, and taxes were increased, so that the peasants who could not produce their own seeds could not afford to buy them either. That meant that one poor harvest led to others.

Much the same could be said of the twentieth century, too!

The poor weather was a terrible misfortune for the Scottish people during the union of England and Scotland in the 1690s, as described by S. M. Trevelyan (1942) in Hubert Lamb's book: [22] During the 1690s, in the last several years of William's reign, there were six consecutive years of poor weather when the harvest did not ripen. "The country had not the means to buy food from abroad, so the people had laid themselves down and died," Trevelyan says. In many parishes, one-third to one-half of the people perished.

> This sombre experience, from which the nation was slowly emerging during the years when the Treaty of Union was under debate, coloured the North Briton's outlook, deepened his superstitions and darkened his political passions, especially in relation to the hated English who had watched the kindly Scots die of hunger, and had moved no finger. . . . Fortunately a cycle of fat years . . . followed.[23]

Lamb explains:

> We see from this the complex psychological factors of which the historian must take account. The Union was however the decision of those in power rather than of the grumbling rank and file, the majority of whom probably opposed it in both countries. It seems to have been a case of dire necessity since there were further harvest failures later—for instance, in 1709. This is also the background to the "Dickens' Christmasses", to some of Keat's poetry and possibly to some of the fearful descriptions of mountain landscapes at home and abroad in the 18th and early 19th centuries.[24]

The winter of 1709 is another well-documented example of climate-induced famine that led to riots and sickness. In March the poor Frenchmen in Anjou, Paris, and Rouen rioted to stop merchants from exporting the sparse supply of wheat. Hunger and epidemics that winter were probably as deadly as a major war in Europe.

The last great famine in France occurred during the years 1740 to 1743. Again social unrest occurred, as Ladurie describes:

> The price of wheat in Liège in 1740 went up again to the astronomical heights it had reached in 1709. The poor people of the town menaced the canons and others who had well-rounded bellies, threatening to ring their carillons to a tune they would not find at

> all to their taste. Prince Georges-Louis, the Governor, told the more prosperous citizens to "fire into the middle of them. That's the only way to disperse this riffraff, who want nothing but bread and loot. . . ." [25]

Ladurie points out that since that time improvements in the farming, transportation, storage, and trading of food have helped to avert starvation in France (and elsewhere), even during times of unfavorable weather and harvests. Technology and organization have begun working their changes on society.

Lamb points out that the ends of the sixteenth and seventeenth centuries were periods of significant volcanic activity, when a layer of volcanic dust was probably spread throughout the upper atmosphere, reducing the amount of solar energy reaching the earth's surface. But, before we conclude too quickly that volcanic dust was the sole cause of the cool periods just described, we must also consider that part of this cool period was marked by an absence of sunspot observations by astronomers, who both knew of their existence and had telescopes adequate for observing them. Jack Eddy, an astronomer at the National Center for Atmospheric Research, postulates that this apparent absence of spots for about fifty years could provide an alternative explanation for the Little Ice Age, assuming that the sun emits less radiant energy when missing its spots, a theory advanced some fifty years ago by Charles Greeley Abbot of the Smithsonian Institution and supported more recently by the observations of two Soviet scientists. (I shall explain in the next chapter the dangers of ascribing cause-and-effect links to climate variations, given the present state of our knowledge.) But let us return to the effects of the Little Ice Age period on western civilization.

Utterström quotes from a Swedish parish record: "In that year (1601) the sun did not have its right natural shine or heat, but in a clear sky shone as though through smoke until 9 or 10 in the day and in mid-evening lost its shine again." [26] Utterström also says that from 1596 to 1598 there were several famines in Stavanger, Norway, and people were forced to eat bark and make bread out of hay, straw, and chaff.

Like 1316 and 1675, 1816 was wet and unpleasant in Europe. That summer Mary Shelley, not yet nineteen, wrote *Frankenstein*, and John William Polidori wrote *The Vampire*. Both writers spent the summer near Lake Geneva, Switzerland, with Lord Byron and Percy Shelley. In the introduction to *Frankenstein* Mary Shelley wrote, "At first we spent our pleasant hours on the lake or wandering on its shores. . . . But it proved a wet, ungenial summer, and incessant rain often confined us for days to the house. . .'We will each write a ghost story,' said Lord Byron . . . ." [27]

With the wet winter, the cold and late spring, and the wet and late summer, wheat harvests were extremely poor in 1816. The wine grapes were ruined, and the bread was damp and sticky, according to a peasant in Brie, a French province rich in wheat.[28] Another Brie peasant described the following year, "In the month of May wheat had got so dear everyone thought to die of hunger. In June came revolutions in all the markets and towns, for bread was no longer to be had at the bakers." [29]

These examples should serve to illustrate the dependence people have had on the variations of the weather in pre-eighteenth century Europe. We turn now to more recent examples; but first a digression to Iowa.

A Two-Hundred-Year Drought in Iowa. Reconstructing past climatic events in America is more difficult than tracing the climate history of Europe or China for no written accounts can be brought to the task. Instead, climate historians must rely on paleoecology and environmental archaeology to reconstruct the past. In the still "undiscovered" United States, there was a culture that archaeologists call Mill Creek, near the boundary between the tall-grass prairie region now known as the corn belt and the short-grass prairie plains now called the spring wheat region or Great Plains. These native American people were corn farmers who moved across the plains to the base of the Rocky Mountains and established small towns about 900 A.D. They settled in an area with tall grass on the hills where they hunted deer, and woods in the valleys where they raised corn.

Then a remarkably long drought began, and within about twenty years the tall grass gave way to shorter varieties and the forests dried up except for some willow and cottonwood trees along the streams. The deer that had comprised about 97 percent of the Mill Creek people's diet disappeared, and bison, a grazing animal, became the meat staple. Food was less plentiful, and farming settlements were forced out of existence to the west. "That drought lasted 200 years," says

Reid Bryson. "Clearly two hundred years of drought in the 'bread-basket' of North America is possible." [30]

More Recent Climatic History. As we approach our discussion of climatic events during the past century, we will first develop an overview of long-term temperature trends (see Figure 2). Then we will show how modern civilizations, despite their technologies, can still be dangerously vulnerable to climatic variability, particularly when several good years make it easy to forget that bad years will inevitably recur, as the ancient Egyptian Pharoah needed to be reminded by Joseph in the seven fat years to prepare for the inevitable seven lean years.

The climate history of Europe has been marked by periods of relative warmth and coldness (Figure 2b). But even in the Little Ice Age, when climatic extremes were experienced and suffering was abundant, the average cooling was no more than 1 to 2°C (1.8–3.6°F), a seemingly small change. The dramatic effect in some places of the small change in average temperature can be seen by comparing a recent photograph of a mountain glacier and a picture of the same scene a century earlier.[31] One of the most spectacular comparisons is provided by the Argentière Glacier. Around 1855, an etching (Figure 5) was made of Argentière, a town in the French Alps. At that time, the glacier was spreading down into the town; but a photograph (Figure 6) of the same scene taken in 1966 shows that within a century the glacier had retreated well above the town. One explanation for this dramatic change is the fact that the northern hemisphere was several tenths of a degree Celsius (0.5 to 1°F) cooler in 1855 than 1966. (Extraneously, it is noteworthy that the citizens of Argentière saw fit to eschew the modernity of contemporary architecture; the charming village of the 1855 etching looks much the same in the 1966 photo. Argentière seems to have found a formula that preserved its character and staved off the bulldozer.)

To offer another comparison, if we were to set our home thermostats down by a degree, we'd hardly notice the difference, but our heating bills would be reduced noticeably. The analogy is applicable to these "small" shifts in climate. Even during the celebrated Little Ice Age, human civilizations were not wiped out. Most people had enough food to survive, though many had just barely enough. And the few extra degrees of cold certainly did not threaten the survival of the whole human race, although it did force the relocation of some living in the high country. However, the cold also caused frequent reductions in the expected size of many harvests—reductions that could not be tolerated in regions both depending on local harvests and subsisting dangerously close to the food-production levels that had predominated in the good harvest years of the past.

Although modern technology (e.g., transportation, food storage, irrigation) has multiplied crop yields and often reduced the dependence of its users on the vagaries of the climate, not all people presently enjoy its benefits. Many still grow most of their own food, but must often supplement it with imports. In fact, so many people have become so dependent on others for their survival that reduction or termination of food imports—because of serious droughts in the world's breadbaskets, for example—pose a threat to larger numbers of people now than the Little Ice Age endangered in days when people had no choice but to grow locally what they needed for themselves. (Of course, there were only about 15 percent as many people then as now.) The threat would be reduced, however, if adequate food reserves were maintained to insure against climatic variability, and this point will be treated in greater detail later.

Returning to the recent climate record (Figure 2a–b) the most remarkable feature besides the Little Ice Age period is the warm peak that has characterized the middle of the twentieth century. Today, by international agreement, the climatological norm is defined as the period between 1930 and 1960. (This will be replaced soon with the 1940–1970 period, and about ten years later with the 1950–1980 period, etc.) Reid Bryson is fond of pointing out that what we now term normal is perhaps the most abnormal period in a thousand years—that is, abnormally mild. The optimum (warmest) years occurred in the 1940s, and a cooling trend set in subsequently, a trend much celebrated in the media and often blamed for the weather troubles in 1972 and 1974. Actually, the cooling trend has been greatest in the Atlantic Ocean region of the high latitudes of the Northern Hemisphere and no cooling at all has been proven for the Southern Hemisphere.[32]

In fact, I am far from certain that even the Northern Hemisphere has, on the average, been experiencing a continuation of this well-known trend beyond the late 1960s (although much of the North

Atlantic sector has been cooling).[33] It is not evidence against the cooling that makes me so dubious, but rather a lack of evidence in general. The reason the climatic normal period still in effect in 1975 is the 1930–60 rather than the 1940–70 span tells much of the story: It is a vast job to collect and analyze all the climatic records (and only a few people work at it). Thus, climatologists are often frustrated with a situation that ironically finds them not very knowledgeable (statistically speaking) about the climate of the five to ten years they have just lived through!

The Recent "Cooling Trend": Does It Mean More Climatic Variability?

Recall that Bryson has found paleoclimatological evidence relating monsoon rain failures in northwest India to cooling trends in the northern latitudes. He also has produced evidence that the frequency of drought occurrence in northwest India in the twentieth century is correlated to cooling in the high latitudes. The chance of drought was nearly one in four in the first part of this century when high latitude temperatures were lower (Figure 7). But the threat was reduced to as little as one chance in twenty during the period from the mid-1920s to the mid-1960s. Bryson argues that the middle of the twentieth century marked the coincidence of not only the introduction of medical and other technologies and high population growth rates, but also unusually favorable weather, the isolated cases of severe famines of 1940 and 1942 notwithstanding—although there is evidence that these famines, the result of a rice disease, were partially weather-induced.[34] The population of India has almost tripled in the seventy-five years since the turn of the century (Figure 7). However, in 1966 a severe drought occurred, and after five more good years monsoon failures occurred again in 1972 and 1974. Bryson points out that while temperatures in the northern latitudes began to increase, the frequency of drought in Northwest India decreased; and, conversely, as the frequency of drought has been increasing since the mid-1960s, temperatures have been decreasing (or at least some evidence suggests this). If this line of reasoning is correct, the near-term outlook for India may be quite tenuous. Monsoon failures devastate crops sorely needed to support the 600 million people now living there (nearly three times the U.S. population); and there are some 10 to 15 million new mouths to feed *each year*, a number roughly equivalent to the combined (but static) populations of New York City, Boston, and Washington, D.C.

Bryson has also looked over parts of the historical record for other periods of cooling to see how many years it usually takes before a cooling trend reverses itself. His "epoch analysis" shows that over the last thousand years cooling trends of the magnitude of the recent one tended to last from forty to eighty years;[35] and since the "present" cooling trend began in the mid-1940s, he reasons that it should continue another ten years at least. If his supposition that monsoon failures are more likely to occur in cool periods proves to be true, and if the cooling trend continues, which is not at all certain, a grim prospect could be forecast for India and other monsoon-dependent lands over the next decade—especially if the current shortages of fertilizer, low levels of food reserves, and high prices of food and energy continue. However, the statistical significance of his assertions are often challenged, as discussed later on and in greater detail in chapter 5.

Hubert Lamb and Tadashi Asakura, a Japanese climatologist, have also argued that cooler periods are marked by higher crop-damaging climate variability—droughts, floods, and temperature extremes. Yet, the summer temperature and rainfall trends in the five major U.S. wheat-belt states since 1900 seem to show the opposite, if anything (Figure 8). In chapter 4, it will be shown in detail why summer temperatures slightly below normal and rainfall just above normal are ideal conditions for growing grain crops (particularly corn and soy beans) in the central states. On the other hand, above-normal temperatures and below-normal rainfall are the characteristics of drought.

The darkened parts of the graph in Figure 8 indicate above- or below-normal values of temperature or rainfall, depending on their location about the midline. The most obvious feature is the nearly unbroken ten-year streak during the 1930s, which was characterized by above-normal temperatures and below-normal rainfall. That period marked the great American drought that eventually caused this once fertile area to be called the dust bowl. Literature and history are filled with accounts of the devastating effects of the drought, how blowing sand and endless stretches of dry, hot, dusty summer days ruined crops, wiped out family farms, and spawned massive migrations—from

Oklahoma to California, for example. The reinforcing coincidence of the dust bowl and the Great Depression of the 1930s created as menacing a threat to the stability of the American democracy as any since the American Civil War, and although the unfavorable climate did not cause the depression, it undoubtedly aggravated an already grim situation. Thus, although some climatologists contend that cooler periods are marked by higher variability in climate, high variability (at least in the United States) can also occur during warmer periods (see Figure 2a).

Other droughts (i.e., periods characterized by below-normal summer rainfall and above-normal summer temperature) have occurred in the 1910–15 period and the early 1950s (Figure 8). In fact, longer records indicate a remarkable periodicity of about twenty to twenty-two years for high plains droughts over the past hundred years or so. Walter Orr Roberts, who has long studied these seemingly cyclical droughts, suggests that they may be connected to sunspot cycles observed on the face of the sun.[36] The main point, he often repeats, is that high plains droughts seem to recur at regular intervals and thus, regardless of their cause, we should be prepared for them and their effect on crop yields, soil erosion, and food security. (He also points out the next such drought is "due" in the mid- to late 1970s!)

Returning again to Gilman's findings (Figure 8), one more important feature must be recognized. That is, directly following the 1950s drought the next fifteen years or so in the wheat belt saw nearly uniform summer climatic conditions, slightly higher than normal rainfall and below-normal temperatures—conditions that are ideal for high crop yields in the plains. Also, since 1955 the use of technological aids, such as chemical fertilizer, certain hybrid food strains, and further developments in mechanical farm equipment, mushroomed. Crop yields have also risen steadily, and at the same time the variability in crop yields per acre from harvest to harvest has decreased since 1955. Technology has been given the credit for both blessings, according to a statement from officials of the U.S. Department of Agriculture, but until very recently, nearly everyone has overlooked the fact that the weather during this period was also unusually favorable.

Many seem to have been lulled into happily believing that our technology has pretty much overcome nature's variations. However, in the next chapter there will be a detailed discussion of a study that proves how critical the fifteen nearly consecutive good growing years were to the maintenance of high yields and (particularly) to the low variability of yields. This study suggests that the chances of enjoying such a favorable growing climate for another fifteen years are quite low. Meanwhile, the U.S. government followed Secretary of Agriculture Earl Butz's advice and got out of the farming business, an action that has contributed to a situation that permitted our food stocks to reach their modern-day low levels as we entered 1975. Reserves may not have seemed necessary in the recent past because some believed that technology had significantly reduced the chances for weather-related crop failures, thus rendering the need for reserves, at least by implication, somewhat superfluous. Regardless of the validity—which I challenge later—of this belief in the developed nations, let us examine the effects of recent climate variations in countries unable to afford, or unwilling to use, modern technology to offset weather variations.

Sahelian Drought: The Desert Advances

Perhaps the most dramatic examples of the continuing influence of the weather on human welfare are the recent situations in Africa and India. On both continents the stress of drought, coupled with large populations occupying relatively fragile, highly variable environments, has been disastrous.

In the drought-prone area bordering the southern fringe of the Sahara Desert, known as the Sahel, human populations had increased by a third over the previous thirty-five years, and livestock numbers doubled.[37] This rapid growth has been possible mainly because of Western intervention in the economic, political, and social practices of the Sahelian nomads and sedentary farmers. In recent years the nomads have experienced the breakdown of their traditional way of life—a life that Nicholas Wade, a writer for *Science* magazine, has called "a remarkably efficient adaptation to the semidesert environment."[38]

A major part of the Western influence in this region has been technological: medical and mechanical. The latter came partially in the form of deep water wells, designed to encourage the proliferation of the nomads' herds.

The growing herds of cows, donkeys, camels, and goats trampled the ground as they traveled from well to well, eating everything in their path. With continued overgrazing, trampling, and a lasting drought, vegetation became scarce, and the animals were forced to compete more intensely for the remaining bit of forage. They, especially the goats, desperately tore up the indigenous plants by their roots, destroying the ability of much of the vegetation to reproduce itself.

At the same time, the increase in the human population led to even more Sahelians farming still more land. In addition, the best lands were often used to raise cotton and peanut crops for foreign markets, and the farmers began to cultivate increasing amounts of marginal (ecologically fragile) lands to feed the growing local populations. Traditionally, the Sahelian farmers had wisely given their marginal lands fifteen- to twenty-year fallow periods, but recently they began to allow the land a meager one to five years to rest. By disregarding the proven grazing and farming techniques of their ancestors, the Sahelian peoples were helping to bring about desertification, a process that would make much of the Sahel a wasteland.

Desertification is extremely difficult, but not impossible, to reverse. Its first signs are a slow loss of fertility in the soil, which is followed by a rapid degradation process. The soil eventually loses its structure and becomes unable to hold sufficient moisture or vegetation.

As the desert moved southward, so did the people and livestock, stressing the retreating fringe more and more, creating further denudation and deforestation, and perhaps (as will be discussed in chapter 5) creating a self-reinforcing process. And so the degradation of the social and ecological systems of the Sahel continued until they finally collapsed, along with the starving humans and animals whose swollen bellies contained only water from the wells drilled by Western technology. By the late 1960s desertification had set in, and with no end of the drought in sight, a major international effort was required to keep millions of Sahelians alive. The first large-scale public awareness that a food crisis had begun thus actually arose only several years after the Sahel had, in fact, been devastated. It is reasonable that climate variations were not solely responsible for the disaster. Clearly, thoughtless applications of technology and poorly examined social and economic priorities must also be deeply implicated.[39]

The nomadic survivors are suffering not only from physical hardship, but also from the shock of having to give up their traditional way of life. Until Western technological and economic ways were bestowed on them, these peoples lived as their ancestors did. They and their herds followed the rains, even across political borders. Now some of the survivors live a humiliating, parasitic existence in refugee camps, robbed of their herds, their possessions, and their traditions. The governments of the drought-stricken nations usually want them to assimilate and adopt the life of the modern African, often only for the reason that nomadic ways do not easily adjust to taxation, census, and government controls. "Properly run, the Sahel should be able to raise enough beef on the hoof for half this hungry continent," says writer Claire Sterling. "Yet after almost fifteen years of bountiful international aid and more than bountiful advice, the Sahel is closer than ever to becoming a wasteland." [40]

Although it is true that normal rains at last returned to the Sahel in 1974 and 1975, the happy future of the region is by no means assured. Political scientist Michael Glantz, who is engaged in a multidisciplinary study of the Sahelian drought, has referred to the return of the rains as "The Paradox of Good News," [41] pointing out that even worse suffering will accompany future droughts if the Sahelian peoples return to their "normal" ways now that good rain years have come for awhile. Land management of this fragile ecosystem and stringent population control of both animals and humans must be practiced, Glantz argues, if the ugly spectacle of the recent drought is not to repeat itself on an even grander scale when the next drought period begins and provide a regional example of Boulding's Utterly Dismal Theorem.

Similarly destructive processes of desertification, which are probably caused by a combination of human and natural factors, have also been occurring in India, Chile, Peru, and in southern Africa.

In India desertification is only one contributing cause of that nation's serious food problem. Thus, in the state of Rajasthan, population pressure and drought on a fringe of the Thar Desert is destroying thirty thousand cultivable acres each year [42] and has already taken half of the state's area.

According to Robert R. Brooks in an article in the *Saturday Review,*

> A classic illustration of large-scale destruction is afforded by the spectacle of wind erosion in Rajasthan. Overgrazing by goats destroys the desert plants which might otherwise hold the soil in place. Goatherds equipped with sickles attached to 20-foot poles strip the leaves off trees to float downward into the waiting mouths of famished goats and sheep. The trees die and the soil blows away two hundred miles to New Delhi, where it comes to rest in the lungs of its inhabitants and on the shiny cars of foreign diplomats.[43]

Some people think that the likelihood of unusual weather events (like droughts) is connected with large-scale, long-term climatic trends, as we explore next.

Unusual Weather May Be Increasing

The Sahelian drought is not the only manifestation of unusual weather. A number of recent incidents of unusual weather are now given, to point out some of the dependencies people still have on the weather. Tadashi Asakura has documented the very recent history of climate with colorful examples.[44] The winter of 1962–63 was a record breaker, he reports. "The first wave of cold began on November 19 . . . in Europe and in England. A man was killed while riding on his bicycle on a bridge and was blown away along with his bicycle by a gust of wind. The cold wave was that intense. Thereafter England was overlaid by high atmospheric pressure which caused formation of bad smog. It is well known that 136 people died as a result." By New Year's Day six meters (nineteen feet) of snow had fallen since autumn in London, which was the greatest snowfall there in over one hundred years. As a result, 152 kilometers of highway were closed, he reports.

In Italy and Germany railroads shut down. There was so much drifting ice on the coasts of the Scandinavian countries that the harbors were closed. The Danube, Rhine, and Thames rivers were frozen, as was the Lake of Constance between Switzerland and West Germany, which froze for the first time in 130 years. About a meter of snow accumulated over all of Yugoslavia. In Poland twenty-seven thousand schools were shut down, and the soldiers distributed coal. Damage to agricultural products, especially oranges and olives, was so extensive that Italian farmers demanded a tax exemption from the government. "Florida, which is known as a sunshine state," Asakura notes, "also was attacked by severe cold waves causing oranges to fall from their trees." [45] He also cites the winters of 1967–68 and 1968–69 as "unusual." In Japan 1967–68 was an exceptionally cold winter. In Niigata the trains stopped running, and ninety thousand households were without electricity. Canned juice froze at Maibashi, and on December 30 Tokyo had sixty fires, which broke the previous record of fifty-three fires in one day. (Whether this means anything for the climate is addressed shortly.)

One year later, however, the winter was unusually warm. Asakura writes:

> However, the unusually warm winter was welcomed by housewives. Vegetables grew too well and their prices fell. Five pieces of large radishes sold for ten yen [a few cents], four heads of Chinese cabbage sold for 20 yen, and the sale of kerosene was sluggish and sold at about one half the price of the previous year. However, the business community felt that it was a warm-winter recession. The sale of heating stoves fell by 20 percent, and winter clothing was difficult to sell which contributed to bankruptcy of some textile merchants. Department stores promoted bargain sales of their winter clothing as early as December, but their customers bought only raincoats and polo shirts. Ski resorts did not have enough snow and travel bureaus remained apprehensive through the month of January. Thus the unusually warm winter was helpful to the consumers but it was a cold winter which was helpful to producers.[46]

Several researchers at the Institute for Environmental Studies at the University of Wisconsin detect emerging climatic patterns that they feel are similar to those that prevailed during the nineteenth century. One result of such a shift could be a return to heavier rainfall in the western plains and Rocky Mountain States. Reid Bryson noted that "Many of the forty-niners who made the trek to California recounted that a hazard of crossing the plains was losing sight of the main party amid endless seas of head-high grass growing in regions that are practically desert today." [47]

Nor have we escaped the vagaries of the weather during the early 1970s; 1972 in particular was a year in which climate variations damaged agriculture considerably and caused other disruptions as well. In early May a freak snowstorm dumped seven inches of snow on the golf courses in Halifax, Nova Scotia, Canada, shortly after the province's tourism department had advertised, "In Nova Scotia, spring often seems like an early summer. This being so, we recommend an early summer vacation." [48]

1972 and 1974: Food Production Drops, Food Prices Soar

After two successive years of good grain yields, the U.S.S.R. experienced unfavorable weather in 1972 that led to a harvest 12 percent below expected goals.[49] That winter, a combination of insufficient snowcover (snowcover provides ground moisture for spring growth and insulation from the harsh winter temperatures) and severe cold in the month of January destroyed almost a third of the winter grain crop. To make matters worse, the European part of the U.S.S.R. was beset by a hot, dry spring and summer that worsened the winter grain situation and affected the spring grain crop as well.

The celebrated wheat deal between the Soviet Union and the United States and Canada was one result of the poor Soviet crop. In the transaction, the Soviets bought up about one-quarter of the U.S. wheat crop for that year. A prestigious group of people have even blamed the deal in part for "triggering world food shortages, production cost increases, massive shifting of foreign exchange expenditures, violent responses in prices, and a new nationalistic policy of food export controls that resulted in anguished reactions from France, Japan, and elsewhere." [50] So what began with bad weather may have had a profound impact on the political and economic underpinnings of today's world, in which (as with climate) every part of the complex system is related to every other part.

The year 1972 was actually something of a climatic anomaly around the world. In addition to the Soviet drought, there was a delay of several weeks in the onset of the Indian monsoon, continuing droughts in Central Africa, and floods in Pakistan. A change occurred in the currents offshore of Peru, and the cold upwelling waters that are vital to fishing disappeared. (Incidentally, ocean currents are primarily wind driven and thus even fishing is dependent on the climate.) After more than twenty years of a steadily increasing world fish catch, the devastation to the Peruvian anchovy fishing industry was so severe that the total world fish catch declined sharply; and Peru, which until 1972 had led the world in total fish catch, dropped to a poor fourth place.[51]

However, the anchovy may not have disappeared only because of natural occurrences. The human influence was probably another factor. As in the Sahel, where overgrazing in a fragile ecosystem exacerbated the impact of a drought, overfishing may have added to the demise of the fish populations. In 1972 and much of 1973, the anchovies seemed to disappear from the usual offshore fishing areas. This situation did not seem serious initially because slight changes of temperature and the occasional change in the course of the Humboldt current, known as *El Niño* (the "Christ Child") since it often occurs near Christmas, had caused temporary moving of the anchovies in the past. However, there is evidence that overfishing may have been an additional culprit; a Food and Agriculture Organization study in 1970 concluded that 9.5 million metric tons was the maximum sustainable yield of the Peruvian fishery.[52] However, catches in the late 1960s and early 1970s were apparently between 10 and 12 million metric tons. When *El Niño* comes, the tongue of cold upwelling water the fish need apparently [53] shrinks slowly in width from some tens of miles to a few miles; and the fish then crowd into a narrow zone, making fishing extremely easy—for the first wave of fishing boats at least. "Presumably, the Peruvian government will allow marine biologists to exercise much closer control over fishing efforts in the future so that fishmeal production can be restored to sizeable—and sustainable—levels within a few years," [54] counsels Lester Brown.

One more unusual climatic event of 1972 was noted by George and Helen Kukla of the Lamont-Doherty Geological Observatory of Columbia University; they found that the minimum brightness of the earth observable from space from an earth satellite went up sharply in 1971 and leveled off in 1973. They interpreted this as a sharp increase in the amount of snowcover and floating sea ice in the Northern Hemisphere—some 13 percent, they reckoned.[55] Whether this coincidence was responsible for, or even related to, the series of climatic disruptions of 1972, is not yet known with certainty. However, although some observers believe that figuring minimum brightness may be a good method for detecting large changes in snowcover or sea ice extent, it is possible that one is really observing a variation in the number of low-level, small-scale clouds; that is, clouds that are individually smaller than the satellite scanner can resolve. It is to be hoped, however, that this technique can be refined and combined with others to provide some forewarning of unusual years like 1972.

What do these many stories portend for the future climate? First let us examine a few more specific examples of unusual weather conditions in 1973 and 1974.

Despite the continuation of the Sahelian drought, 1973 turned out to be an unspectacular year for the global climate, especially when compared to the wrenching food-production problems caused by the climate in 1972 and 1974. The year 1973, however, did turn out to be a record-breaking one for tornadoes in the United States. By September the previous record of 930 was broken, and by the end of the year the figure had climbed to a remarkable 1,107, including new peak numbers in eleven states. The 1973 U.S. tornado season was also the longest on record, it involved more states, and produced more "supertornadoes" than any year since records of tornadoes began.

Because of highly variable weather 1974, like 1972, was a disappointing year agriculturally. The worldwide food harvest in 1974 was nearly the same as in 1973, but more seeds were planted in 1974—and there were some 80 million new mouths to feed, as there are each year now.

In Canada, as in the United States, seeding was hampered by a late spring and heavy rains. A hot, dry summer and early fall followed, stunting growth and reducing crop yields. Canadian wheat production was about 13 million tons, 3 million tons less than in 1973.

The Soviet wheat crop also suffered because of the weather. Yields were about 10 percent lower than in the previous year.

Despite increased plantings of rice in Japan, yields there were down slightly because of a blight that affected 40 percent of the cultivated area. In India the production of corn, barley, rye, and oats was down about 15 percent and wheat yields were 10 percent lower, smashing that country's hopes of self-sufficiency. The monsoon rains were late again, and during June and July they were light in many parts of the subcontinent that are important to grain production.

But the major agricultural news in 1974 was the United States, the major world granary. The year began with a drought that substantially reduced the winter wheat crop in Texas and Oklahoma. Most of the Coloradan wheat crop was saved in early June only by an unusually heavy storm. The storm dumped two inches of rain—which is all that is normally expected in the Colorado plains the entire month of June—on parched wheat that would have been lost had the drought that began in early April continued for one more week or so.

Meanwhile, the spring of 1974 proved to be one of the wettest on record in the most important region, the highly productive Midwest. The excessive moisture made it difficult to plow and plant seeds. Even though 13 percent more corn eventually was planted than in 1973 (which was a record-breaking yield year), crops in many cases had to be replanted a number of times and were weeks behind the normal growing schedule.

Then, as if to counteract the wet spring, July was remarkably dry in the Midwest. The midwesterners who had bemoaned the wet fields in June were praying for rain by mid-July, when sunny, dry 100°F days prevailed for days on end. The seeds that had been planted and replanted late in the spring became starved for moisture during their critical growth period and thus grew more slowly than usual. And, up to near the end of the harvest, the USDA insistently predicted a record harvest and feared probable surpluses [56] that would have depressed prices—and profits.

Since August was more normal, and if anything a bit on the wet side, it appeared that if only the crop could be allowed to ripen a little later than normally planned, 1974 could still be a record year. But in mid-September, with a major portion of the soybean and corn crops still unharvested, record-breaking cold weather struck the plains. Despite an increase of more than 10 percent in acreage planted, the total corn harvest dropped many percent below the 1973 yields, and the yields per acre were reduced tens of percent. Soybeans were also badly hit.

Louis Thompson, an agronomist and Dean of Agriculture at the Iowa State University, and James McQuigg, a climatologist at the University of Missouri, had quietly issued (for the Department of Commerce) a report [57] in 1973 that warned the government that this sort of reduction in yields was quite likely; their warnings were belittled and challenged by Agriculture Department officials. Department reviewers wrote:

> A comprehensive study published by this Department in 1965 evaluating the effect of weather and technology on corn yield in the Corn Belt for the years 1929 through 1962 concluded that through

> the use of better varieties and improved cultivation and fertilization practices, man has reduced variation in yields in both good and bad weather. It seems logical to assume that continued progress has been made since that date, particularly in the use of fertilizer, improved cultural practices and the increased benefits of mechanization.[58]

Shortly after this exchange, McQuigg and Thompson, who have no special interest beyond the natural desire to see agricultural planning more firmly based in reality, had their worst fears confirmed as 1974 corn yields were reduced some 20 percent—by the weather. The plants just couldn't live up to some people's repeated statements that technology had all but eliminated the dependence of crop yields on the weather.

What Does It All Mean?

In this chapter, I have attempted to demonstrate that human civilizations are still quite vulnerable to climatic changes, particularly through setbacks in food production. Historically, local climatic changes have usually created only local consequences and they still do today, but to a lesser degree than before agricultural technology was widespread. For example, because most parts of today's world have access to food imports, locally bad weather can often be overcome by food importation. But today, weather disruptions in just a few of the world's major granaries can cause food price inflation and some shortages in even these well-fed countries, while creating famine conditions in the poor areas that have allowed themselves to depend on food imports to subsist.[59] In 1972 and 1974 the highly variable weather led to increases in hunger on a massive enough scale to get the attention of the world—for a few months, at least. I have cited many examples of recent climatic variability and repeated the warnings of several well-known climatologists that a cooling trend has set in—perhaps one akin to the Little Ice Age—and that climatic variability, which is the bane of reliable food production, can be expected to increase along with the cooling.

But a skeptical reader might ask: "What does any number of individual examples of recent unusual weather mean unless a smaller number of unusual events can be proved to have occurred before the present 'unusual' period?" In the same vein, what do a few isolated records of temperature trends mean with respect to global conditions? We need the average of many separate records to establish overall climatic trends reliably. In mathematical terminology we must ask, for example: "How *statistically significant* is the assertion that climatic variability is increasing now, when it is based on evidence such as the demise of a bicyclist in an unusual storm or the coincidence in the shapes of two graphs taken over only one cycle?" To draw an analogy, I could flip a coin a hundred times and might expect some fifty heads and fifty tails. Of course, it would be surprising if precisely that many occurred, but the exact chance (i.e., the probability) of a certain number of heads and tails occurring in a given number of flips can be computed for an "unloaded" coin. It would not be an extremely remote probability to get, say, five heads in a row. If a "sucker" walked into the room at that instant, I might offer this magical "loaded" coin to him for a price well above the face value of the coin. However, if many more flips were made, the total number of heads and tails would soon tend to equalize.

The point of this analogy is that dramatic examples of unusual weather reported in the press (or from time to time in historical chronicles) may not necessarily be indications of an increasing trend of unusual weather variability; rather, they might simply indicate the increased vulnerability of society to weather disruption, or even a growing interest on the part of the media (or chronicler) to report the continued subservience of society to the elements. A few examples of unusual weather do not necessarily imply a climatic change (just as five heads in a row do not necessarily indicate a loaded coin). The same sort of question can be asked concerning the correspondence between the recent cooling trend in the Northern Hemisphere (which needs further documentation itself) and the increased frequency of monsoon failures. Since the apparent correlation is given only over one "cycle" (compare Figure 2a to 7), there is admittedly a significant probability that the apparent correlation of these separate trends has occurred by chance. Regrettably, much climatic inference is still based on a very small sample of data or on tenuous reconstructions from records of phenomena indirectly related to climate, or on stories of unusual weather conditions that are difficult to validate and, in mathematical terms, whose statistical significance is unestablished.

Unfortunately, this sort of statistically incomplete evidence is often all we have to go on. But ignoring it or waiting for more certain proof may be foolhardy, given the dangers that could follow if the contention that unusual weather is increasing proves accurate. Thus, it may be most prudent to follow Reid Bryson's advice and at least admit that there is an uncomfortable probability that the worst of climatic history might repeat itself. More on this shortly, but first let us explore a promising way to improve the accuracy of the historical record.

Tree rings, whose size and shape are indirect indicators of local climatic conditions, can be used to reconstruct some aspects of climate history. (A new ring is added annually to most species of trees.) The fine work being done by H. Fritts and V. LaMarche in the Tree Ring Laboratory at the University of Arizona is beginning to show that considerable information on the character of past climates can be obtained by analyzing the widths and characteristics of individual rings. Unfortunately, climate history cannot yet be reconstructed from tree rings in enough detail to be useful for agricultural planning, such as the determination of a safe level of food reserves based on the probable fluctuations in year-to-year crop yields as derived from year-to-year weather variability. Regardless, tree-ring analyses (which also provide a precise dating system for past events related to trees—building shelters from felled trees or indicating local climatic changes, for example) have already been invaluable in determining the age, building patterns, and habits of native American cultures in the southwestern United States; and they have revealed periods of climatic stress that corresponded closely to convulsions in these societies.[60]

The study of the behavior of a certain biological species provides another example of the reconstruction of climate history from climate-influenced phenomena. One celebrated example is the date each spring of the blooming of the Japanese emperor's cherry trees. Earlier bloom dates would seem to indicate favorable climatic conditions and later dates, poorer conditions. Thus, the trend toward later bloom dates (Figure 9) seems to fit the general pattern of recent Northern Hemisphere cooling. The cherry blossom records go back hundreds of years, but originally were for blooms in Kyoto rather than Tokyo, where the emperor formerly resided. It is rumored that some people have even found correlations between cycles in bloom dates and sunspots. But, the business of climatic reconstruction from nonclimatic indicators can be perilous. For example, perhaps the trend since 1950 toward delayed spring blooming of the Tokyo cherry blossoms is not really a reflection of weather trends, but rather of an increase in the quantity of biologically damaging air pollution—certainly a distinct possibility in Tokyo. As for the longer-term bloom record, kept over hundreds of smogless years, it turns out that it is not a record of the dates that the cherry trees bloomed, but rather a chronicle of the emperor's excursions to see them, a highly unlikely event to be correlated with sunspots! Of course, it could be argued that the emperor always went to see the blossoms at their peak or that air pollution is less important in this connection than climate. But I have brought up these problems merely to point out why I feel it is still desirable to exercise a considerable degree of caution in making use of quantitative statements about past climatic trends—let alone future ones.

The Actuarial Approach to Climate Prediction

On the other hand, many inferences about recent climatic trends can be drawn, albeit not always with a high degree of certainty. A record of Asakura's [61] showing a doubling in the number of extreme temperature minima since 1930 (Figure 10), Bryson's data indicating unusually favorable monsoon occurrences in northwest India from 1920 to 1965 (Figure 7), or Gilman's graph (Figure 8) showing a remarkable string of good growing seasons in the U.S. wheat belt from the mid-1950s through 1973 do point to a possibility that climatic conditions may be less benign in the immediate future than they have been in the recent past—at least in these areas. Obviously, more records of this nature must be assembled from every possible location and compared simultaneously to determine, with some statistical significance limits, just how climatically typical the recent past has been and whether or not there is convincing evidence that the weather will be highly variable in the immediate future. Dendrochronologists (tree-ring specialists) have just begun to analyze a vast set of tree rings to help establish a quantitative record of past climate variability. The analysis will probably take many years because they are still in the process of calibrating tree-ring characteristics against known climatic variations. As a result, we may in the meantime be forced to rely on statistically insufficient, but growing bits of evidence that suggest climatic variability is increasing or at least is likely to do so soon. And we may need to do this for no better reason than that the cost of ignoring such evidence—if it eventually proves to be correct—could be terribly high. In my judgment the evidence suggests there is at least an even chance that the probability of climatic variability will increase in the near future. Who is willing to gamble with the weather when the odds are at least even?

One more complicating factor should be mentioned in connection with the actuarial approach to climatic change taken so far (by "actuarial" is meant analysis of past records *per se* to provide insights into the probabilities of future climatic variations). This kind of analysis from past records is a viable approach, *provided* the physical and chemical factors that cause climatic variations are essentially the same now as they were when the historical fluctuations occurred. However, the constancy of these causal factors over the long term cannot be easily determined; in addition, human activities have been complicating the otherwise natural system. Therefore, the extent to which one can infer future climatic variability from analyses of past records alone is somewhat uncertain. Nevertheless, from studying the magnitude of human influences relative to natural forces, it seems unlikely that human influences on climate have (or will in the next decade or so) become sufficiently dominant to change the characteristics of the climate system enough to render historical climatic fluctuations irrelevant to the actuarial study of future climate variations. An actuarial approach, which is imperfect now and will perhaps become inapplicable by the next century, is still the best way to estimate the probability of potential climatic threats to our crops and livelihoods over the next decade or two.

Actions Before Certainty

Despite the agonizing uncertainties in estimates of future climatic trends, we should not ignore the lesson of climatic history that Reid Bryson so frequently relates: "What *has* happened *can* happen; climatic history can repeat itself."

Perhaps, you are reading these pages right after learning of a bumper harvest; you may be listening to an interview with an Agriculture Department official or reading an editorial that is castigating the prophets of doom, whose dire predictions of food shortages, price inflation, and famine have once again and finally been repudiated by the latest bumper crop and "surpluses" it has generated. To anticipate this kind of attack—which will certainly happen if the climatic dice give us a good year or two soon—I must caution that I am not predicting famine (or feast) with *certainty*. Rather, I am warning that the worst possibility is at least as likely to happen as the best possibility, and since the worst is accompanied by well-known horrors, it seems logical to insure against its possible impact. To that end we should not let a few good years—and I have little fear that all of the next ten years will be bad—dull our sense of preparedness for the bad ones. We should not forget the Genesis Strategy taught by Joseph in Egypt. If we are browbeaten by Pollyannas opposing a food reserve following the good news of one or two fine harvests, remember that in 1966 when the Green Revolution was widely introduced, many optimists proclaimed that famine had been banished from earth. Continuous "surpluses" did, in fact, mark the next half dozen years in North America. Yet, that unrealistic hope was soon shattered by the combination of bad weather in 1972 and 1974 and the increasing demand for food because of growing populations and affluence.

We will need several good years after 1975 simply to return to the safety levels of reserves we had a few years before (unless, of course, food consumption patterns change); the chances of several more bad years afflicting us before the reserves are built up cannot be glibly dismissed by attaining a bumper harvest or two in the interim. Food supply situations can turn around very rapidly, as they did between 1972 and 1974, and as the heavy 1974 rains in the U.S. corn belt gave way to a severe drought a month later. And, if the seemingly real twenty-two-year drought cycle in the Great Plains continues, drought should occur in the mid- to late 1970s, and there are strong indications that it has arrived.[62] Those who belittle our incomplete understanding of drought must not be allowed to prevent preparations against its effects. Nor should we let the soothing talk of special-interest groups or short-sighted optimists, bolstered momentarily by a good harvest or two, weaken our resolve. We must now begin urgent actions to correct the long-term conditions that render us vulnerable, sooner or later, to extended periods of unfavorable weather, which, as history has taught us, are inevitable.

Much research remains to be done even to determine accurately what the recent climate has really been and to assess quantitatively the chances that various climatic events might recur and affect food production. But, to my mind, the important question is, Can we afford to be unprepared for a recurrence of known types of climatic fluctuations (e.g., several years like 1972, the dust bowl, or even a two-hundred-year corn belt drought), recognizing the suffering any of these would cause if we were unprepared for them? To ignore such possibilities is tantamount to gambling with the lives of those most heavily dependent on the weather.

4

The North American Grain Drain

North America today controls a larger share of the world's exportable surplus of grains than the Middle East does of current world oil exports.

—LESTER R. BROWN WITH ERIK P. ECKHOLM
By Bread Alone (1974)

The conventional wisdom in agriculture since 1960 has been that we are experiencing an explosion in technology and that we must remove about sixty million acres of cropland from agricultural use and convert this land to other uses such as parks and recreational areas by 1980. The purpose of my presentation is to challenge the conventional wisdom of agricultural policies based on an explosion in technology.

—LOUIS M. THOMPSON
Associate Dean of Agriculture, Iowa State University, addressing a conference on agriculture in 1965.

The probability of getting fifteen consecutive years that good is about one in 10,000 [*referring to the unusually good weather for agriculture in the United States from 1957 to 1972*].

—JAMES MCQUIGG
University of Missouri and National Oceanic and Atmospheric Administration (NOAA) of the Department of Commerce (1974)

In the previous chapter I sketched out an historical perspective of climatic change and its often devastating effects on society. It was pointed out that the influence of climatic variability on food production is by far the climate's most important long-term impact on people (short of dramatic changes such as an ice age). It was also shown that the weather problems of 1972 and 1974, despite the mitigating effects of our powerful transportation and food-storage technologies, proved many regions still are vulnerable to weather variations, and in some cases are even more vulnerable because of technology (as in the case of the Sahel). Simultaneous reductions of only several percent in the harvest of the few major world granaries can threaten at least as many people today (who are dependent on food imports) as local, weather-induced shortages did before the Industrial Revolution. This chapter will examine the relationship between climate and principal crops in the world's major exporting granary: namely, cereal grains produced in the central part of North America.

North America: The Food Giant

Before examining the important influence of weather variability on North American agriculture, it is useful to understand the role that this granary plays in the world food system. We can begin by looking at the dominant role of North America in world grain trade. Grains, in their many varieties, account for as much as 70 percent of the food consumed throughout the world. Table 1 shows the changing pattern of world grain trade, as compiled from U.S. Department of Agriculture data by Lester Brown, an economist and food expert. Brown divides the world into a number of very large (primarily continental) regions, although individual countries in the several regions may experience import-export behavior different from their region's average. The striking fact is that since the 1930s, when most regions except Western Europe were net *exporters* (shown by plus signs on Table 1) of grain, all large regions except North America and Australia had become net *importers* (minus signs) by the early 1970s, including the regions containing the important Soviet and Indian wheat granaries and the Southeast Asian rice granaries.

From the table it can also be seen that, although Australia is an important exporter, North America has been the only really substantial

Table 1
The Changing Pattern of World Grain Trade

Region	1934–38	1948–52	1960	1970	1976*
			(million metric tons)		
North America	+5	+23	+39	+56	+94
Latin America	+9	+1	0	+4	−3
Western Europe	−24	−22	−25	−30	−17
Eastern Europe and U.S.S.R.	+5	—	0	+1	−25
Africa	+1	0	−2	−5	−10
Asia	+2	−6	−17	−34	−47
Australia and New Zealand	+3	+3	+6	+12	+8

* Preliminary, fiscal year.
Plus sign indicates net exports; minus sign, net imports.
Note. Based on U.S. Department of Agriculture data, compiled by Lester Brown.[1]

grain exporting region in the 1970s. Total world grain production in 1976 was about 1,200 million metric tons, and North America exported 94 million metric tons. Thus, most grains produced in the world are consumed within the regions of production, and less than 10 percent is traded. More than 90 percent of this interregional trade comes from the "surplus" of crops in North America. Small wonder that, as Brown pointed out, Canada and particularly the United States "today control a larger share of the world's exportable surplus of grains than the Middle East does of current world oil exports." [2] Although much of these grain exports have recently been sold for foreign currencies to Eastern Europe, the U.S.S.R. and Japan—regions that desire and are able to afford importing food for higher standards of eating—an important proportion of exported American grain is shipped to countries where hundreds of millions live at marginal or subsistence levels. It becomes distressingly evident that any serious cut in the level of imports from grain-rich areas or a drop in food production in North America can mean increased levels of malnutrition and even famine for many people throughout the world.

World Food Security

To assess the vulnerability of the world to fluctuations in food production, Brown compiled what he called an index of world food

Table 2
Index of World Food Security, 1961–76

Year	Reserve stocks of grain	Grain equivalent of idled U.S. cropland	Total reserves	Reserves as days of annual grain consumption
		(million metric tons)		
1961	163	68	231	105
1962	176	81	257	105
1963	149	70	219	95
1964	153	70	223	87
1965	147	71	218	91
1966	151	78	229	84
1967	115	51	166	59
1968	144	61	205	71
1969	159	73	232	85
1970	188	71	259	89
1971	168	41	209	71
1972	130	78	208	69
1973	148	24	172	55
1974	108	0	108	33
1975	111	0	111	35
1976*	100	0	100	31

* Preliminary.
Note. Prepared on the basis of U.S. Department of Agriculture data, compiled by Lester Brown.[3]

security for the years 1961–1974, shown here in Table 2. The entries (in millions of metric tons) represent the actual reserve stocks of grain, the grain equivalent of idled U.S. cropland, the sum of these (total reserves), and the number of days the reserves could feed the world population at the annual consumption rate (index of world food security).

The "landbank" program that idled U.S. cropland in the 1960s and early 1970s was originally implemented to keep grain production from becoming too high and creating "surpluses" that would depress domestic grain prices and reduce farmers' profit incentives. Despite a world situation in which hundreds of millions went to bed hungry each night,[4] Brown explained in 1974 that "for the past dozen years or so, the U.S. government has paid its farmers to keep roughly 50 million acres idle. Although this source of supply cannot be tapped so quickly as grain reserves, most of the acreage can be brought back into production within twelve to eighteen months once the decision to do so has been made."[5] After the global weather troubles in 1972, grain reserve stocks did indeed fall precipitously, and the idled U.S. cropland was brought back into production. Fortunately, no payments were made to prevent food production in 1974, a year punctuated by the specter of famine, food shortages, and food price inflation.

It is sobering to consider how valuable in both human and monetary terms the grain that farmers were *paid not to grow* from 1970 to 1973 would have been if we had had it in 1974. For example, the cropland that was idle from 1970 through 1973 could have produced an incredible total of 214 million metric tons of grain! Table 1 shows that the entire North American grain exports in 1976, worth tens of billions of dollars, were only 94 million metric tons (only 40 percent of what could have been available if cropland hadn't been deliberately idled after 1970). In human terms the figures are perhaps more startling. Delegates at the World Food Conference in Rome in November, 1974, determined that about 10 million metric tons of grain were needed urgently to alleviate famine in the wake of the disastrous weather-induced shortfalls in 1972 and 1974. Yet, Secretary of Agriculture Butz, head of the U.S. delegation, bitterly disappointed the conference when he said the United States could provide only a small part of the needed grain because the poor weather in 1974 had impaired the United States' "ability to deliver"[6] that year. (The Food Conference is examined in detail in a later chapter. To put these figures in perspective, the conference called for 10 million tons of grain to check famine conditions, which was less than 5 percent of the amount that could have been grown on the purposely idled croplands from 1970 through 1973.)

Returning to Table 2, we see that the number of days of grain reserves fluctuated considerably over a fifteen-year period, but the trend has clearly been downward since the big U.S. "surplus"[7] in the early 1960s. In the year 1964, for example, note that the total grain reserves were 223 million metric tons, equivalent to some eighty-seven days of food security at that time. In 1970, a year with eighty-nine days of food security, that level of security required 259 million metric tons of total grain reserves. Thus, for almost the same number of days of world food security, it took about 30 million metric tons (about one-third of the 1976 North American grain exports) more grain in 1970 than in 1964; that is, our reserves must rise several million tons (India's grain shortfall in 1974, for example) each year just to keep the level of food security stable. The reasons for this ever larger basic food requirement are well known: rising population, particularly in underdeveloped countries, and rising affluence, particularly in newly rich countries.[8]

However, in 1972 there was a pattern of worldwide climatic disruptions, and food reserves plummeted. World food production dropped about 1 percent, the first such drop since the widespread adoption of Green Revolution methods in the mid-1960s, while population increased by 80 million. Despite a record world harvest in 1973, food security continued to drop and had plummeted to about a month's supply in 1974. Thus, for the second time in only three years and despite the elimination of the landbank program and the restoration of millions of acres to production, world food output fell disturbingly, signaling trouble ahead.

Thus, the world is entering a final quarter of the twentieth century suffering from dangerously depleted food reserves, low levels of fertilizer stock, high energy prices, and 80 million new mouths to feed yearly. And if the historical pattern of twenty-two-year drought cycles continues in the American Great Plains, some part of that region will be visited again by drought. Thus, since the vast bulk of the world's exportable grains comes from the surpluses of that region, the prospect of a serious world food crisis is not at all remote.

Food supply and demand projections differ on whether the world will produce enough food each year to prevent severe famine over the next decade. (The background of these uncertain estimates will be examined in a later chapter.) But there are a few conclusions that seem to be widely agreed upon and often bemoaned by those who study supply and demand questions: (1) The few major granaries will continue over the next decade at least to distribute their surpluses to food-deficient regions, which will remain unable to achieve self-sufficiency in food; (2) food production simply must be increased at a faster rate than population growth—unless the rich countries are willing to reduce their accustomed high standards of eating, and even this unlikely prospect would merely postpone briefly the impending disaster.

If I read correctly the state of the world's political consciousness, which is motivated primarily by the generally narrow short-term self-interests of its many nation-states, there is regrettably little likelihood that population growth rates will diminish enough during the next decade or two; nor does it seem likely that affluent nations will consume less wastefully. I believe that populations in general—and perhaps particularly in the less developed nations—will probably not voluntarily curtail their growth, nor will those in grain-rich countries sacrifice their diets to prevent immediate famine in countries that are not self-sufficient in food production. And, since the surpluses in the North American granary are likely to continue to be a major source of emergency food for countries unable to feed themselves, it is imperative that the vulnerability of this granary to crop-yield variations be understood. In particular, I will try to show why James McQuigg of the University of Missouri, a noted American agricultural climatologist, warned at a recent conference that *"world grain reserves are now roughly equivalent to (or smaller than) the difference between production in a favorable weather year worldwide and an unfavorable year."* [9] Or, why at the same conference a respected Canadian meteorologist, Gordon McKay, cautioned that "famines appear inevitable" if we accept global population projections. Both of these men are familiar with the recent climatic patterns in their respective countries, and have constructed methods to estimate the effects of weather variability on crop yields in their regions.

Now let us take a closer look at the influence of weather variability on cereal grain yields in the North American granary.

What Makes High Yields

Consistently high crop yields depend on many factors: water, good seed stock, maintenance of genetic variability of crops, fertilizer, high productivity of the soils, pesticides, the pest-control services of natural ecosystems, good management skills of the farmer, capital to acquire and maintain technology, and stable climatic conditions.

Food contains energy, the energy of the sun that is (along with the atmospheric gas, carbon dioxide) captured and transformed to plant tissue during plant growth. Technology helps to increase production by augmenting the quantity of sunlight absorbed, and hence the weight of crop yield per unit area and sometimes the nutritional value per unit crop weight. In recent years, it has been possible by selection to obtain an array of hybrid strains that substantially improve plant growth in many local environments. Because of the large, unsatisfied needs for more and richer foodstuffs, the development of high-yield rice and wheat varieties especially suitable for growth in tropical regions, and of hybrid soybeans and high lysine (a building block of protein) grains have been notable achievements.[10]

The major technological boost to food production in North America, however, has been the use of chemical fertilizers, especially on fields planted with high-yielding cropstrains (see Figure 11). Wayne Decker, an agricultural climatologist at the University of Missouri, points out that "since 1900 the annual amount of fertilizer applied in the U.S. has increased by 50 times, while the number of acres in production has remained virtually constant (land harvested in crop production actually *declined* by 11 percent from 1930 to 1969).[11] Crop yields up to 1973 have also increased dramatically, as the case of corn yields in Missouri demonstrates (Figure 12). Decker indicates that the coincidence between increased yields and increased use of fertilizers is not accidental.

The actual annual yields of corn do, however, vary considerably from one year to the next. "The primary cause of this variation is weather," Decker suggests, "although the influence of pests (insects and disease), economic considerations and government control programs are confounded with the effects of weather."[12] The scatter of the individual annual yields away from the long-term trend line (the solid curve in Figure 12) indicates that before the mid-1950s the deviation of yields from the long-term trend was considerably larger than for the two decades after the mid-1950s. That is, after about 1956 the actual values of yield each particular year (represented by the dots or squares in Figure 12) are closer to the trend line (the solid line drawn through the middle of the dots). Thus, we can see that after the mid-1950s not only did yield per acre go up dramatically, but the variability of yield went down (represented by the diminishing scatter of the dots away from the trend line). It is tempting to conclude that improved agricultural technology has also accounted for the reduced variability in corn yields since 1956. In fact, the U.S. Department of Agriculture has often pointed to the reduced yield variability since the 1950s as an implicit justification for its policy of reducing the government's food reserve holdings nearly to zero. The USDA has argued that since technology has reduced the vulnerability of crop yields to weather variations large food reserves are now less necessary. However, their argument fails, as we shall see, to consider that weather played a significant role in the almost twenty-year period of high crop yields and low yield variability.

Predicting Crop Yields with Mathematical Models. Through photosynthesis (the conversion of atmospheric carbon dioxide and water to carbohydrates and oxygen in the presence of sunlight) solar energy is "bound" by the plants and later converted into plant tissues. Humans require some of these plant tissues in their diets. Geographic location and the nature of regional weather continue to control the amount of sunlight that reaches plants and to affect the efficiency with which they absorb and utilize solar energy. Local (so-called micrometeorological or canopy) weather conditions determine the rates at which carbon dioxide is taken up, water vapor is transpired, and solar energy is utilized. The lengths of the day and growing season, the average cloudiness, temperature, humidity, soil moisture and nutrients, and soil flora and fauna all affect the rate at which plants grow. Attempts to build mathematical models of the complex interactions among micrometeorological conditions, soil moisture, and plant growth have been made. These models, based on theories derived from knowledge of plant physiology, have been useful, Decker reports, "in predicting the effect of different genetic characteristics" on the production of plant tissues in individual cases and for carefully controlled conditions. "But, generally," he continues, those sophisticated models "have not proved successful in simulating the total seasonal growth." [13] However, extrapolating the experience gained with theoretical plant models to large crop-growing regions (e.g., the U.S. corn belt) has led to the use of less rigorous, so-called semi-empirical, statistical models. These models have been remarkably successful in predicting trends and variability in corn, wheat, and soybean yields in the North American granary. The only data required to use the models successfully are monthly rainfall and temperature information, and a factor to account for improvements in technology. In mathematical terms, empirical coefficients are determined by the statistical technique known as regression analysis, and the resulting simple algebraic equations that make up the model can then be used to estimate with surprising reliability the influence of weather fluctuations on crop yields.

Henry Wallace, the renowned American agriculturalist and vice president under Franklin Roosevelt, was one of the first investigators to work out relations between crop yield and weather as part of his thesis work published in 1920.[14] The most successful work in that field to follow Wallace has probably been that of Louis Thompson, dean of agriculture at Iowa State University. Thompson correlated weather data (monthly temperatures and precipitation) and corn, wheat, and soybean yields for major U.S. grain-growing states. He is able to explain (statistically speaking) over 80 percent of the variability in crop yields on a state-by-state basis, using only the monthly weather data for each state. For those readers with a statistical bent, Table 3 from Decker summarizes the weather data Thompson uses for each crop, and the fraction of the yield variations that can be accounted for by weather data alone. (A value of 1.0 on the table would mean the model predicted crop yields perfectly.) July rainfall is the most critical factor for grain production in the U.S. granary, as the table shows, although other months are also influential. The technique is actually conservative, since by using monthly averages of weather it does not include the major crop losses that would occur from severe events such as hail or frosts in the late spring or the early fall. Thus, Thompson's technique gives the general pattern of crop yield variations with weather variations, but actually *underestimates*

Table 3

Crop	Best prediction climatic variables	Fraction* of yield variability accounted for by weather variations alone in the Thompson model
Wheat	April, May, June, and July precipitation May, June, and July mean temperature	.81–.86
Corn	July precipitation July and August mean temperature	.86–.89
Soybeans	July and August precipitation June and August mean temperature	.90–.97

* 1.0 implies the model perfectly accounts for the yield variability of each grain crop.
Note. From Decker.[15]

yield variability by not including factors such as frosts, hail, and pests—which in some cases could be related to weather, although knowledge of these relations is still meager. Next, Thompson applied his model to major cereal grains grown in the United States to determine how influential weather fluctuations have been on recent crop yields, and to test the U.S. Department of Agriculture's contention that technology was primarily responsible for the very low yield variability in U.S. grains from 1955 through 1973. The results are described next.

For Consistently High Yields, The Weather Still Counts. A team of scientists headed by James McQuigg and Louis Thompson prepared a *Report to the Administrator of the National Oceanic and Atmospheric Administration of the Department of Commerce.*[16] The report, dated December 14, 1973, concluded that a "straightforward analysis of agricultural and climatological data" that "can hardly be questioned" indicates "very strongly that the production of grain in the United States has been favored by extremely good weather in recent years. Any national policy that does not take into consideration the fact that less favorable weather is far more likely than recent nearly optimum conditions, is likely to place us in most unfortunate circumstances." [17] This conclusion was reached from the use of Thompson's crop yield–weather model in the following manner.

The authors first calibrated their model by comparing yields of corn, wheat, and soybeans with weather data in the five major production states for each crop. They next assumed the influence of technology to be fixed (i.e., technically speaking, the technology parameter in the regression equation was constant) in the model at 1973 levels, and then made new calculations employing the historical weather records. The result of this model simulation for corn (see Figure 13) shows that the average yields under these assumptions are around one hundred bushels per acre over the entire eighty-year period, whereas a curve of the actual yields (see Figure 12, for example) shows clearly that yields in Missouri increased from some forty bushels per acre in the 1940s to nearly one hundred bushels per acre in the 1970s. But the model simulation (Figure 13) is not a simulation of *actual* yields each year since 1890, but rather a hypothetical case in which it is hypothesized that the benefits of 1973 technology were available each year since 1890. This explains why the high yield per acre of the simulation (seen in Figure 13) persists throughout the entire hypothetical eighty-year period. By deliberately holding the influence of technology fixed at 1973 levels, McQuigg, Thompson, and their co-workers were in fact isolating the influence of weather on crop yield variability. Since technological effects were fixed and pest or economic influences were absent, weather effects were isolated by this model simulation. Furthermore, because the kinds of weather variation experienced in the corn belt from 1890 to 1970 are as good a guess as any we can make now about the weather that will occur over the next eighty years, it is more than an academic question to ask the model to predict how past weather might have influenced crop yields if present technology had been available.

Several prominent features that show up in the simulation correspond roughly to the patterns that were also noted by Gilman (Figure 8). For example, the period of severe drought in the central United States during the 1930s, which was characterized by above-normal temperatures and below-normal precipitation, also shows up (in Figure 13) as several years of low corn yields. If there were a recurrence of these drought years, yield reductions of as much as thirty bushels per acre would occur, even with modern technology (since only the weather was allowed to vary, and thus only its year-to-year fluctuations influence the crop yields). The model also shows poor crop yields in the early 1950s, late 1910s, and at the turn of the century, times coincident with the controversial twenty-two year drought cycle.

As with Gilman's wheat region graph, the most remarkable feature on the simulation is the period of almost two decades, stretching from the mid-1950s through 1973 (characterized by a tight clump of points in Figure 13), indicating both high yields per acre and low variability in yields from year to year. Before the most optimistic technologists among us quickly cry that technology accounted for these favorable circumstances, it should be repeated that the variation in corn yields (Figure 13) assumes a *technology factor fixed at 1973 levels,* with weather as the only parameter allowed to vary from one year to the next. Thus, the inescapable conclusion reached by McQuigg and Thompson in their report is that "there is strong evidence that it is, indeed, 'good' weather, as well as recent technological gains that has produced these high yields" and that there is "evidence that technology has not influenced the susceptibility of crop yields to weather" [18]—at least as long as most U.S. grain continues to grow on unirrigated lands.

The McQuigg and Thompson report went on to warn again that "the message is strong that we have been unusually fortunate in recent years to have experienced such high grain yields. It is imperative that we not be lulled into a dangerous and unjustified expectation that such fortunate circumstances will continue." [19] The message is worth repeating.

Their analysis for soybeans and wheat showed similar results, although wheat yields have not been as unusually consistent as corn or soybeans in the past fifteen years, perhaps because much of the wheat grown in North America is "winter" wheat. That is, it is planted in the fall and grows until the cold weather makes it dormant, only to revive and grow to maturity in the spring. Thus, winter wheat depends on weather conditions in two seasons, fall and spring, and is less dependent on the vagaries of the weather in any one month (e.g., July) than corn or soybeans.

Facts Versus the USDA. Using the actuarial approach of calculating the chances for drought from past weather records, McQuigg and Thompson computed that "the probabilities of one or more drought years in the next three for soybeans is 7%; but for corn it is 26% and wheat it is 29%." [20] Interestingly, in 1974, the year following the publication of their report, the corn and soybean yields were actually reduced sharply by poor weather.

Yet, despite the strong message of these respected analysts, the U.S. Department of Agriculture responded to the Department of Commerce report with their familiar line: "The analyses fail totally to measure the gains in technology—extension of irrigation, increased use of fertilizer, improved varieties and strains, mechanization, cultural practices—which clearly decrease susceptibility of yields per acre to adverse weather." [21] Although USDA critics admitted that "a substantial shift to adverse weather has the potential to reduce yields per acre to disappointing levels," they still insisted that "favorable yields during the 1960s and early 1970s are largely a result of technology." [22]

Louis Thompson replied incredulously on February 13, 1974, to Secretary of Agriculture Earl Butz: "Whoever criticized the study does not understand the method of analysis. I do not attribute the big increase in yield since 1960 to weather." [23] Thompson explained that his main point was that the *consistency* in the high levels of yield per acre after 1955 was largely a result of an unusually long string of good (unvariable) weather years; he was not suggesting that technology was irrelevant to the dramatic increase in yields since the 1950s, nor that his model accounted for it perfectly. Rather, the yield records of this post-1956 period should not automatically lead to the conclusion that technology had also reduced the year-to-year variability of crop yields, because this era was also one of very consistently good growing weather; and his analysis suggested that the weather was the dominant influencing factor on the consistency of the yield variability. Thus, he was recommending that agricultural policy not be based on the premise that technology leads to consistently high yields, a dangerous premise that implies large food reserves are no longer needed to hedge against weather variability.

Deputy Administrator Bruce Graham of the USDA replied to Thompson on March 1, 1974, that the USDA reviewers "not only studied and understood the NOAA report, but had additionally consulted your published articles to which the report referred and are satisfied with their remarks already forwarded to you." [24]

In still another response, Thompson pointed out that in some cases technology could make things even worse (see Figure 38). "Fertilizers help a corn plant to extend its root system to tap subsoil moisture in dry weather, while unfertilized corn suffers in comparison. However, when the subsoil moisture is depleted . . . fertilized corn yields less than the unfertilized corn." [25] Thompson recalled a visit to the Sanborn plots of Columbia, Missouri, on August 2, 1954, "when the only corn plots still green were the unfertilized control plots." [26] Thompson ended the series of correspondence by repeating his earlier warnings.

Who was right? Thompson or his critics? The weather and crop yield statistics of 1974, for example, which showed corn yield some 30 percent [27] below USDA predictions speak for themselves. However, I must raise some questions here about the significance of the numerical results of the McQuigg–Thompson simulation. Although I have little doubt about the validity of the McQuigg–Thompson conclusion that weather variability can, despite modern technology, still ruin agricultural yield expectations, it should be pointed out that the Thompson crop model is an admittedly simple empirical approach to crop yield forecasting, and it cannot be expected to yield more than approximate results. Furthermore, although the model does account for technology *trends*, it does not, in technical parlance, appear to include a technology factor in the weather terms of the empirically fitted regression coefficient. Thus, it may then miss some of the mitigating (or enhancing) effects on yields that technology change might produce when the weather varies. However, this inadequacy is probably not too serious since the model is, nonetheless, still able to explain more than 80 percent of the crop yield variability by the input of weather variability data alone. Thus, more research is, of course, urgently needed to test and improve the performance of crop models, but I strongly believe it was not premature for McQuigg and Thompson to use this qualitative tool to remind us once again of the proven wisdom of the Genesis Strategy.

Not long after the unpublicized debates between the USDA and Thompson and the disruptive climatic events of 1974, a USDA publication *The World Food Situation and Prospects to 1985* appeared. It clearly recognized the climatic vicissitudes of 1972 and 1974 and admitted that "policies and programs for expanding food production should recognize the possibility that weather conditions could be either less favorable or more favorable than normal. This underscores the need for flexible world food policies to adapt to changes in conditions and to provide a margin of security against sudden or unexpected changes." [28] It seems especially unfortunate to me that a crisis is needed before reestablishment of a margin of safety becomes fashionable enough to be recognized, even if only by a sentence or two buried in the middle of an obscure technical publication. Yet, Secretary Butz still argues that the U.S. government should not be in the business of holding food stocks. Who, then, will perform this essential service? Perhaps there is a way to solve this dilemma even within the Secretary's philosophical viewpoint of maintaining profit incentives. The U.S. government could, for example, consider tax or tariff concessions to those grain traders who held certain minimum quantities of reserve stocks, or who withheld sales until a certain level of national stocks were achieved. No doubt, others can and will propose and work out many such suggestions. But first, we need some encouragement from Washington.

Canadian Grain Production

But what of Canadian grain yields? Are they influenced by weather in the same ways as crops in the central United States? The answer is, Yes, but. . . . As Thompson points out, "Wheat grown anywhere in the United States is adversely affected by higher than normal temperature from flowering time until the crop is mature." [29] Canada, however, faces the additional problem of a more northerly location, and thus severe climatic fluctuations (such as late spring frosts) that do not show up clearly in monthly average weather data (input to simple empirical models) are particularly important there. Thompson explains that spring wheat grown in Canada "is also favorably affected by lower than normal temperature, but the problem in Canada is that lower temperatures are related to a shorter growing season. The fluctuations in yield of wheat from year to year in Canada are greater than in the United States and are highly related to weather." [30] The severe spring weather in 1974 reduced the Canadian wheat crop by more than 10 percent and contributed to the rise in wheat prices through mid-1974.

Canadian meteorologist Gordon McKay reported at a conference in Sterling Forest, New York, in December, 1974, that Canada shared in the hemispheric warming trend early in this century (see Figure 2a), but that a cooling phase followed, starting in about 1940 in the west, and in the 1950s in the east. "The resulting shorter growing season has adversely affected agriculture, and the heavier sea ice has hampered fishing and coastal transportation," McKay noted.[31] In Canada (as in the United States) "the present area of economic agriculture has shrunken from that defined by the normal period statistics for 1921–50 that are commonly used in planning. The extensive opening of new lands is unlikely" in Canada he suggests, "since these lie generally to the north where the risks are greater. At the same time urbanization is diminishing the supply of prime land in 'safe' areas." [32]

In addition to weather factors, McKay contends that for Canada "technology has not had a major impact on grain production in the past and there is little reason to expect major advances now in light of the energy and fertilizer supply restraints." [33] In regard to both these restraints and to weather factors, McKay concluded—with a dramatic understatement, I suspect—that "whether or not these conditions will persist, and the nature of any change are questions of great importance." [34] In the uncertain business of crop yields and weather relations, the seemingly obvious importance of the entire subject is often not easily accepted by some in the agriculture establishments. Thus, the need for a dramatic understatement, absurd as it seems, is very real indeed.

Another Canadian-based meteorologist with broad interest in the effects of climatic change on food production is Derek Winstanley of Environment Canada. Although Winstanley has specialized in the analysis of climatic data pertaining to the Sahelian drought, data to be discussed in more detail later, he stresses the impact of climatic variability on the high-latitude, arid granaries of Canada and the Soviet Union. "When only 1.1 percent of Soviet agricultural land lies in areas with an annual rainfall as much as 28 inches, compared with 60 percent in the U.S.A.," Winstanley argues, "it is to be expected that Soviet grain production will continue to be highly vulnerable to rainfall fluctuations." [35] A recent example of this vulnerability was the severe drought in 1972 that led to the Soviets' purchase of 20 million metric tons of grain from the United States and Canada,[36] and marked the beginning of the recent period of rising grain prices and shrinking grain reserves. And, again in 1975, the media began to tell of further troubles with Soviet wheat production and their negotiations for another major purchase of U.S. wheat, this time some 10 million metric tons. Thus, those who belittled the seriousness of the food crisis in 1975, and complacently pointed to a bumper U.S. wheat harvest as a harbinger of long-awaited food price decline will have to wait another year for such price relief, now that the Soviets have bought up most of that 1975 cushion. And in 1976 or 1986 will it be India, the Soviets again, or China who buy off our surpluses; or will bad weather produce no such surplus at all? Then, what will happen to food prices? Perhaps the lesson learned by Russian Premier Khrushchev is instructive to all of us who expect increased agriculture to materialize quickly. The premier built Soviet expectation for vastly increased grain production on a big push to open new lands for production in Siberia, lands in climatically marginal areas. When frequent crop failures followed,

Khrushchev absorbed much of the blame.[37] The rest of this story is now history.

USDA's Long-Term Outlook: Extrapolation from Good Years

The importance of stable weather to stable production levels in the North American breadbasket has been emphasized here because the vast majority of world grain exports come from U.S. and Canadian surpluses. As Wayne Decker points out, when a drought "occurs over a major portion of the grain producing area, the drought incident has a major impact on food production." But he adds, "Even when drought occurs in a region that is not part of the 'world granary' the impact may be felt worldwide because of the increased demand for the available export grains." [38]

What, then, is the outlook for world food production and food demand up to 1985? There is no obvious answer to this complicated question, though some important components will be discussed in chapter 8. However, since we are concerned here with the impact of climatic variability on food production, I will touch only briefly on this aspect of the recent U.S. Department of Agriculture publication entitled *The World Food Situation and Prospects to 1985.*

The USDA has been frequently optimistic about the food situation and outlook. That department's optimism was evidenced in the 1960s, for example, when Paddock and Paddock predicted correctly that world demand for food would catch up with the U.S. surplus supply by 1975,[39] but the USDA put the date off till 1984; and in 1974, when Thompson and McQuigg warned that crop yields were still very vulnerable to climatic variability despite improved technology, and the USDA minimized their warning. Thus, we might expect that the USDA outlook to 1985 would suggest that "there is sufficient world grain production capacity to allow for continued improvements in the developing world's per capita consumption levels." [40] This conclusion is based on four alternative future scenarios, each founded on different assumptions about population, economics, and food-production growth rates. The report says that at best the growth rate of food production will be considerably higher than that of population to 1985, and at worst only moderately higher. Without critical analyses of their technological or economic assumptions here, and several criticisms were made at the Sterling Forest Conference,[41] the optimistic prediction that *at worst* the world will muddle through the food–population problem to 1985 can be strongly challenged on climatological grounds. In referring to the USDA projections, Louis Thompson commented that their "optimism is not widely shared because their projections are based on coefficients developed from a period when weather was generally favorable, when there was a significant improvement in varieties of wheat and rice, and there was a significant uptrend in the use of nitrogen fertilizers on all cereal grains throughout the world." [42] Thompson's point is that projections based on extrapolations of trends during a period of very favorable weather (1961–73 in this case) or rapid technological growth can be dangerous if we take too much comfort in them. The absurdity of long-term extrapolations from unrepresentative short periods was recognized by Mark Twain who wrote, with typically biting sarcasm and wit:

> In the space of one hundred and seventy-six years the Lower Mississippi has shortened itself two hundred and forty-two miles. That is an average of a trifle over one mile and a third per year. Therefore, any calm person, who is not blind or idiotic, can see that in the old Oolitic Silurian Period, just a million years ago next November, the Lower Mississippi River was upward of one million three hundred thousand miles long . . . seven hundred and forty-two years from now the Lower Mississippi will be only a mile and three-quarters long. . . . There is something fascinating about science. One gets such wholesale returns of conjectures out of such trifling investment of fact. [43]

"As we look to the future of grain production," Thompson cautions, "we must recognize that weather is becoming a limiting factor as well as technology . . . If the weather should become as variable from now until the year 2000 as it was in the corn and soybean belts from 1890 to 1955, the average yield would be reduced about 3 percent." [44]

Weather Variability Belongs in Long-Range Planning

Given the earlier evidence that population growth and rising affluence require an increasing food reserve each year just to maintain a constant level of world food security, it seems clear that more serious consideration by the policy-makers of the interactions between food production and climatic variability is long overdue.

Thus, it is not so much the celebrated cooling trend since the mid-1940s that threatens the productivity of the American granary (although its continuation—if it still exists—could cause a gradual southward shift of the grain belt boundaries), but rather the unknown variability in climate that might accompany any future climatic trend.

Perhaps the immediate reality of the impact of weather variability on the food-producing system is well demonstrated in a routine newspaper story (by H. J. Maidenberg and appearing on page 53 of the February 13, 1975, *New York Times*) that reminds us that winter is a time of critical choices for farmers—choices such as how much seed to plant, what kind of seeds to plant, how much fertilizer or herbicide or insecticide to buy, and speculation as to how much money a bushel of harvested crop will bring at market time. Also, the Department of Agriculture officials have to decide how much encouragement they should offer farmers to invest money in a variety of crops and how much food should be stored, conserved, or sold abroad. However, despite "all this planning by farmers and Government officials, the weatherman may have the last word," Maidenberg concluded, for "a repetition of last year's uncommonly wet spring, dry summer and early frost would nullify the best laid plans of both farmer and bureaucrat."

Is it really too much to hope that the food security planners will soon account for this contingency, too, lest it becomes too late to put into practice the age-old Genesis Strategy adopted by Pharaoh after Joseph interpreted the monarch's nightmare? If only the President of the United States would have a dream . . .

5

Everything Is Connected to Everything Else: A Review of the Theory of Climate

> *The general aim of a theory of climate is to predict the global or large-scale climate of the atmosphere . . . Climate is determined by a balance among numerous interacting processes in the oceans and the atmosphere and at the land surface . . . If we are to assess the possibility and nature of man-made climatic change, we must understand how the physical processes produce the present climate and also how past changes of climate, clearly not man-made, have occurred.*
>
> —Report of the International Study of Man's Impact on Climate (1971)

"The Message Is the Medium"

The earth is round. But that's nothing new, you say. Despite the fundamental effect that Columbus's mission of discovery had on the evolution of history, the belief in the roundness of the earth still has a profound political message in the twentieth century. The message, simply put, is that the air that passes over each location on earth is a medium that eventually blows over every other location; everyone, like it or not, is sharing the very same air. Of course, many things may happen to a parcel of air in its travels: It may pick up and give off various amounts of energy, water, dust, salt, minerals, bacteria, pollen, spores, smoke, pollution, and even radioactive wastes, among many other chemical constituents. The concentration and distribution of these substances in the atmosphere depend on a variety of factors: the amounts injected; the time it takes them to react chemically with each other, with other substances in the air, or with sunlight; the speed with which they are removed from the air by falling to earth or being washed out by rain or snow; and where they go before they are removed, if they are removed. Thus, the atmosphere is a resource shared by all peoples and places on earth, and is endowed with the potential to transport itself and some of the materials it has incorporated anywhere. Not surprisingly, there are many political implications attached to the sharing of such finite, continuous media by the competitive interests we call nation–states, as we shall explore in later chapters. Here, however, I shall confine myself to a basic outline of our present understanding of the climatic system. Several popular theories of climate will be discussed along with a brief description of how researchers gather data on a system as vast as the earth's entire climatic system, which consists of the atmosphere, oceans, land areas, and snow and ice masses. Finally, I will offer a short explanation of how climatic theories evolve and how some "additional" research "efforts" (in science "additional effort" is often a euphemism for financial support) promise to improve substantially our understanding of how the system works. This material is written in generally nontechnical terms, but attempts to include many of the basic ideas—and disagreements—of the "art" of climatology. Thus it may appear unnecessarily technical for the tastes of some readers, but my constant purpose will be to prove that despite the present inability of climatic theory to account adequately for each of the cause-and-effect links of the climate system, *every place on earth is connected to some extent by the climatic system to every other place;* that is, a kick in one spot will cause a bulge elsewhere.

The message of this chapter having been stated, the reader may now follow his or her inclination and either venture briefly into the toddling science of climatology or skip to the next chapter without cutting the thread of the argument.

Climate and Weather Are Different Things

A simple definition of climate can be found in the dictionary: the composite or generalization of weather conditions of a region, such as temperature, pressure, humidity, precipitation, sunshine, cloudiness, and winds, throughout the year, averaged over a series of years. That is, the climate is fundamentally a grand statistical mean of large numbers of weather events with account taken of the average variability or standard deviation. For example, the thirty-year average or climatological mean temperature for, say, New York City in July is about 77° F. But that number by itself is not very descriptive. The average departure from the normal temperature during specific Julys over this thirty-year period is also important. A measure of the magnitude of the variability of the temperature from its long-term mean is called its standard deviation. In New York, where I used to live, the city is near the climatically moderating influence (or thermal inertia) of a large body of water, the Atlantic Ocean. Thus, the standard deviation of temperature is much less than where I live now, in Boulder, Colorado. Although the mean temperatures for Boulder and New York are similar, since both cities are at about the same latitude, the climates are very different. This difference is largely due to the cities' difference in proximity to an ocean; Boulder is in the center of the North American continent, far from the stabilizing effects of the ocean's thermal inertia; also, the mile-high altitude of Boulder contributes to large variability. Thus, the standard deviation of temperature is much larger in the middle of the continent than on the coasts. A New Yorker knows pretty much what to expect from the weather just by looking at the calendar. January is cold, the light clothes are stowed away; July is hot, the woolens are in mothballs. Although it could get as warm as 45 or 50°F sometimes in January or as cool as 55 or 60°F in July, by and large the weather is relatively consistent with past experience.

In Boulder, however, the weather is much more variable. My first experience there was a visit in mid-September, 1971. The first day it was 80°F in the afternoon. Twelve hours later, it was snowing. Eighteen inches of snow fell! Trees hadn't begun to turn colors yet, and the weight of the early snow on the leaves caused branches to break all over town. Many of the plants and trees cultivated lovingly by Boulder residents—plants that are not naturally adapted to the highly variable climate of Colorado—perished in the snow and cold. Tomatoes and other garden vegetables, many of the flowers, and a fair number of newly planted trees and older fruit trees were badly damaged.

On the other hand, when I returned to Boulder in January, 1972, the temperature was in the sixties in the afternoon (and in the twenties at night)!

An average taken over many years shows that Boulder's mean temperature patterns from month to month are similar to those at other locations at the same latitude. Even though tulips often bloom in March, Boulder's climate is relatively harsh for agriculture compared to that of New York, because the weather is so variable and the likelihood of frosts in late spring or early fall is much greater than in places closer to an ocean or at lower elevations.

Of course, variability in climate has its virtues for nonagriculturalists. One is rarely bored with the weather! In Colorado, although there are a number of extremely cold winter days, with temperatures much lower than in New York, there are also many warm days, when one can bicycle, play tennis, hike, or ski without a jacket. This comparison is not meant to point out the virtues of one clime—particularly since these depend on value judgments as to virtue—but rather to emphasize two important facts about climate in general: (1) Climate is the long-term average of highly variable, short-term weather events; and (2) the variability itself is an important characteristic of the climate in a particular location, especially where agriculture or water supply are concerned. Therefore, we can see that any change in the climate of a place involves a variation in long-term means and/or standard deviations of one or more of the important climatic variables of temperature, precipitation, winds, air pressure, humidity, cloudiness, haziness, etc.

The Sun: The Prime Mover

The climate system derives nearly all its energy from the sun. Only a negligibly small fraction of energy comes up from the earth's interior and other sources. These geothermal sources create, at most, short-term influences on only a small fraction of the earth. For example, tornadoes have been spawned by the intense heat densities of large volcanoes.[1] Every object—the sun, the moon, the polar icecaps, your own body—emits radiant energy in an amount proportional to its surface type and temperature. All radiant energy travels at the speed of light, about 186,000 miles per second in a vacuum. This radiative energy, depending on its wavelength, occurs over a spectrum that includes gamma rays or x-rays, which have extremely short wavelengths; visible and ultraviolet light, which have short wavelengths of about one five-hundred-thousandth of a centimeter; thermal or infrared-heat radiation, which covers a wavelength spectrum some ten to one hundred times greater than the wavelengths of light; microwaves or radar, which are radiation typically with wavelengths of centimeters; and finally, radio communications, which can have wavelengths greater than a few meters.

The wavelength at which the strongest radiation is emitted from an object decreases as the temperature of the object increases. And, the sun, which has a surface temperature of about 5,500° Kelvin (10,000°F), emits radiant energy over a spectrum of wavelengths from very short (x-rays through ultraviolet radiation) to short (visible radiation) to somewhat longer (near infrared radiation) to long (microwaves). Most of the energy emitted from the sun, however, is in the "visible" and near infrared wavelengths; but the ultraviolet is also a critical component since it contains several percent of the total solar energy and, more important, because it interacts strongly with some biological tissues. For example, ultraviolet radiation in large doses can damage plants and animals and will cause sunburn or even skin cancer in humans, particularly light-skinned people. The total radiant energy output from the sun is often called the solar constant, but this is a misnomer since instruments that have been put to the task are not yet sensitive enough to tell whether the energy output of the sun is really constant to better than about 1 percent. The power density of the sun's radiation reaching the earth's orbit is about 1,350 watts per square meter, and variations by as much as ten to twenty watts per square meter cannot yet be ruled out in the short run. Even larger long-term variations in the output of the sun are also possible, but now these can only be studied by examining theories of how the sun works.

Because the earth possesses an absolute temperature in relation to absolute zero—the lowest possible temperature an object can have, corresponding to the temperature when even molecular motions cease—it gives off radiant energy. But since the earth's absolute radiative temperature is only about 255°K, it emits much less energy per square meter than the hot surface of the sun. This terrestrial radiation occurs in an invisible part of the spectrum of wavelengths, and is called terrestrial infrared radiation.

Humans can see radiation at the wavelengths of peak solar emission, and this visible portion of the spectrum of solar radiation is visible to us precisely because our eyes have evolved biologically over the eons in the presence of solar radiation. Thus, at the present evolutionary stage, humans are very well adapted to see the middle part of the spectrum of solar radiation. Some nocturnal animals (e.g., vipers), on the other hand, have evolved under different radiation environments, and can see better at the longer wavelengths that predominate at night. Such animals have a distinct advantage hunting at night, since they can see infrared radiation better than daytime creatures, who often fall prey to nocturnal predators.

The Radiation Balance, the Albedo, and the Greenhouse Effect

Part of the sun's radiation that is intercepted by the earth is immediately reflected back to space by clouds, air molecules, dust, and the earth's surface and thus does not heat the earth. This reflectivity of the earth, also called the planetary albedo, returns about 30 percent of the sun's energy to space. The remaining 70 percent is absorbed by the earth-atmosphere system and warms the planet. As the planet warms up, it in turn emits terrestrial infrared radiant energy proportional to (the fourth power of) the earth's radiative temperature. Usually the amount of terrestrial infrared radiation emitted by the earth to space closely balances the amount of solar radiant energy absorbed by the earth; when that occurs, the earth is in planetary radiation balance, or radiative equilibrium. Should something disturb either the albedo (reflectivity) of the earth or the proportionality factor (emissivity) between the earth's temperature and the quantity of terrestrial infrared radiation emitted, the earth's radiative temperature changes appropriately to correct the imbalance and restore planetary radiation balance. For example, the albedo may be disturbed when smoke and dust are dumped into the air, and the emissivity can be disturbed by this (or by the burning of gas, coal, or oil), which in turn increases the amount of carbon dioxide in the air. Because disturbing the albedo or the emissivity of the earth is already within the power of humans, these factors are often referred to as climatic leverage points at which people can—advertently or inadvertently—modify the climate.

Returning to the planetary radiation balance, it is important to know that the earth's radiative equilibrium temperature (the point at which it emits radiation to space) is very low—about 255° Kelvin (−18° Celsius or 0° Fahrenheit), while the average surface temperature of the earth is much warmer, about 288° Kelvin (15° Celsius or 59° Fahrenheit).

It is well known that the difference between the warm surface temperature of the earth and the relatively cold planetary radiative temperature is due to the presence of the earth's atmosphere. The atmosphere contains such gases as carbon dioxide, water vapor, and ozone, and similarly, such particles as the dust, smoke, and water droplets that comprise clouds and hazes. These gases and particles absorb both solar and terrestrial infrared radiation and emit infrared radiation proportional to the temperature of the gases or particles. The gases are much more transparent to solar radiation than to infrared radiation, and thus about half the solar energy that hits the top of the atmosphere makes it down to the earth's surface, where it can be absorbed. The solar energy warms the surface, which, in turn, emits infrared radiation upward. Next, since the gases and particles are not as transparent to infrared radiation, most of these warm surface emissions are then absorbed by atmospheric constituents, only to be reradiated by them. However, the earth's atmosphere is heated primarily from below by the vertical convection of heat originating as absorbed solar energy at the surface. Thus, the average temperature of the atmosphere *decreases* with height, at a rate—called the lapse rate—of about 6.5° Kelvin per kilometer up to about twelve kilometers.

Perhaps this seeems contrary to what one might first think, since the higher one goes up in the atmosphere the closer one is to the sun. Thus, one might expect it to get warmer. But the sun is roughly 93 million miles away, and a few miles up or down are not going to change the amount of solar heat in any significant way. Thus, the ancient Greek legend of Daedalus and Icarus is technically out of line. Icarus's waxed-on wings could not have melted as he flew higher in the atmosphere, since it gets colder as one moves upward. However, the legend could be both scientifically accurate and preserve his demise as well; the story should be revised so that the wax would freeze and crack, and his wings would fall off just the same!

Above twelve kilometers or so, in a region called the stratosphere, the temperature increases again with height, mostly because of absorption of incoming solar ultraviolet radiation there by ozone gas. Ozone is concentrated most heavily in the high stratosphere, where it is produced by the action of ultraviolet on oxygen molecules in the presence of other constituents. Because the ozone layer prevents most of the solar ultraviolet radiation from reaching the earth's surface, where it can cause biological damage, the latter's integrity is of crucial import, and the recent struggles against the SST and chlorofluorocarbon aerosol propellants attest to this concern.

Earlier it was said that the radiation from the warm surface of the earth is largely intercepted by atmospheric particles and gases, and that these constituents re-emit infrared radiation—both up toward space and back down to earth—proportional to the temperature of the particles and gases. Furthermore, since the temperature of the lower atmosphere up to twelve kilometers (called the troposphere) decreases with height and since 90 percent of the gases and particles are contained in the troposphere, the terrestrial infrared radiation that eventually finds its way out of the atmosphere to space comes mainly from tropospheric particles and gases. Because these are colder on the average than the earth's surface, the earth's radiative equilibrium temperature ($-18°$C) is some $33°$ colder than the earth's surface temperature ($+15°$C).

The blanketing, retentive effect that the combination of the atmospheric gases and particles has on the terrestrial infrared radiation is, as I have said, much stronger than the atmosphere's effect on solar radiation. This is popularly known as the greenhouse effect, by analogy to the well-known phenomenon in which solar energy penetrates the glass of a greenhouse and becomes trapped inside the glass, ultimately making the greenhouse much warmer than the air outside. The term, greenhouse, is actually a misnomer in reference to the earth, since the effect of the glass in blocking the escape of terrestrial infrared radiation from the greenhouse is small, compared to its effect in inhibiting the warm air in the greenhouse from mixing with the cold air outside. Nevertheless, the phrase, greenhouse effect, is widely used to refer to the atmosphere's trapping of infrared radiation (heat) and is instructive as an analogy if not taken too literally.[2]

So far, the important point to remember about the factors that cause climatic change is that the average surface temperature of the earth is determined by two principal factors: the earth's albedo, and the quantity and nature of gases and particles—mainly the water droplets in clouds—in the atmosphere that can absorb, scatter, or emit infrared radiation.

The Weather Machine

Thus far, I have been talking about the world as an entire closed sphere. But this global average approach is not very useful if we consider the important climatic differences that occur among various parts of the earth. Therefore, I shall discuss how the spherical shape and rotation of the earth give rise to the climatic differences among the tropics or torrid zone, the temperate or mid-latitude zone, and the frigid or polar zone.

The equatorial, or tropical, regions of the earth receive the bulk of the incoming sunlight because they lie more directly perpendicular to the path of the rays (Figure 14). On the other hand, the mid-latitudes are heated less than the tropics because they receive a more glancing incidence of solar energy, while the poles of course receive the weakest dosage of solar radiation since the incoming rays are nearly parallel to the surface. Because of the relatively high albedo of the icy poles, most rays skip back out of the atmosphere like flat stones across a pond. Thus, the spherical shape of the earth makes the tropics torrid, the mid-latitudes temperate, and the polar zones frigid. Like the warm air that can be seen rising from a hot radiator or over a hot blacktop road in the summer, warm air rises in the tropics and moves poleward at high altitudes in the atmosphere. At the same time, a return flow of air from the colder latitudes moves equatorward under the poleward-moving streams.

All points on earth make one rotation around the earth's axis each day; the closer a point is to the equator, which is the longest latitudinal circle, the faster it has to travel to make a complete rotation in twenty-four hours. If this is difficult to visualize from the two-dimensional diagram in Figure 14, try spinning a globe and observe how much faster a point at the equator moves than one near the poles. The warm air that rises in the tropics carries with it the fast rotational speed of the equatorial regions into the mid-latitudes providing momentum and energy, which help to establish the belt of winds around the mid-latitudes known as the circumpolar vortex, or jet stream.

Since a parcel of equatorial air that rises and moves poleward was originally (before rising) moving eastward at the rapid speed of the equator, the parcel tends to deflect eastward (relative to the underlying surface) as it moves toward either pole. This deflection, called a Coriolis deflection, accounts for prevailing eastward winds in the mid-latitudes (eastward winds are called westerlies in meteorological jargon for the directions *from* which they blow).

On the other hand, a parcel of colder air starting out near the surface in the midlatitudes flows equatorward to replace the warmer rising equatorial air that flows poleward; but the colder parcels do not move fast enough to keep up with the eastward speed of the equator and thus they deflect (relative to the surface) toward the west, forming the renowned trade winds that blew the European explorers to the West Indies and perhaps the Egyptians to Mexico centuries earlier. Similar winds blow equatorward from the midlatitudes in the Southern Hemisphere and collide at the equator with trade winds from the Northern Hemisphere to create a region of storminess called the intertropical convergence zone (ITCZ); in the ITCZ the colliding air streams rise, creating clouds and rain before moving poleward again to create the midlatitude westerlies and the return-flow trade winds. Clouds are thus created by rising motions, because rising air experiences a decrease in atmospheric pressure with height. This decrease in pressure is accompanied by an expansion, which causes a cooling. But the rising, cooling parcel of air contains moisture it picked up near the surface, and eventually the cooling is sufficient to cause the water vapor in the rising parcel to condense into liquid droplets. The condensation process releases the latent heat of vaporization to the air that was needed in the first place to evaporate liquid water to form water vapor. When the condensation process accumulates a sufficient quantity of liquid water (in a cloud), it precipitates out as rain and snow.

The bulk of the rising warm air that flows away from the ITCZ later sinks in a region between the humid tropics and the temperate midlatitudes, a zone called the subtropics, the region in which most of the world's great deserts exist. Sinking air has the opposite effect on clouds to that of rising air. The sinking air evaporates the tops of clouds and inhibits rainfall; thus, the great deserts generally exist in the subtropics under perenially sinking air. During the summer, however, the rising air associated with the intertropical convergence zone moves up to the equatorward fringes of the deserts and creates a few months of often heavy rainfalls called monsoons. These rains bring life-giving water to India, Southeast Asia, central Africa, and central South America, for example. If the monsoons fail or do not move far enough poleward, the desert encroaches equatorward, as the Sahara did into the Sahelian zone of central Africa for several years before the return of normal rains in 1974.

The strength of the midlatitude westerly winds depends to a large extent on the volume of hot air that rises in the tropics and moves poleward, and that volume of air, in turn, depends on the contrast in the temperatures between the tropics and the poles. When this contrast—also called the equator-to-pole temperature gradient—is great, as in winter, the circulation of the atmosphere becomes vigorous and the midlatitude westerlies are fast (i.e., in the winter the poles cool down but the tropics remain more or less steady in temperature). In the winter, when there are strong westerly winds, it often takes much more time to fly from New York to San Francisco than vice versa. But the flying times are often quite similar in summer when the equator-to-pole temperature gradient is smaller and the westerlies are weaker.

Furthermore, the midlatitude westerly winds at high altitudes, called the jet stream, tend to meander and break down into eddies (Figure 14), which are connected with the highs and lows the weather person refers to on television weather broadcasts. The breakdown into eddies is much more likely when the winds are fast than slow. The eddies transport heat and moisture from the warm, wet tropics to the cold, dry poles and bring the winter storms that recharge the soil moisture in the midlatitudes. Some may not like the often cloudy skies and stormy weather in winter and spring that come with a fast-moving, frequently unstable jet stream, but the moisture brought in by such storms is needed to grow the crops that feed billions of people.

The most important concept that emerges from this discussion is that much of the world's weather patterns are caused by the differences in heating between the tropics and poles, and that a change in the climate in one region is likely to be related to a change in the conditions thousands of miles away. The climate is really a gigantic weather machine, driven by the sun and controlled by the radiation balance of the earth. Some of the political implications of the fact that every place on earth is connected by the atmosphere to every other place will be drawn in the next two chapters.

Earth Is the Water Planet

The description given so far is barely more than an oversimplification of the workings of the climate. For example, I've hardly mentioned the oceans or the ice- and snow-covered regions, which play crucial roles in shaping the climate.

The oceans are massive compared to the atmosphere. The entire weight of the atmosphere (which creates a surface pressure of some fifteen pounds per square inch) is equal to the weight of only the first ten meters (thirty feet) of the oceans. The mass of the oceans permits them to store vast quantities of heat compared to the atmosphere; and the oceans can release that heat to the atmosphere days or perhaps even hundreds of years later. Thus, climatic variations can have their origins in events that occurred decades, and perhaps even millenia, earlier.

Because the oceans cover 70 percent of the earth's surface and have relatively low albedos, they absorb most of the solar radiation that reaches the earth's surface. But wet surfaces that absorb heat convert only about 10 percent of the heat absorbed by the surface directly into warming that surface; the other 90 percent of the absorbed heat is used to evaporate water, a process that takes a lot of energy. Think how long you must run your stove to boil away a kettle of water! And when the water vapor that has been evaporated from the ocean surface later recondenses to form clouds, it releases in the atmosphere the latent heat that originally went into the evaporation process.

The bulk of the sun's heat absorbed at the earth's surface is given up to the atmosphere in tropical rainstorms, which are caused by rising, hot, moist air. In the process of evaporation, condensation, precipitation, and cloud formation, the composition of the atmosphere is substantially modified, at least locally. Recall, the composition of the atmosphere determines the earth's albedo and emissivity, which in turn determine the earth's temperature. But the equator-to-pole temperature gradient also influences the amount of hot air that rises, and that air, in turn, influences the chemical composition of the atmosphere. Thus, the climatic system is comprised of processes that feed back on themselves—particularly through the hydrological cycle, in which the processes of evaporation, condensation, cloud formation, and precipitation are driven by temperature differences between high and low latitudes; but the cycle itself determines those temperature differences by influencing the earth's albedo and emissivity.

The complexity of the climatic system, however, goes far deeper than that. The oceans can retain heat for weeks and longer and act as a "thermal flywheel" on the climate system. But the temperature patterns of the oceans are influenced by the ocean currents, which in turn are driven by the frictional force of the winds. We all have seen how the wind kicks up waves and whitecaps on water. These waves influence the ocean currents, which in turn alter the sea-surface temperature patterns, which influence the atmospheric motions, which to close the circle again, create the waves. Again, a feedback mechanism.

No Polar Bears in England, Thanks to the Gulf Stream

The importance of ocean currents to the world's regional climates is easily seen on any globe. Notice that England and Scandinavia, for example, are very far north of the equator, at latitudes between 50 and 65°. Even "sunny" Spain is at 40°N. Now look at North America. London is at the same latitude as Hudson Bay in Canada. And while London's winters can be quite chilly (often near the freezing point of water), Hudson Bay is frigid in winter; polar bears can be seen on the ice in the bay! Sunny Spain is at the latitude of Iowa, and who vacations for a suntan in Iowa during the winter?

Northern Europe is much warmer than many other places at the same northern latitude to a large extent because of the warm Atlantic Ocean current known as the Gulf Stream. Relative to the present the northern Atlantic Ocean was quite cold during the Ice Age seventeen thousand years ago (Figure 3). At that time the Gulf Stream curved far to the south, and northern Europe was buried under a glacier.

If one considers that ocean currents cause regional climates (i.e., over areas of a few hundred thousand square kilometers) at the same latitude to differ greatly from each other, it becomes clear that important differences in climate occur not only from one latitude zone to another (north-south), but from one longitude to another (east-west) at the same latitude.

The Ice-Feedback Mechanism

Another important feedback mechanism identified by climate theory is known formally as the ice and snow-temperature-albedo feedback, nicknamed ice feedback. Simply stated, the ice feedback theory says that when it gets cold, snow falls or ice forms on bodies of water and since ice and snow are very bright relative to water or snowless land, a decrease in temperature usually leads to an increase in the brightness (i.e., albedo) of the earth's surface. This means that more sunlight is reflected away from the surface, and less solar heating occurs. In turn, the surface is cooled further, and that leads to more ice and snow and thus a higher albedo.

This self-amplifying or positive feedback mechanism has led many scientists to fear that the polar regions are much more climatically sensitive to changes in energy inputs than are other regions. My colleague at the National Center for Atmospheric Research (NCAR), William Kellogg, has extensively studied the role of the poles in shaping the global climate. He claims that climatic changes at the poles are, in fact, relatively larger than climatic changes at lower latitudes. For example, it is clear that the Northern Hemisphere has cooled several tenths of a degree Celsius since 1945 (see Figure 2a). And, Herman Flohn, a widely respected German climatologist, has pointed out that since 1960 most of the surface cooling trend in the Northern Hemisphere has occurred at very high latitudes, greater than 50°N, and that the magnitude of the polar cooling has been several times greater than the hemispheric average cooling.

Kellogg believes that the poles are important to global climate concerns even though they cover a relatively small fraction of the earth's surface, and attributes their importance to two basic factors: (1) The poles have high albedo because of their snow and ice cover, and thus they play a relatively important role in the earth's radiation balance; and (2) the huge continental glaciers on Greenland near the

North Pole and the Antarctic continent at the South Pole have enough water locked up in the ice phase to raise the world's sea level tens of meters—if they were partially melted by a polar warming or were to slip into the sea (in as short a time as a century) in a glacial surge.[3] Although little is known of the causes or likelihood of major glacial surges, if one were to occur today the effects would be swift and stunning. Obviously, these phenomena deserve more study.

However, the high albedo of snow cover on the continents or on floating sea ice depends not only on sufficiently cool temperatures to have freezing, but also on the availability of sufficient moisture to create snowfall. The latter condition is part of the hydrological cycle that is connected with the atmospheric motions, which are in turn driven by horizontal heating gradients that themselves depend on the albedo of the earth's surface. Therefore, if a disturbance to the climatic system creates a change, that change can feed back to modify the effect of the original disturbance. Some of these feed-back mechanisms tend to stabilize the climatic system, and some tend to destabilize it.

The point is that the atmospheric motions that create the winds, clouds, rain, temperature patterns, ice masses, or the particular climate of a given place on earth are driven by the heating contrasts between the tropics and poles, the land and sea, and even between lands of different surface characteristics (e.g., forested land and cleared land).

Causes of Climatic Change

What, then, causes climatic changes? The most obvious candidates are called external causes, since they are not themselves influenced by the climatic state (e.g., the solar energy output of the sun; dust from volcanic eruptions; or the carbon dioxide, dust, or heat put into the atmosphere by human activities). Also, the varying periods of time that oceans or glaciers can store energy, only to release it days to centuries later, suggest that variability in the climate can have internal causes; that is, the climatic system can distribute and store energy, moisture, and motion among the atmosphere, oceans, or glaciers in a way that creates self-fluctuations over spans of months to millenia, even if external factors such as solar input remain unchanged.

A brief summary of a few possible causes of climatic change follows, beginning with the most external factors.[4]

Natural Factors

Since the sun is our major energy source, an obvious theory of climate change concerns fluctuations in the output of the sun. Fluctuations at very low and very high wavelengths have been determined, but these wavelength regions account for only a small part of the total radiant energy output of the sun. Unfortunately, we do not know to better than a few-percent absolute accuracy what the total output of energy from the sun is and how it varies in time. Therefore, it is very difficult to determine whether previously observed planetary warmings and coolings are caused directly by changes in solar output. Accurate observations of the solar output are finally being contemplated; it is a difficult measurement, to be sure, but it can be done if given impetus commensurate with its importance.

Earlier, a possible relationship was mentioned between sunspots and drought in the U.S. plains, the so-called twenty-two year drought cycle (graphically shown in Figure 15). The sunspot number cycle indicates the average number of spots that appear across the face of the sun in a year. The lower part of the cycle (that is, numbers below the zero line) does not represent negative sunspots, but is an indication of the second eleven-year part of the twenty-two-year cycle, in which the magnetic polarity of the sunspots is reversed from the first eleven-year cycle. For some reason, the sunspot-drought relation (Figure 15), indicates that droughts often occur near the minima of the double cycle (that is, the point when there are the fewest number of sunspots). That occurs usually every twenty-two years, rather than every eleven years. Unfortunately, no satisfactory explanation of the physical processes that might lead to such a relationship is available, assuming the apparent correlation is more than a mere statistical quirk.[5]

One hypothesis, that the total energy output of the sun is modulated by the sunspot cycle, has been advanced a number of times. Clifford Mass, a graduate student at the Department of Atmospheric Science at the University of Washington in Seattle, and I recently made a computer calculation [6] with a climate model (models are discussed later on). We constructed this model to investigate the possible influence of sunspots on the earth's temperature. Although we were able to conclude that it is feasible that the postulated sunspot effect on total solar energy output could have caused many of the climatic fluctuations that occurred over the past few hundred years, we were continuously frustrated by our inability to verify conclusively or deny our calculations against observations. The reason for this frustration stems from the poor quality of current knowledge of the historical variability of both factors we were testing: variations in solar output and the long-term climate history of the earth. Quantitative records of both are poor indeed.

To measure to high precision (about 0.1 percent) the solar energy output, the so-called solar constant, one would ideally want to place a high accuracy instrument in space where it would be outside the contaminating influence of the atmosphere. Yet, no long-term high-precision measurement of the total energy output of the sun has been attempted from space, despite the fact that the earth's climate is most fundamentally driven by the input of solar energy. Moreover, the measurement will be fiendishly difficult and costly, but I think justifiable. By way of analogy, this situation in climatology is akin to a family or business trying to prepare a budget without precise knowledge of their economic income. A rough knowledge of income permits the budget-maker to have a general idea of expected spending patterns, but adjustments to individual budgetary items come only after the variations in income are accounted for precisely. In climatology, we still do not know even if these smaller, yet important, variations exist. In my opinion, this measurement of solar variability could turn out to be the most important single geophysical observation that could be made (from space or earth). The lamentable oversight that permitted the world to watch men skipping, planting flags, playing golf, saluting presidents, and race driving across the moon without quietly placing an instrument capable of shedding light on the debate as to whether variations in the sun cause variations in the climate seems one of the major scientific omissions of the 1960s! To be sure, it is not an easy measurement to make; but if a very tiny fraction of the 5 to 10 *billion* dollars spent on this television odyssey were dangled in front of prospective instrument builders, I suspect we would now be considerably closer to knowing if we should be worrying about whether the Great Plains will experience yet another devastating drought in the mid- to late 1970s, as the double sunspot cycle comes out of its possibly ominous minimum.

Another theory of climate change is related to the fact that the earth's orbit varies slightly with respect to the sun; for example, the relative location of the poles changes from summer to winter over periods of roughly 10,000 to 100,000 years. This theory has been used by some [7] to explain long-term ice ages but it certainly cannot account for climatic fluctuations such as the Little Ice Age and the shorter-term variations that wreak havoc on crops. Changes in the concentrations of atmospheric dust and carbon dioxide and in the complexion of the land surface are also thought to cause changes in climate. One clear cause of climate change is the moving continents, which have drifted around for millions of years; but that factor is not significant for climatic changes over periods less than a million years.

However, there is another natural external factor that merits a closer look.

Great volcanic eruptions spew vast amounts of debris into the atmosphere. The biggest are even powerful enough to inject material into the stratosphere, where it can remain for several years. This debris, known as volcanic dust veils, can block out as much as a few percent of the sunlight from the earth's surface for several years in a row. Many have speculated that eras of high volcanic activity should coincide with cool climatic periods and vice versa.[8] Until the big blast in 1963, when Mt. Agung erupted in Bali, there had been remarkable little volcanic activity since about 1915. It is well known (Figure 2a) that the temperature of the earth also warmed up since 1915, and a number of scientists have postulated that this warming was connected to the lack of volcanic dust veils. A computer calculation [9] by Clifford Mass and myself tended to support this conclusion, but our calculations also implied that the apparent cooling that set in after 1945 cannot be explained by volcanic activity. Thus, although the role of volcanoes in forcing climatic change may prove critical—in which case I should be trying to learn to predict volcanic eruptions and not spend most of my time building atmospheric models—as yet we cannot be sure the extent to which the potential climatic effects of volcanic blasts are detectable in climatic records.

Another major cause of climatic changes could be the interaction between the atmosphere and the ocean. The oceans are truly a critical component of climate. They have, as I described earlier, a vast capacity for energy storage and can release their energy on time scales of days to hundreds of years. This process may very well be responsible for short-term climatic fluctuations. Internal oscillations in the climate system—which consists of the atmosphere, oceans, land, and glaciers—have also been postulated; [10] although it is possible to perceive short-term climate changes as being forced by external causes, they probably represent redistributions of energy among glaciers, oceans, and atmosphere.

Human Impact on Climate

In addition to natural causes, the effects of human activity must be considered in relation to recent climate changes. A number of these are briefly summarized here, and treated in detail in the next chapter. Carbon dioxide affects the transfer of radiation in the air: The greater the amount of carbon dioxide in the air, the warmer the surface of the planet (the so-called greenhouse effect). Carbon dioxide (CO_2) has been controlled in the atmosphere by natural processes over geological time. However, it also is a by-product of the burning of fossil fuels, an activity that has already increased the atmospheric concentration of carbon dioxide from about 290 parts per million (ppm) before the Industrial Revolution to about 320 ppm today. The rate of change of carbon dioxide from natural processes is slow (except for a minor seasonal fluctuation due to the growth and decay of plants) relative to the human input. Thus, this is an important problem. Based on current growth curves, the CO_2 concentration could be as high as 400 ppm by the end of the twentieth century—less than twenty-five years hence! Although the present increase in carbon dioxide may be causing only a slight warming of the climate (on the order of a few tenths of a degree according to current estimates), the projected CO_2 increase would certainly exert a decided impact on the system—unless our model estimates of this effect are completely in error. According to our best, but still crude, estimate,[11] an increase in CO_2 to 400 ppm would bring about a temperature increase on the order of 1° Celsius, which is as large as any global change observed in the past thousand years.

The pollution of the air by dust particles created by human activity provides the basis of another popular theory of how humans may have contributed to the recent cooling trend. (The need of some people to blame even climatic changes on modern technology while other explanations are possible has been dubbed the "masochism of the over-affluent"[12] by one irate technological optimist. Later we shall see if his backlash is any more justified than that of the pessimists.)

By looking at astronomical records, climatologists have learned that air in the continental regions may be hazier now than it was fifty years ago;[13] atmospheric dust has long been a nuisance to astronomers and they have kept good records of it over time. Various kinds of dust interfere with the transfer of radiation reaching the earth's surface. Whether the dust warms or cools the climate depends on whether it is relatively more white or black than the medium it overlies; blacker dust tends to warm and whiter particles tend to cool the atmosphere. The whiteness or blackness of dust is determined by its chemical composition, and the dust now surrounding the earth is highly nonuniform.

Determining the climatic effect of dust generated by primitive agricultural practices, such as slash-and-burn agriculture, and by industrialization requires much more study. The current consensus of published literature is that pollution caused by human activity could result in a cooling of the climate by preventing sunlight from reaching the earth; some have already proposed this theory as an explanation of the earth's cooling since 1945. But consensus is a poor way to do science. I do not rule out the validity of the theory, but want only to point out that it is difficult to establish a global effect when we suspect that industrial pollution covers only about 20 percent of the earth's surface; and it is yet to be established that such pollution cools the earth. On the other hand, dust pollution associated with soil erosion from overgrazing may be more widespread than industrial aerosol pollution and perhaps could have contributed to the cooling since 1945.[14] But, in my view, there is not yet an adequately evolved understanding of the climate system, one that would allow us to assign cause and effect with certainty. However, if current trends of human activities persist through the year 2000, the magnitude of the climatic forcing function—the radiation effects owing to increasing carbon dioxide concentration, and perhaps increasing dust pollution, for example—would be sufficient to provide a clearer indication of the relative impact of humans on climatic change. But at what cost? A closer look at the serious implications of these trends is offered in the next chapter.

Another element of the human impact on climate to be considered is the heat generated to produce electricity, to turn our wheels, run our industries, and heat our homes. The total use of energy around the world equals about one ten-thousandth of the solar energy that reaches the earth's surface. However, unless a no-growth philosophy prevails or some intervening catastrophes prevent further growth, the world population may reach 10 to 20 billion before the end of the next century with each person consuming about twenty kilowatts—nearly twice the current per capita standard in the United States. Should that situation occur, the total energy released around the world would average about half of 1 percent of the surface solar input and, if our model predictions are correct, could produce a global temperature change of several tenths of 1° Celsius. (This effect would be added to the even more serious warming from increased carbon dioxide.) Moreover, since most of that energy input would be concentrated over land masses with unnatural energy densities many times the global value, such localized anomalies as regional heat islands, severe storms, and other deviations from the common rule would probably occur.[15] This process is already apparent in the cities, where it is considerably warmer than the surrounding countryside, particularly at night. This condition, which is called the urban heat island effect, stems from several factors.

Where the phenomenon is found, heat is consumed and generated directly in the city, and the large masses of stone tend to retain the heat that builds up in the afternoon. The paving-over of vegetated fields also reduces evaporative cooling of the surface, which contributes to the heat island effect. It is becoming clear that when such heat is concentrated in an area, the urban heat island effect can spread out and cause downwind anomalies in weather. For example, there is good statistical evidence that rainfall up to fifty miles downwind of industrial areas has been increased by as much as 15 percent as a result of the heat that is released in the city and the dust that accumulates in the clouds over the city. If industrial centers continue to grow rapidly, regional climates may change significantly over the long term, as explained in some detail in the next chapter.

Finally, the destruction of vegetation by grazing animals and the use of land for agriculture or urbanization affect the brightness of the earth's surface, the amount of solar energy that is absorbed, and the rate at which water evaporates from the surface. Some climatologists claim that goats of ancient times ate away the vegetative cover of the Fertile Crescent in the Middle East and helped to create the desert that exists today.

Space satellites provide many dramatic examples of overgrazing. One photo taken by the U.S. National Aeronautics and Space Administration's (NASA) Earth Resources Technology Satellite (ERTS) shows part of the modern-day Fertile Crescent, in this case some tens of miles around the Gaza Strip (Figure 16). The darker region to the right is the Israeli Negev Desert and the lighter region to the left is the Sinai Desert, which until the 1967 war was part of Egypt. The 1948 armistice boundary between Israel and Egypt is clearly seen—from space! But political boundaries are not physical and should not be visible from space. As Joseph Otterman, a satellite expert at Tel-Aviv University, has pointed out, since the 1948 armistice that fixed the boundary, the Bedouin Arabs' goats in the Sinai have defoliated enough land on the Egyptian side to make the political boundary visible from space.[16]

Another fascinating space photo (Figure 17) is a NASA Skylab shot of an extinct volcanic mountain in New Zealand taken in 1973. The camera was looking straight down on the top of cone-shaped Mount Egmont, and the snowcap can be seen as an irregular bright area at the center. The most remarkable feature, however, is that the mountain slopes, which take on darker tones in the photo, are surrounded by a light-toned area that makes an almost perfect circle around the bottom of the mountain. I was told by Robert MacQueen, a Skylab astronomer from NCAR, that the dark area is the forest of a New Zealand national park, and the circle is the border of range land on which a fence has been placed to keep the local sheep from straying inside the park boundaries and destroying the vegetation cover. In other words, the sheep have grazed up to the fence, which encircles the mountain, and have changed the complexion of the land so drastically that the border can be easily spotted from space.

The next chapter will discuss the relative seriousness of these various theories of climatic modification.

Some Theories on the Sahelian Drought Disaster

A final example of how overgrazing can change the brightness and perhaps the climate of an area is the case of the Sahel, where a 1973 ERTS satellite photo (Figure 18) of the drought-devastated land revealed the remarkable presence of a geometric, darkened shape (indicating that vegetation cover is heavier within the pentagonal shape than without). The existence of this area prompted an investigation,[17] which turned up a large ranch surrounded by a fence to keep out the hungry, foraging herds of the Sahelian nomads. Even in that severe drought year, there was enough rain so that some vegetation could grow in the Sahel—as long as heavy grazing pressures were absent.

Since the Sahelian drought and its devastation to a number of African nations was discussed earlier in some detail, a few comments here on possible theories for the drought are worth summarizing.

The oldest theory is that the recent drought was merely another episode in a region whose history has been punctuated with droughts. There were droughts in 1941 and 1942, in 1910–1914, and on back into time. But such a theory explains nothing of possible causes and merely suggests that whatever caused the previous droughts caused the recent one, too.

Another theory is that the drought in the Sahel—a region whose precipitation falls at the extreme northern edge of the summer African monsoon belt—is related to a shift of the monsoon belt toward the equator as a result of the apparent Northern Hemisphere cooling trend that set in after 1945. Reid Bryson, the leading proponent of this theory, also suggests that the dust kicked up from human activities (which he calls the "human volcano") have helped to cause this cooling, which in turn enhanced the vigor of the atmospheric circulation, thus displacing the jet stream equatorward. Bryson argues that the jet stream's movement also pushed the monsoon belt south toward the equator and thus away from the Sahara, causing the encroachment of the desert.[18] (At the same time, there is some supporting evidence that the northern fringe of the Sahara has been enjoying excess rainfall.) Such a theory is fraught with political implications, for if human activities have indirectly ravaged central Africa the victims may well demand compensation from the perpetrators.

Bryson has been repeatedly attacked by his fellow meteorologists for this theory of aerosol cooling leading to monsoon failures. Sometimes, the arguments over Bryson's public statements warning of possible monsoon failures have grown so bitter and personal that some enraged "responsible" scientists ostensibly suggested that Bryson's conduct has been unethical for a meteorologist. Perhaps a personal story may help to explain why Reid Bryson speaks out so vehemently and often on the question of monsoonal droughts.

In chapter 2, I recounted how in June, 1974, Bryson and I had briefed a White House policy group on the prospects for weather-related crop failures, including in our presentation a discussion of the situation in heavily populated monsoon-dependent regions of Africa and India. After the session we were stopped outside by a newspaper reporter who somehow had heard of the briefing and wanted to know from us what went on. We related in general terms what happened at the White House meeting, and also discussed in some detail the question of a food–climate crisis. A few cocktails later the reporter interrupted my conversation, apologized in advance to Bryson, and, as I recall, asked me a blunt question: "No offense to Dr. Bryson intended, but you, Dr. Schneider, seem to be agreeing with most of what Dr. Bryson has said. But I have heard that he is known in meteorological circles to hold an extreme opinion of the seriousness of a food crisis. Do you associate yourself with his position?"

I was, admittedly, a bit shocked by the candor of the question, and was also acutely aware of Bryson's presence and the reporter's pad and pen. All I could think to say was how I felt: "I very much share Reid's sense of urgency, but I can't always share his sense of certainty."

Bryson then looked me in the eye and asked, "Steve, have you ever had to make a forecast that someone's life depended on?"

I knew the question was rhetorical, even though I offered the obvious, "No."

"Well, I have." He told a story something like this: "Back in the war I had the responsibility to predict the landing weather at an airbase several hours later. Those forecasts often determined whether certain aircraft should take off on their missions, since bad weather at the airport upon their return could be dangerous. I did all right as a forecaster in those days, but one time I was wrong—dead wrong—and nine men died in a crash because of heavy rain and low clouds at the airport I earlier thought would be clear. From that day on I swore to myself if ever I could issue a forecast to save someone's life, I'd stick my neck out as far as possible to get the message across."

I know Bryson feels he has a good enough chance of being right in his predictions that monsoon failures will increase over the next decade or two, and that he has tried as hard as possible to get the message across, despite the abuse he has taken from his fellow scientists for stating his position so forcefully.

Although I am less convinced than he is about the likelihood that his *theories* are correct, I fully agree he could well be right about famine in India from increased monsoon failures, on actuarial reasoning alone—and the human consequences of that possibility might be counted in units of hundreds of millions of lives.

Therefore, to listen to some of the vicious remarks about Bryson's *motives* made by some members of the discipline makes one wonder just who in the profession is being most unethical.

A variant of Bryson's theory, which also relates the drought to large-scale changes in atmospheric circulation far away from the Sahel, is espoused by Canadian-based meteorologist Derek Winstanley, who agrees with Bryson that the jet stream (also called circumpolar vortex) has indeed moved equatorward recently; but Winstanley also maintains that the vigor of the circulation has *weakened* at the same time, not increased.[19] Such decreased vigor of atmospheric circulation, Winstanley contends, would be accompanied by a decrease in the volume of tropical air that rises and moves poleward; that is, that the monsoon, which is caused by rising tropical air, would be weakened. He presents evidence of a remarkable coincidence between the incidence of westerly-type weather over Britain—a measure of the strength of the upper level westerly winds at England's latitudes—and the rainfall patterns in the Sahel. This relation suggests that the recent decrease in the frequency of westerly-type weather in the British Isles correlates with Winstanley's record of a long downward trend in rainfall in the Sahel (Figure 19). However, since the Sahelian rains come only in the summer, and westerly weather in the British Isles may or may not occur in the summers, the correlation between these data may not be for the same season. Furthermore, as was pointed out earlier, just as one of Bryson's suggestions (that cooling in the polar regions may be connected to drought in northwest India) could be a mere coincidence, it is likewise not improbable that the high degree of correlation between the two separate trend curves for Sahel rainfall and westerly winds in Britain occurred merely by chance. Thus, the caveat that applies to the correlation in Figure 19 also applies to the correlation between Figures 2a and 7. This is not meant as a repudiation of Bryson's or Winstanley's interesting ideas, but rather as an attempt to point out that climatic cause-and-effect linkages are very difficult to demonstrate with full confidence, because there is still no adequate quantitative theory of climate to explain if any of the possible causes we suspect is responsible for a particular climatic event such as the Sahelian drought.

Bryson also raises a "local origin" theory in the case of the Rajasthan Desert in India and suggests that overgrazing there has led to desertification.[20] In semidesert regions, he says, overgrazing leads to erosion and a loss of the vegetative cover. This, in turn, allows the wind to pick up large quantities of soil no longer held down by vegetation and distribute them in the atmosphere. These windblown dust particles or aerosols block out sunlight, thus preventing some of the sun's heat from reaching the earth's surface. But surface heating is one of the driving forces behind rainstorms. With less surface heating there is less rain produced, and desertification can begin. Other scientists have applied Bryson's dust-desertification theory for India to explain recent desertification in the Sahel, and some have extended it to account for the potential role the windblown dust may play in directly altering the precipitation of the clouds themselves.

I turn now to two other theories of drought that have only recently been proposed.

Jule Charney of the Massachusetts Institute of Technology, a pioneer in the development of a theoretical basis for meteorology, espoused a quantitative version of the theory that local factors can contribute to the formation of deserts.[21] Charney has devised a local-origin theory based on a different mechanism from dust, but which is also related to overgrazing. Charney's idea was sparked by a satellite observation that showed the Sahara Desert to be a region of "negative radiation balance"; that is, more terrestrial infrared radiation apparently escapes to space over the Sahara than is absorbed in the form of sunlight. The main reasons for this condition are the relative absence of clouds, which block the escape to space of warm infrared radiation from the surface of the desert, and the high surface albedo of the desert, which reflects as much as 40 percent of the incoming solar energy back into space. Because of this negative radiation imbalance, energy must be imported into the desert region to fill the deficit and maintain the temperature of the surface. This extra energy comes from the atmosphere. The winds that descend into the Sahara region rose first in the tropics—e.g., the monsoons—and subsequently sink over the sands, supplying the missing heat. The air grows warmer as it descends, thus inhibiting cloud formation; and the desert rapidly becomes a stable geographic and climatic feature that is able to perpetuate itself.

Charney reasons that the high surface albedo of desert lands helps to preserve the aridity by first causing the radiation deficit, which then favors the sinking air that inhibits cloud formation; the sinking air also brings in the needed energy to fill the deficit created by the high surface albedo. Thus, since the Sahel is at the northern fringe of the rising summer monsoon air and is also just south of the sinking air over the Sahara, Charney speculates that the overgrazing and chewing away of the (dark) vegetation cover by the animals (as in Figure 18) causes enough additional reflection of solar energy to create an increase in the Sahel's surface albedo, and thereby a radiation deficit. The deficit in turn causes the warm monsoon air to sink farther south into the Sahel and thus create desertification at the region's northern fringe. Charney, not content with a qualitative theory, made several calculations (with a mathematical model of his own and one from NASA's Goddard Institute for Space Studies in New York City) to test his theory. The results (shown in Figure 20) indicate a large decrease in rainfall predicted for the Sahara region, using a model with the surface albedo increased from 14 percent to 35 percent.[22] Although a number of details of his analysis (beyond the scope of this chapter) are still being argued among theoretical meteorologists, Charney's work certainly demonstrates quantitatively the possibility that local causes may have contributed substantially to the Sahelian drought.

Still another local explanation of the drought has been proposed by Russell Schnell, a scientist concerned with the physics of ice formation in clouds. Schnell argues that the microphysical properties of clouds—that is, the individual droplets, ice crystals, and tiny floating particles that act as nuclei for a cloud's constituent ice crystals or water droplets—are important factors in influencing the amount of rain that falls from the clouds. In particular, the tiny ice crystals are believed to enhance the precipitation efficiency of a cloud as they drift and tumble through it, gathering liquid water until they are heavy enough to fall toward earth. Formation of an ice crystal requires the presence of a very tiny floating particle in the atmosphere called an ice nucleus; and the size and composition of the nucleus determines the temperature at which a crystal forms. These temperatures are usually well below freezing, and thus the presence of efficient ice nuclei—those that activate ice formation at temperatures close to freezing—can significantly increase the freezing rate of cooling water droplets. In the Sahel, Schnell says, most of the *efficient* ice nuclei are of biological origin.[23] That is, they are small bacterialike particles or decayed bits of organic matter found in the litter of decaying plants. He postulates (so far no atmospheric measurements have been made) that they are carried upward into the atmosphere by the winds and enhance precipitation efficiency. If most of the vegetation is eaten away by animals, little material remains to decay and be carried up to provide efficient ice nuclei. Thus, Schnell believes that when the efficient ice nuclei needed to form crystals in clouds become scarce, the amount of precipitation is proportionately reduced. With less rain, the already sparse vegetation becomes sparser, forcing the large herds of grazing animals to compete more fiercely for food and thereby decreasing the supply of biological ice nuclei even further. This is another example of a "positive" or self-amplifying feedback mechanism, but in this case the coupling encompasses biological processes as well as meteorological ones—a so-called biogeophysical feedback mechanism.

Both Charney and Schnell postulate that a local biogeophysical feedback mechanism may have caused the Sahelian drought. In short, overgrazing depletes the vegetation cover, which triggers mechanisms that bring on or enhance drought, which necessitates further overgrazing and sustains the drought. Only after the grazing animals die (or are removed in some other way) does the vegetation recover slightly, thereby increasing the number of efficient ice nuclei and decreasing the surface albedo, which then encourages the return of the rains. Thus, the cycle begins anew and the vegetation springs up once more (if the soil hasn't already blown away) until a new generation of animals again overgrazes the land—assuming no controls over the numbers of grazing animals are implemented to interrupt this vicious cycle.

Actually, it is a remarkable coincidence that the bulk of the Sahelian nomads' herds were decimated in 1972 and 1973, and that 1974, the first year of reduced grazing pressures, was a year of sufficient rainfall to create extensive vegetation cover again. Perhaps these theories of drought were then, at least, partially correct in predicting that the drought would end only after the vegetation was allowed to regenerate somewhat.

My own view is that too little is known to single out one of these theories as clearly superior to the others. Perhaps a combination of them ultimately may provide the most plausible explanation of the recent drought. For example, a reasonable synthesis might be that in years of plentiful rainfall, there is sufficient plant growth to support many animals and still leave an ample covering of vegetation. But, with the introduction of technology—medical science to cure diseases and civil engineering to provide the wells needed to sustain larger populations of both people and grazing herds—a dangerous stress is introduced into the fragile ecosystem of the Sahel, where drought has been a common feature.

In summary, when animal populations are high, causing substantial grazing pressures, changes of the type proposed by Bryson or Winstanley in the atmospheric circulation system (or whatever has caused the seemingly random occurrence of Sahelian droughts before) rapidly reduce rainfall. Consequently, the vegetation cover diminishes until there is not enough food for the animal population that grew when rain was adequate. The animals then quickly overgraze the land, causing a rise in the albedo, an increase in windblown dust, and perhaps a reduction in the number of efficient ice nuclei. The local mechanisms suggested by Bryson, Charney, or Schnell then could exacerbate and perpetuate the drought until either the animals are gone or the atmospheric circulation patterns elsewhere change sufficiently to bring back more rainfall—or, as I suspect, both occur.

One lesson is clear regardless of the validity of the various theories of drought. Because the Sahel is a region of frequent droughts, the animal population must be conscientiously held in check so that, especially in the worst years, overgrazing doesn't occur. Once the ground cover is gone, even for a single year, it is nearly impossible to avoid tragedies. The recent Sahelian experience must not be allowed to repeat when the droughts inevitably recur. Furthermore, the analogy of the Sahel case to other fragile ecosystems is instructive: *We must be careful not to overtax fragile ecosystems in an attempt to raise their carrying capacity, for we may be in danger of ultimately reducing that capacity even below present levels.*

Returning again to the general question of the possible climatic consequences of various land-use practices, there is no doubt that these can indeed cause local climatic changes. The chief question is whether the scale of the effects is ever sufficiently large to cause large regional changes. One might argue that altering the land surface is an activity that has been going on for many years and that the system has therefore already accounted for whatever changes in climate these activities may have induced. However, a large-scale alteration in our land-use practices is particularly likely to occur if food and population pressures continue; such a change cannot help but produce new weather phenomena on a local and perhaps regional scale at a much faster rate than in the past. If this causes a significant change in climatic conditions, much more serious attention to adjustments in agricultural practices may prove absolutely necessary. It should be noted that in the past, when climate shifts (for whatever reason) caused hardships, people often moved; but now large populations are, for the most part, severely constrained by political boundaries.

Mathematical Models of the Climate

How does one judge the relative importance of these various factors in climatic change? It is a difficult task to separate quantitatively the cause-and-effect linkages from what seems a plethora of factors, all of comparable magnitude, yet tending in opposite directions. As with any science, the process of developing a quantitative theory of climate begins first in careful observations, from which useful hypotheses are eventually generated. The next step is to design experiments to test the elements of these hypotheses, which in turn demand further observations to verify the experiments.[24] In the real atmosphere, only a limited number of small experiments can be carried out and, moreover, it would be nearly impossible to devise a laboratory experiment that could capture even the barest fundamentals of the complex interactions that produce the earth's climate. Fortunately, one can learn a good deal about some of the important individual processes by studying other planets, which after all obey the same laws of physics as the earth. But this is not enough, for without a twin earth the only way to simulate the behavior of the earth's atmosphere and oceans effectively is through the use of mathematical models. Since the climate system is too complex for even the largest computers, the model must break the system up into a finite number of places or points at which information is obtained. All events or processes that occur in the atmosphere or oceans on a scale larger than those points can then be treated explicitly in the model, although smaller occurrences cannot be added to the data bank or predicted directly. For example, knowledge of the wind, temperature, humidity, and pressure at Denver and at Salt Lake City does not provide sufficient data to predict exactly how the clouds change between those places, because the sizes of the clouds are on the order of a mile or so, though our data were taken on a scale of a few hundred miles. Yet, whenever the clouds, or whatever element or process we are trying to represent, occur on a scale smaller than the scale of collectible information, we must still make the statistical assumption that these small-scale (yet important) phenomena can be related to events occurring on the larger scale. This is the canon of faith on which climate modeling now operates. But climatologists are constantly struggling to determine the validity of these approximations by testing models of different complexity against each other and against the real world, and undoubtedly our perceptions and understandings of the weather machine will be greatly refined in the coming decades as these tests proceed.

There are many different kinds of models, ranging from simple one-dimensional forms that focus solely on vertical parts of the atmosphere to highly sophisticated representations of the entire atmosphere and the world ocean system. Any detailed analysis of long-term climate variations requires a coupling of the models of the atmosphere, the oceans, and even the cryosphere (the ice and snow parts) since their interactions clearly affect climate as thoroughly as do their internal processes. In addition, while many processes are included in such models, there is no certainty that they have been included correctly.

We can, for example, include the "positive" feedback mechanism of ice and snow-temperature-albedo feedback in a mathematical climate model. This can be done by assuming a relationship between the temperature of the earth's surface and the albedo of the earth on an assumption based on both theory and observations. But, because the theory is not completely developed and the observations are not extensive, there is a discomforting amount of uncertainty in the formulation. To the extent that we can define which feedback processes are represented correctly, these models can have some utility for climate studies in the near future. Unfortunately, *for the task of estimating the potential impact of human activities on climate, the models are just about the only tools we have.*

All sorts of processes have a place in climate modeling, and many of them feed back on each other; e.g., the sun's energy heats the earth's surface, which causes warm air that can create clouds. Clouds in turn block out some solar energy, thereby reducing the surface heating, and so on. The problem is that although we may know accurately the effect of individual processes on the climate to, say, 10 percent, many of them may operate in opposite directions; thus, when we attempt to average the collective effects of all of them, the overall impact on the climate is difficult to determine.[25]

Should we ignore the predictions of uncertain models? Despite the limitations implicit in present-day climate modeling, the predictions climatologists make with their models can still have important implications for policy-makers. To cite one example, climate modelers understand that increases in the amount of carbon dioxide in the atmosphere affect the transfer of infrared radiation required to warm the earth; in addition, they can compute with a fair degree of certainty the effect of an increase in carbon dioxide on the global radiation balance. Since, under certain assumptions, this increase can be related to a change in the earth's temperature, how then do we respond politically to a predicted temperature change of, perhaps, two degrees, a change that could easily result from the doubling of carbon dioxide in the atmosphere? Do we ignore this forecast as an example of the "masochism of the overaffluent"[26] in the belief that feedback processes improperly included or entirely omitted from our model may cancel it out? I think not—a political judgment, of course. It is just as possible that such feedbacks could amplify this prediction, and it is clear that: *Uncertainty is not biased toward optimism.* Unknown feedback processes can operate in either direction; thus, clouds might reduce the temperature change that results from a doubling of carbon dioxide to two-tenths of a degree or, alternatively, they might increase it to four degrees. My view is that once we know reasonably well how an *individual* climatic process works and how it is affected by human activities (e.g., CO_2-radiation effect), we are obliged to use our present models to determine whether the changes induced by these human activities could be large enough to be important to society. If they are, then the most rudimentary ethic requires that we decide as a society whether we are willing to risk expansion, or even continuation at all, of that activity, recognizing that its ultimate effects may be both global and irreversible. At the same time, we must also recognize that those effects might prove either totally insignificant or more disastrous than we originally imagined, if our models are poorly constructed.

Furthermore, there is more than theoretical evidence to show the validity of some of these theories. Carl Sagan, astronomer and author from Cornell University, often points out that the great Martian dust storm of 1971 significantly altered that planet's climate for months and, in addition, that the inhospitably high temperatures on Venus are existing proof of the carbon dioxide greenhouse phenomenon.[27]

The processes at work in determining climatic change are simply not yet fully understood. Despite the uncertainty in measurements and in theory, estimates must be given and difficult decisions may have to be made on the basis of the available knowledge, as I will discuss in the next two chapters. In any case, efforts to develop better estimates and models of potential effects will be absolutely necessary to help us reduce the uncertainties in decision-making to a tolerable minimum. Improvement of the quality of these estimates is the responsibility that atmospheric scientists and their funding agents owe to long-range planners, for the climatic effects of human activities are self-evidently the outer limits to growth. The real problem is: If we choose to wait for more certainty before actions are initiated, then can our models be improved in time to prevent an irreversible drift toward a future calamity? And how can we decide how much uncertainty is enough to prevent a policy action based on a climate model? This dilemma rests, metaphorically, in our need to gaze into a very dirty crystal ball; but the tough judgment to be made here is precisely how long we should clean the glass before acting on what we believe we see inside.

6

Weather and Climate Modification

Our great-grandchildren will live in an extraordinarily vigorous and highly technological society, and by 2050 A.D., or at least 2100 A.D. climate changes due to mankind will be very evident, especially at high latitudes.

—WILLIAM W. KELLOGG
National Center for Atmospheric Research (1974)

Well-planned weather modification efforts should be a strong part of total water resources management concepts and be fully organized and implemented before *drought conditions reach critical level.*

—North American Interstate Weather Modification Council, Resolution No. 75–5 (1975)

The predictions of Wofsy and McElroy are, perhaps, typical. They find that even if the manufacture of halomethanes were to be halted immediately, the depletion of the ozone layer would reach 5 percent by 1990 . . . A 5 percent ozone depletion (10 percent increase in radiation) might thus cause anywhere from 20,000 to 60,000 additional cases [of skin cancer] in the United States alone.

—ALLEN L. HAMMOND AND THOMAS H. MAUGH II
Research News, *Science* Magazine (1974)

Climatic Limits to Growth

Early in human history, before people banded together in social units, built shelters from the elements, or tilled the land to assure a more stable source of food, the prevailing climatic conditions largely controlled the style of life. Since the advent of social organization, people have increasingly collaborated on ventures designed to free them from a slavish dependence on the whimsies of the weather. Modern technology has helped societies to build the complex structures, shape and improve the frequently unforgiving landscape, and live in increasing physical comfort, often in environments too hostile for survival without technology.

A key, but imperfect, indicator of high material standards of living is the per capita energy consumption figures of a society.[1] Energy is an essential element of modern, technological agriculture and industry, which provide high levels of living to the peoples that create technology. At the same time, the majority of the world's peoples do not share the high per capita energy consumption levels of the industrialized nations, and the relative inequities in living standards cause friction between the rich and the poor. As a solution to the serious problems arising from technological inequity, the poorer countries often propose that the rich nations foster a rapid expansion of industrialization throughout the developing world. On the other hand, the solution most often proposed by the richer industrialized nations is a substantial cut, or even a reversal, in the soaring (around 2.7 percent per year) birth rate typical of poorer countries. Of course, the inevitable answer is to do both, but the important point for this discussion is that each side agrees on one thing: Higher material standards are desirable ends. And, since raising those standards often correlates with increased per capita energy consumption, the total world use of energy is projected to rise with increasing population and/or increasing development.

There are, as discussed in chapter 2, numerous and bitter debates about how fast energy resources can be developed and transferred to less developed parts of the world and on just how long the resources will last. A resolution to these debates is crucial if we are to solve the problems of, first, how many people can ultimately be supported from our known resources at a given per capita standard of living and, second, how rapidly those standards can be increased in the meantime.

It is becoming clear that industrialization and modern agriculture are limited in their growth potential by factors other than the obvious constraints on resource supplies. Technology and high material standards of living are generally very energy consuming, as I already said, but it is to be stressed here that energy production is often accompanied by the disagreeable by-products popularly known as pollution. Pollution is not merely an offense to the aesthetic sensibilities of people disgusted with litter, dirty water, or hazy brown skies. Some types of pollution also affect the atmospheric and oceanic systems and can even influence the weather and climate. In addition, it is well known that some pollutants can seriously damage essential ecosystems, the aspect of the pollution problem that focused public consciousness on the "environmental crisis." Here, the climate aspects of pollution will be stressed.

Furthermore, the technological equipment used to abate pollution, such as recycling processes or antipollution devices on automobiles, often require additional energy inputs and thus increase the amount of heat released into the environment. This heat is called thermal pollution. Although such pollution may seem secondary because its impact is not always immediately apparent as the more visible effluents, it may ultimately be more dangerous precisely because of its seemingly low priority as a pollution problem. I will outline some ways in which the rapidly growing use of energy may harmfully modify the climate and give order-of-magnitude estimates of climatic disruptions that could occur if present growth rates continue. In essence, I will try to quantify *the climatic limits to growth.*

The Energy–Growth Dilemma—Economic Development Versus Climatological Disruption

A particularly difficult aspect of this problem is that energy production is often a key to improving living standards and is thus universally desirable. However, if global energy consumption expands greatly in the next few decades, and this proves to be climatically dangerous, a great dilemma will befall the world. It will present a choice between curtailing energy production and thus curtailing a vital prop to the world's economy and food-producing systems, or continuing to use energy at high growth rates and risking serious climatic disruption (such as the melting of the polar icecaps, for example).

Society's adjustment to changes as fundamental as energy use takes decades to accomplish smoothly. And, at the same time, the potential for widespread climatic disruption from increased energy production could also occur as soon as the year 2000; thus, present estimates of the climatic effects of energy production take on far greater importance than might appear at first glance. Unfortunately, as we shall see, present estimates are mired in uncertainty.

Urban Scale Effects: The Heat Island. The magnitude and scale of any possible climatic impact from human activities depend on the degree of interference with the natural processes that determine the climate—primarily the energy flows in the land–sea–ice–atmosphere systems described earlier. Since human activities are not uniformly distributed across the earth, we might expect the largest and most apparent human impact on climate to be felt in regions where industrial activities and high population densities are concentrated. In fact, this is precisely what is found.

The urban heat island is a phenomenon well known to city dwellers, even if most haven't heard the name. Because of their intense energy consumption and unnatural physical characteristics, as pointed out earlier, cities are often hotter downtown (especially at night) than the surrounding countryside. The heat island is most noticeable at the center of cities, and the mean annual isotherms (lines of constant temperature) of Paris, for instance, are as much as 2°C (3.6° F) warmer at the city center than out of town (Figure 21). Very similar conditions can be shown for other cities, as summarized in an excellent article [2] by Helmut Landsberg, a noted senior climatologist at the University of Maryland.

Table 4 [3] shows the energy consumption (EC) statistics for a number of industrial and urban areas. To gauge the possible local climatic consequences of energy consumption, one can compare the value of average net radiation (last column on Table 4) with the EC density (next to last column on the table). The natural average net radiation is merely the difference between solar energy absorbed by the surface, a net gain of energy, and infrared radiation lost from the surface. In the exceptional case of Manhattan (i.e., it is a most intense heat island) where the energy consumption density is many times higher than the natural average net radiation, the city is considerably warmer than before the days of high energy consumption density. We can also see from the table that the growth of interconnecting urban

Table 4

Energy Consumption (EC) Densities [4]
(in watts per square meter)

Place	Area (km^2)	Population (in millions)	EC density (W/m^2)	Natural average net-radiation (W/m^2)
Manhattan	59	1.7	630	93
Moscow	878	6.42	127	42
Los Angeles:				
City	3,500*	7.0	21	108
County	10,000	7.0	7.5	108
West Berlin	234*	2.3	21.3	57
Fairbanks	37*	0.03	18.5	18
Bosnywash †	87,000	33	4.4	90

* Building area only.
† 21 metropolitan areas between Boston, New York, and Washington, D.C.

areas, called megalopoli, is spreading high energy consumption densities over large areas. For example, the twenty-one metropolitan areas between Boston, New York, and Washington, D.C., comprise a region of eighty-seven thousand square kilometers, affectionately called "Bosnywash" by meteorologists. Even over such a large area, a regional heat island of some 5 percent of natural average net radiation at the surface seems to be emerging from energy consumption densities.

Before we leave the discussion of urban scale, a few related sources of urban climatic modifications need mention. To recap, the paving-over of regions covered with vegetation and the consequent construction of buildings have several influences: (1) water that would have been absorbed by the soil before the paving is free to run off after a heavy period of precipitation; (2) some water that would have been stored by the soils and evaporated by plants has run off and is not available for evaporation, and much of the subsoil moisture that remains is prevented from evaporating by the pavement; (3) the roughness of the land surface is altered by the presence of rigid structures; and (4) a number of chemical effluents—gases and particles—are released into the air in high concentrations.

Water runoff is of primary concern because it threatens areas downstream with flooding after heavy storms, unless sufficient runoff and storage capacity is built into local sewer systems. The evaporation of moisture directly from soils, as well as through the vegetation cover, is an important factor in keeping surface temperatures down when the heating rate is high. The evaporation process returns to the air water vapor containing much of the heat absorbed at the surface; that latent heat is later released high in the clouds when the evaporated water recondenses to liquid drops. As a result, a severe reduction in the evaporation rate in an urban area will allow a larger fraction of the city's heat input—from either absorption of sunlight or energy consumption—to appear simply as a surface temperature increase. Thus, the reduction in evaporation contributes very directly to the heat island effect.

Furthermore, there is a positive feedback (i.e., mutual reinforcement) between the higher nighttime temperatures in cities in the summer and the very energy consumption densities that created the heat island; that is, the hotter temperatures themselves result in the greater use of air conditioners, which, in turn, consume energy and increase the quantity of urban heat still further. On the other hand, during winter the warmer night temperatures produced by the city's heavy energy consumption (particularly space heating) are beneficial since less energy is necessary to heat buildings. However, in both seasons energy consumption could be substantially cut by the use of better thermal insulation in buildings and by reducing the wasteful uses of heating and air conditioning.

The modification of the winds by large rigid structures in urban areas has been convincingly demonstrated by Landsberg and others, who found that wind speeds near the ground in cities were often more than 10 percent lower than unimpeded upwind speeds. This reduction in wind speed continues many miles downwind into the lee of the cities, causing a concomitant reduction of the normal evaporation rates, a change in the natural temperature patterns, and a substantial alteration in the dispersion of smoke and other air-polluting effluents.

Urban Air Pollution. But perhaps the most serious "climatic" disturbance caused by highly industrialized and populated concentrations is air pollution itself. In urban areas, air pollution has many effects; the primary one for people is a general degradation in the quality of the air,[5] particularly during atmospheric temperature inversions. An inversion occurs when a body of warm air hangs over a mass of cold air, creating a stable pattern that inverts the more normal situation in which warm air lies below cold air. In the more normal pattern, warm air near the ground enhances the process of vertical mixing since it is buoyant and rises. The ascending drafts of warm, dirty air carry up and disperse locally heavy concentrations of pollution. On the other hand, when an inversion is present the winds are calm, precipitation is absent, and warm air aloft descends to trap cold air at the surface; the pollutants generated near the surface can accumulate, and a situation called air stagnation occurs.

Air pollution can occur in the form of noxious gases or particles or, more commonly, in some combination of the two. Exposure to any kind of pollution can be not only harmful to life, but also can affect the climate in the city and many miles downstream as well. In fact, certain effluents remain in the atmosphere long enough to create regional and even global air pollution levels that are potentially significant to climate. For example, carbon dioxide (CO_2), a gas that accompanies the burning of fossil fuels (oil, gas, coal, etc.—our primary energy source), has a very long life in the atmosphere, and its concentration in the air has been increasing exponentially since the beginning of the Industrial Revolution. Fortunately, CO_2 poses no health hazard in the dilute concentrations that now exist in the atmosphere, but it is of tremendous importance to climate.

Sulfur oxides, on the other hand, SO_2 (sulfur dioxide) in particular, are major by-products of industrial activities and are quite hazardous to health in the high concentrations that are now often found in city air. In an increasing number of urban districts, strict emission controls have been imposed on the sulfur content of fuels burned by both industries and power plants, necessitating the use of more expensive low-sulfur coals or oils and sometimes forcing certain industries to move to more sparsely populated suburbs. Although such relocations certainly reduce the danger of unhealthy concentrations of SO_2 in the cities, they scarcely reduce the total amount of SO_2 injected into the air. Worse yet, the total injection rate may ultimately be increased, since the dispersal of industries that emit SO_2 encourages a return to the buring of cheaper, high-sulfur fuels that do not seem as dangerous when they are diluted.

Concern about the total emission of sulfur dioxide is often expressed because SO_2 is a gas convertible into sulfate particles by reaction with sunlight, often in the form of various sulfuric-acid droplets. Such photochemical gas-to-particle conversions account for a large fraction of the tiny particles injected into the atmosphere by human activities; and particles can interfere with the transfer of solar and terrestrial radiation as well as become nuclei for cloud droplets, thereby influencing cloud and precipitation patterns.[6] But sulfur oxides are not the only significant source of urban-produced pollutants that can react photochemically and be converted to particles. The nitrogen oxides from automobile exhaust are also considered[7] a major source of urban aerosol pollution, and a major component of the smog problem in Los Angeles, Denver, Tokyo, Mexico City, and other places with extensive automobile exhausts.

Particles can be recognized by an observer on the ground as hazy air, a condition that often afflicts most major metropolitan areas. Tiny particles are more important for large-scale climatic effects because their lifetime in the atmosphere is much longer than that of larger particles, and they travel much farther downwind from their source than the larger particles. Therefore, smaller particles can cover a larger fraction of the earth's area. And, since climatic processes are driven by the distribution of heating over large areas, the pervasiveness of particles that disturb the radiation is important. Small particles, which we will call optically important if they have sizes between about 0.1 to 1.0 one-millionth of a meter, it turns out, are much more effective (per unit weight) at disturbing the transfer of solar radiation than larger particles,[8] a point we will address momentarily.

A controversy has been raging between those optimists who claim that technology (e.g., particle-catching filters or sulfur-removing processes) has solved the problem of heavy and increasing particle concentrations in the cities, and those pessimists who claim that more industrialization will simply assure exposure to even greater levels of pollution. Some supporters of industry point to evidence[9] supplied by the Environmental Protection Agency (EPA) that the "concentration" of particles in urban areas has been declining steadily since the widespread imposition of emission controls. (It might prove interesting to find out the extent to which some of those who now praise these controls were vigorously opposed to them when they were first imposed.) On the other hand, others point to data taken in the same study that show the concentration of particles in suburban and rural areas rising steadily through the mid-1960s, that is, precisely during the period that the particulate concentrations in city air were dropping. Thus, they claim, air pollution has merely been increasing in the suburbs, and the long-term particulate trend continues upward. Since the climatic consequences of increasing aerosols depend primarily on how extensive an area is being polluted with optically important particles, and not so much on a few heavily polluted small areas, the rural trend implies that the situation has become far more serious. Thus, while the worriers may concede that controls have improved health conditions in cities, they would also argue that even larger areas have experienced increases in particle concentration, and that the potential climatic risk is increasing from this nonurban uptrend in the number of atmospheric particles.[10] However, more recent EPA studies of nonurban particulate concentrations since the mid-1960s show that emission controls have apparently been checking these increases rather effectively even outside the U.S. cities. This evidence, according to some technological boosters, is final proof that particulate air pollution is rapidly becoming a solved problem.[11]

Although EPA data suggest that particle concentrations are not increasing in the United States, at least for the time being, both sides often overlook a critical point. What is meant by "concentrations"? It was mentioned earlier that the efficiency with which a given total *weight* of particles interferes with the natural transfer of solar radiation through the atmosphere depends strongly on the size of the particles; small particles with the optically important sizes are more efficient than large particles, but large particles weigh more.[12] Since the EPA particle concentration statistics are given as the total weight of particles in a standard volume of air, they fail to distinguish the quantities of small particles—which are relatively most harmful—from the large ones—which are less damaging climatologically. Therefore the weight statistics alone are not very useful for judging the potential climatic effects of particles, and worse, they even obscure one important measure of the air quality over cities.

The particle-emission controlling devices that have been used recently generally remove the large particles much more easily than the smaller ones. But, since the large particles have short lifetimes (hours) in the atmosphere relative to the small ones (days), they do not travel far beyond the urban area and thus do not contribute much to the area-wide concentrations—the important factor for potential climatic influence. Furthermore, since a mass of large particles is less efficient at scattering and absorbing solar energy than the same weight (and therefore greater number) of small particles, catching the large ones in emission control devices has little effect on the large-scale climatic influence of particles. This is not to suggest that trapping large particles is a worthless venture; quite the contrary. But one cannot point to the evidence of diminishing *weights* of particulates in the United States as proof that the potential climatic effects of particles are also diminishing. A much more relevant statistic is the turbidity of the atmosphere, a measure of its haziness. In fact, if emission control devices were selectively trapping large particles, while industrial activities (and thus the total of effluent emissions) were creating an increase in the number of small particles or nitrogen or sulfur oxides in the atmosphere, the turbidity index might increase—although the total weight of particles would have decreased. Furthermore, if overgrazing or deforestation lead to increased aerosol loading, as Reid Bryson often suggests, then these kinds of activities could be as influential on climate as industrial ones, and perhaps more so.

All in all, it is likely that not enough is known from the few measurements taken so far to make definite conclusions [13] about recent trends in turbidity over most of the world; some are up, some down. But I can assert with confidence that it is premature to take comfort, as far as climatic effects are concerned, in the recent reduction in the weight of particulates found in the atmosphere in industrial America, for weight alone is an insufficient measure of the likely climatic effects of particulates.

Before proceeding with an even more uncertain estimate of the significance of particles as large-scale climate modifiers, let's return again to the urban scale, where the human influence on climate is much easier to document than on a regional or global scale.

The Urban Rain Machine. The urban heat island boils air upward at a greater than normal rate. At the same time, some of the small particulates emitted by industrial activities or automobile exhausts are carried rapidly aloft, where they can serve as nuclei for cloud droplets.[14] These mechanisms suggest that urban centers may well influence precipitation patterns both over the city and for some unknown distance downwind.

A celebrated and controversial example of the urban effect on downwind precipitation is the case of the so-called La Porte Anomaly. La Porte is a small town downwind of the Chicago and Gary industrial centers. Stanley Changnon and F. A. Huff, scientists from the Illinois State Water Survey, examined records from a network of rain gauges in and around La Porte and concluded that rainfall in La Porte had very possibly been enhanced by the industry upwind.[15] Their conclusions have been sharply criticized because the natural variability of rainfall in the area is so high that a large sample of precipitation data taken over many, many years would be required to establish the statistical significance of the anomaly. The critics charged that the researchers had neither sufficient nor accurate enough data to make a sound determination.

Changnon and Huff then entered into an elaborate experiment with a number of carefully chosen statistical design criteria to test the possible influence of St. Louis, Missouri, on the rainfall pattern both upwind and downwind of the city. The preliminary results [16] of their study show a significant enhancement of precipitation—as much as 15 percent—for some tens of miles downwind of St. Louis, as seen in Figure 22.

If urban activities really are causing precipitation modification, then the effects should clearly be greater on weekdays, when business is in full swing, than on weekends. Helmut Landsberg shows many examples of this in his review article on urban climates, including a study of Paris (Figure 23). A buildup of precipitation can be noted, peaking on Fridays, dropping off suddenly on Saturdays and Sundays, and then beginning to build up again on Mondays. In statistical parlance, the results are significant "at the 95 percent confidence level" [17] (that is, there is only a 5 percent chance that this result was a statistical accident). However, Landsberg warns that not all data show these trends so definitely. For example, some studies show that summer precipitation values are "affected in a random fashion by the heat island but that the winter excess on weekdays is stimulated, in some fashion, by seeding of supercooled clouds that are more prevalent on workdays."[18] Landsberg is a cautious scientist and wisely warns that much more work needs to be done to demonstrate, let alone understand, these connections with reasonable certainty. Nevertheless, he concludes his article with a statement of considerable force for a cautious man:

> It has been established beyond reasonable doubt that urban agglomerations cause measurable changes in the atmosphere immediately adjacent to them. Temperatures are increased, low-level lapse rates steepened, horizontal winds slowed, and updrafts induced. Turbulence and cloud formation are increased, summer rainfall is enhanced, and possibly, some winter snowfall is stimulated. Snow on the ground is diminished and so are surface–near humidities. Most apparent is the increase in pollutants from one to several orders of magnitude. They reduce solar radiation intensity, eliminate all the short-wave and a substantial portion of the long-wave ultraviolet, and shorten sunshine duration. Their effect on cloud formation and rainfall over and in the vicinity of the cities is still somewhat uncertain, but evidence points to occasional cases of stimulation of precipitation and perhaps some rare cases of inhibition. . . .
>
> As cities grow into large conurbations one can foresee that they will have notable regional weather effects.[19]

Regional Climatic Modification. Although the effects of urbanization on climate are readily demonstrated, urban areas cover a relatively small fraction of the earth. Thus, if regional or global climatic modifications could be caused by human activities, these changes would be more important because of the disruption of large geographic areas. However, even though the importance of potential climatic changes increases with the size of the affected area, the magnitude of potential effects decreases. Thus, urban-scale modifications are larger in magnitude and easier to demonstrate but limited in impact, compared to regional or global-scale effects, which are larger in impact but smaller in magnitude; it is, therefore, harder to demonstrate that a change has actually occurred. Furthermore, there is no proof that all climatic changes are reversible once they occur; thus, changes resulting from human activities might prove permanent even if the causal activities were removed.[20]

One example of an almost irreversible regional change in the environment that could create climatic changes locally as well as thousands of miles away is the large-scale clearing of land (for agricultural, industrial, urban, or even recreational purposes). For example, most of the United States was forested land before the arrival of Europeans. Whether the widespread deforestation activities that followed were partially responsible for the climatic shifts Americans have experienced over the past two hundred years is difficult to prove. Regardless, whatever climatic adjustments deforestation may have brought are now pretty much discounted; that is, our present expectations for average climatic conditions include whatever effects, if any, deforestation had on the weather patterns. But today, the ability of modern machines to clear lands would allow us in perhaps only one generation the alteration of as much land as previous generations took hundreds of years to clear. And, since changing the complexion of the land can affect the climate, societies may now be causing climatic shifts much more rapidly than they have before. One immediate danger of rapid changes in climate is that food production would probably fluctuate. Recall that we showed earlier how such fluctuations can cause food crises when world food reserves are dangerously low—as they are now.

Deforestation of the Jungles: Could It Change the Climate? The consequences of deforestation projects are not merely academic. The optimistic views held by many of the world's food planners rest squarely on the fact that only a moderate fraction of the world's arable land is presently tilled, and particular hope is expressed that the clearing of tropical rain forests, such as those of the Amazon jungle in Brazil, will supply enough arable land to feed the world's expanding population for decades to come (see Figure 33). Because this seems an especially useful example of another possible climatic limit to growth, I will concentrate on the range of side effects that might follow rapid development of tropical forests like the Amazon, which the Brazilians hope to cultivate. The project has already begun, in fact, and every year they push construction of the Amazon highway deeper into the interior. According to media reports, a tragic and growing side effect of the development of the Amazon frontier is that the native Amazon Indians are being subjected [21] to the same kinds of treatment that the North American Indians experienced when immigrant settlers pushed westward into Indian lands.

Let's return to the climatic effects of deforestation. I attended a conference on tropical ecology in Turrialba, Costa Rica, in 1973, and one of the problems I was asked to address was the possibility of climatic side effects of tropical deforestation. Since my comments were incorporated into a conference report that both succinctly summarized the problem and was endorsed generally by the conference delegates (including Latin American scientists from countries contemplating development schemes such as large-scale deforestation), it will be quoted liberally in Appendix A. Here, the problem is summarized briefly.

Deforestation replaces the dark forested surface with lighter agricultural or grass lands, thereby raising the surface albedo and causing less total sunlight to be absorbed. Changing the vegetative cover also affects the water-bearing characteristics of the land. Both of these factors influence the climate. As an example of how strongly deforestation affects both evapotranspiration (the rate at which water is evaporated and transpired from the surface and its vegetation cover) and runoff, we can consider the estimates from Sellers [22] given here in Figure 24. Although these data are for a middle-latitude station (North Carolina) and may not be strictly valid for tropical regions, it shows that deforestation can have a significant effect in decreasing evapotranspiration and increasing runoff. The extent to which similar effects occur in tropical regions would depend largely on the relative amounts of water available over the year to the root systems of native forests and to the crops that replace them. Field research and simulation modeling must first be undertaken to determine the annual and seasonal evapotranspiration values from extensive areas of various types of vegetation in tropical lowlands.

To gain some perspective here, I emphasize again that although land-use practices in tropical regions may affect the local climate significantly, the degree to which such climatic changes might be felt elsewhere is still entirely uncertain—and uncertainty does not necessarily imply that an effect is negligible.

Lastly, it seems appropriate to examine the potential climatic consequences of acting on Roger Revelle's projection published in *Scientific American* that enough cultivable land exists on earth to provide "for between 38 and 48 billion people, between 10 and 13 times the present population of the earth." [23] Revelle quotes a 1967 report to President Johnson to the effect that 3.2 billion hectares of land are potentially arable (about 24 percent of the earth's land area and thus about 7 percent of the earth's total surface area). However, only about 1.4 billion hectares are now cultivated. If one excludes much of the humid tropics and those regions where the supply of irrigation water is insufficient, Revelle suggests that "the total potentially arable land is reduced to 2.5 billion hectares (the present 1.4 billion hectares plus 1.1 billion)." [24] That is, 1.1 billion additional hectares of land may need to be cleared in the next generation or so to feed the world. This is about 8 percent of the world's land area and slightly more than 2 percent of the earth's total surface area. Let us suppose that cleared agricultural land has an average albedo of about 20 percent and that uncleared (tropical) forested land has an albedo nearer to 10 percent; then, if the remaining arable land were cleared, an increase in the earth's surface albedo of roughly 10 percent would occur over about 2 percent of the earth's surface. The order of magnitude of the increase in the albedo of the global surface would then be $0.10 \times 0.02 = 0.002$, a rather small change from a planetary perspective. However, in those regions where a large fraction of forested lands are cleared, the changes to both the surface albedo and the water-bearing characteristics of the land surface can certainly have regional climatic side effects, which in turn may influence climatic patterns thousands of miles away (by creating so-called regional anomalies).[25] In addition, changing areas from forests to agricultural land influences the nitrogen-fixation rates in the soils and may, as we will discuss shortly, alter the concentration of atmospheric nitrous oxide, a gas that can destroy ozone in the stratosphere. Thus, while the notion of cultivating the remaining 1.1 billion hectares of arable land seems to portend little *global* climatic limitation, truly serious meteorological obstacles to such cultivation may occur regionally. And the obstacles may not be only climatic, but may be ecological and economic as well.[26]

That the latter obstacles may prove unsurmountable to the Brazilian's efforts to deforest and develop the Amazon has been predicted by the Ehrlichs. In *The End of Affluence* they write: "The current efforts of the Brazilians will merely net them a few harvests at the cost of irreversible destruction of the region. Amazonia will be turned into the world's largest parking lot." [27] (Displacing the Long Island Expressway, perhaps?)

Climatic Barriers to Energy Growth

Earlier in this chapter, it was pointed out that most of the potential climatic consequences of human activities would be proportional to the relative disruption those activities might cause to natural energy flows in the climatic system. These estimates have political overtones, imply future growth limitations, and raise serious, long-term questions that require candid and perhaps immediate answers. But discussion of these sociopolitical aspects will be deferred to a later chapter. Here, we will estimate the potential amounts of disruption.

Since vastly increased energy production is almost a certainty in our future, the discussion of large-scale inadvertent climatic effects resulting from human activities will begin with the case of energy use. The section that follows is intended to show how energy production can lead to climatic modifications on both large and small scales, and will present and justify some rough estimates of the potential climatic impacts of continued energy use over the next several generations. Some of the arguments that lead to these estimates of climatic impacts may seem a bit too technical at times but the scientific details are presented as background to the estimates that will be offered shortly. Much of the following material follows from a collaboration [28] with an NCAR colleague, Roger Dennett.

Climatic Sensitivity to Energy Production

All processes that mobilize or use energy release heat into the environment. If the generation process were 100 percent efficient, then the bulk of the heat would be released to the environment at the point of consumption (in electric house heaters or steel blast furnaces, for example). However, energy conversion efficiencies are typically closer to 30 percent for electric power plants, for example. Thus, in addition to the heat released to the environment from useful consumption, large quantities of heat (up to 70 percent of that generated) are discharged at electric power sites. It is well established that the potential for modifying the climatic environment from energy use is proportional to the relative magnitudes of natural energy flows and energy released to the environment from human activities. Thus, it is necessary to compare natural power densities with those arising from both waste heat production and useful power consumption, in order to estimate potential climatic effects of released heat from any particular process. (Appendix B defines these energy-related terms.)

Natural energy flows include the incoming solar radiation, the outgoing terrestrial radiation, and the transport of heat by the winds and ocean currents, for example, as summarized in Figure 25 and described in Appendix B.

Climatic Feedback Mechanisms

In Appendix B, estimates are given of natural power densities and temperature sensitivities to energy consumption for average conditions across the planet by using results from a simple globally averaged climate model. But the real atmosphere, as explained previously, is driven by the heating of the globe, which occurs quite unevenly and primarily in latitude zones, but with important longitudinal thermal differences created by the special geometry of the continents and oceans. The heating gradients (i.e., spatial differences) drive the motions we recognize as winds and ocean currents and govern the general patterns of the climate. Mass, moisture, heat, and momentum are redistributed in this process, and cloudiness is generated. Clouds, in turn, largely control the albedo (reflectivity) and affect the radiation balance, which alters the heating gradients; and thus we can see that the driving force behind the climatic motions is modified by the very processes it sets in motion (feedback).[29]

There are numerous feedback processes in the atmosphere/ocean/land/ice systems, which collectively comprise the climate system; some of these tend to enhance a disturbance (i.e., positive feedback), and some tend to dampen one (i.e., negative feedback). One positive feedback that is easy to visualize is the snow and ice–temperature–albedo feedback, mentioned earlier. In this, a general increase in temperature decreases the snow and ice cover of the earth, thus lowering the albedo; this, in turn, permits greater absorption of solar energy, thereby enhancing the original warming (or vice versa for cooling). This ice feedback effect has been modeled by a number of scientists. For example, Mikhail Budyko, renowned climatologist and past director of the Main Geophysical Observatory in Leningrad, and William Sellers, a climate modeler from the University of Arizona, have each constructed climate models predicting that ice feedback increases the sensitivity of the surface temperature to energy inputs by up to four times as much as the result for the simple, one-dimensional model for the temperature-energy sensitivity analysis, which is explained in Appendix B. Furthermore, computer experiments by Syukuro Manabe and Richard Wetherald at the U.S. Department of Commerce's Geophysical Fluid Dynamics Laboratory in Princeton, New Jersey, using a very detailed three-dimensional "general circulation model" of the atmosphere, have provided similar conclusions to the earlier results obtained with the Sellers and Budyko simple models; that is, polar regions are several times more sensitive to changes in energy inputs than other regions of the globe. The more complex, three-dimensional experiments showed that the combination of a surface temperature inversion (cold air prevalent near the ground which inhibits vertical mixing of air) and snow– and ice–temperature–albedo feedback enhanced the model's response in the polar zones, suggesting a definite warming trend if energy production continues to grow rapidly. The degree to which this enhanced sensitivity in the ice-covered poles might imply potential changes in world sea level (e.g., from melting glaciers) is difficult to estimate, but deserves careful and increased consideration.

To gain some perspective, even a one-meter (approximately three feet) rise in the sea level could doom the city of Venice. Further rises could be serious for scores of low-lying places such as Holland, Bangladesh, Florida, etc. Unfortunately, since many feedback processes are operating simultaneously in the real climate system, and the synergism of these is largely unknown (the modelers only approximate some of them), the simple globally averaged model probably continues to provide the most useful estimate of the global climate's sensitivity to energy perturbations. But we must not forget either that the model remains subject to error by factors perhaps as great as ten, particularly in the polar regions; thus, in our major climatic and socioeconomic decisions related to energy we will be plagued by an uncomfortably high degree of uncertainty reflected in the crudeness of present climate theories.

One more point needs amplification here. The discussion has been dealing with global-average thermal energy densities so far. But it must be made clear that even if the total worldwide energy use were to remain a negligibly small fraction of these natural densities, actual energy usage would be highly concentrated on local or regional scales. A dramatic nighttime photograph taken from a U.S. Air Weather Service space satellite clearly shows the lights of the major metropolitan areas in the eastern half of the United States (Figure 26). The earth literally glows at night, and the concentration of energy consumption patterns in urban areas is startlingly portrayed since energy consumption, and thus thermal pollution, occur in similar locations to the light generation. Chicago, Miami, Montreal, and "Bosnywash" are clearly identified. With an atlas you can identify dozens of other cities. On these local and regional scales, the heat released to the environment can be a significant fraction of the natural power densities shown earlier in Table 4 (page 155). That these energy-use concentrations could cause local or regional climatic anomalies is almost certain, but what is more serious (but less certain) is that if present growth trends continue, these regional heat islands could also force changes in atmospheric motions that might be global in scope. Before looking further into this question, we need to examine current values and projections for power densities associated with human use of energy.

Unnatural Power Densities. The total use of power by humanity in the early 1970s was about 8 billion (8×10^9) kW (kilowatts). Dividing by the earth's surface area, some 5×10^{14} m^2, we arrive at a global average power density of 0.016 W/m^2. This is less than one ten-thousandth of the global average value of solar energy reaching the earth (called the solar irradiance, $\bar{Q}$) and is insignificant globally (see Appendix B). However, most energy production now occurs, and is anticipated to continue to be generated, in limited areas near industrial regions, as Table 5 shows. Furthermore, it is possible that future energy production will often be concentrated in even smaller areas called power parks (especially designed for the plethora of projected nuclear plants to improve both efficiency and security). These concentrations would discharge tremendous quantities of heat locally, and thus would have extremely high power densities. Some futurists have suggested possible (although highly optimistic) projections for future energy consumption rates for a steady-state world, expected to have an upper limit [30] of 20 billion people using 20 kW per capita (nearly twice the current U.S. per capita power consumption), as shown in Table 5.

Table 5

Pollution Thermal Power Densities *

	Ultimate steady-state future	Present
Global average	0.8 W/m²	0.016 W/m²
West Germany	5.0 W/m²	1.0 W/m²
Industrial area (Ruhr)	1,000.0 W/m²	17.0 W/m²

Large nuclear power parks 30,000 MWe → 100,000 MWth
20,000 W/m²

* Composite from Weinberg and Hammond [31] and Häfele.[32]

This level of usage would raise the global power density by 0.8 W/m², about one-quarter of 1 percent of the average solar power input, $\overline{Q}$, and half of 1 percent of the natural solar *surface* power density, $\overline{Q}_s$, as described in Appendix B. For thermal energy pollution alone, present (no-feedback) climate theories predict from this experiment a *global* average surface temperature increase of the order of 0.4°K. But this could be a considerable underestimate of *surface* temperature increases if the pollution energy were released during times of vertical atmospheric stability. Such stable conditions commonly occur during the winter in the midlatitudes and very often in the polar regions.

However, the projections of Table 5 imply that unnatural thermal power densities in industrial regions could be as large as several times the natural solar power input. These projections also suggest a 500 percent increase in world population and a 1,000 percent increase in world per capita energy consumption standards. I do not endorse such figures, nor do I think they are realistic, given present world social and political systems. However, they have been put forward seriously by noted scientists [33] as possible limiting values for a steady-state world, and as such I am merely estimating the potential climatic consequences of these projections. Some implications of the dissipation of heat pollution of this magnitude are given in the next few sections, along with a few specific examples of the potential consequences of certain energy production methods. It is not necessary for us to assimilate every technical detail of the following few pages. The message carried by these examples can be simply stated: Climatic disruption from human usage of energy will be proportional to our interference with natural energy flows.

Dissipation of Unnatural Power by the Natural System: Some Examples. Nearly all the unnaturally utilized (i.e., by human activities) thermal energy remains in the earth-atmosphere system, and the way it is dissipated into that system largely determines the climatic consequences of its release.

A principal method of dissipating waste heat (liberated at the power plant site) is discharging it into rivers. Rivers are a primary agent for removing the excess water (precipitation minus evaporation) from the continents, and they also provide much of the irrigation needs for highly technological agriculture.[34] Wolf Häfele, an expert on energy production (and an advocate of nuclear power) from the International Institute of Applied Systems Analysis in Austria, has made some order-of-magnitude calculations. He writes: "The relation between rainfall and runoff thus puts certain limits on power densities, and implicitly, on land use. If all runoffs may be heated by 5 °C, only 0.25 W/m^2 can be dumped." [35] Häfele concludes that dissipating the heat by evaporating water in wet "cooling towers" may help for awhile, perhaps fifteen years, but that future projections of energy use imply that other means of dissipation will have to be found if riverine ecology is not to be significantly altered, or if severe regional effects (such as persistent cloudiness near cooling towers) are not to be generated.

Häfele suggests the oceans as a feasible dumping ground of waste heat "at least insofar as heat capacity is concerned," but he wisely recognizes that "its effect on the ecology, and especially on the dynamics of ocean currents, must be studied further." [36] Let's examine this suggestion.

The placing of power parks in a few oceanic regions where strong currents could dissipate the heat has been suggested informally by Häfele and his colleague Cesare Marchetti. Take the case of the Gulf Stream, for example. The difficulty here is that this current transports enough heat poleward to influence radically the climatic conditions of the North Atlantic region. As said earlier, England and Hudson Bay have vastly different climates despite their similarity in latitude. The effects of heat transport by the Gulf Stream include partial control of the extent of sea ice coming down from the Arctic

Sea. The polar regions, as already pointed out, are known to be especially sensitive to changes in energy inputs, and thus large quantities of heat added to the Gulf Stream from power parks could ultimately find their way into the climatic system in a particularly sensitive region: the subpolar part of the North Atlantic at the margins of ice-covered regions. (Similarly, large quantities of heat removed from the Gulf Stream by special power plants,[37] driven by the temperature difference between the warm surface and colder deep water, could also be disrupting to a natural energy flow in a particularly sensitive place.) While Häfele and Marchetti's basic concept—dumping waste heat into the environment at points of minimum climatic sensitivity—could provide an extra measure of safety and is a valuable principle, it is likely that uncertainties in our present knowledge of climatic cause-and-effect links will make the satisfactory identification and selection of minimum sensitivity regions tenuous for many years.

A second potential problem with the dissipation of these large quantities of power on regional scales (i.e., areas a million or so square kilometers) is that such regional heat islands could influence the position and character of atmospheric patterns known as long waves. John Sawyer, a prominent British meteorologist, suggested that surface power density anomalies of about 20 W/m^2 over an area of a million km^2 (about the size of New England) are theoretically capable of causing changes in these long waves. These, in turn, affect the large-scale circulation patterns of the atmosphere. This kind of reasoning, in fact, has been behind attempts at long-range forecasting of climatic anomalies in the United States [38]—but the anomalies in surface power density used for long-range forecasting studies are natural ocean surface temperature abnormalities, not heat pollution patterns.

Climatologist John Kutzbach of the University of Wisconsin has examined this question further and tentatively concluded that "future patterns of energy use may produce large-scale circulation changes of the order of the natural interannual (i.e., year to year) circulation fluctuations which presently are thought to be associated with such factors as anomalous sea surface temperature or snow cover patterns." [39] As we have seen, such anomalies in atmospheric patterns (e.g., the Soviet drought in 1972 or the peculiar weather patterns in the United States in 1974) may have important consequences for agriculture, even if they imply only a slight shift in the normal position of atmospheric patterns. How can the influences of such anomalies be modeled? The answer, I fear, is "not very well," at least with present tools. General circulation models of the atmosphere (GCMs)[40] are a widely used tool, but they suffer from several deficiencies when contemplated for use in energy production studies. Not least among these are—

(1) fixed (or inadequately computed) values of ocean surface temperature;
(2) inadequate treatment of cloudiness and its variations;
(3) difficulties in establishing the statistical significance of individual numerical experiments, which implies
(4) high usage of computer time, resulting in high costs.

Nevertheless, some efforts have been made to use GCMs to study the response of the atmospheric (long wave) patterns to surface anomalies, and out of these pioneering experiments is emerging, at least, a methodology for the statistical design of such experiments.[41] Furthermore, preliminary results by NCAR scientists Paul Julian and Robert Chervin suggest that ocean temperature anomalies in the tropics could cause larger atmospheric disturbances than midlatitude anomalies. If so, then current suggestions to place power parks in tropical oceans may need to be re-examined. It is reasonable to hope that these gigantic computer models will soon provide elementary insight into the regional climatic problem, but the precise utility of this approach won't be known for several more years. As for the global heat discharge problem, the restriction to fixed oceanic surface temperatures renders most of the present generation of GCMs inappropriate to the task. The exception is one of Syukuro Manabe's models, which does compute ocean surface temperature crudely, but has not yet been used to estimate *surface* heat release effects on climate.[42] Also, this model does not use a realistic distribution of land and ocean and is primarily a research tool; although it still could provide useful estimates of potential heat pollution effects on surface temperature.

Tornadoes and Heat Pollution. Another problem with the dissipation of high pollution power densities in small regions is the possibility of initiating conditions for anomalous cloudiness, or even of enhancing the chances for the occurrence of severe convective storms—thunderstorms, hailstorms, and tornadoes. Since such severe convective storms are dependent on mesoscale (regions of 10 to 100 kilometers on a side) conditions of atmospheric stability, humidity, cloud nuclei, wind patterns, and surface heating, extremely large concentrations of surface heating could, under appropriate environmental conditions, trigger atmospheric instabilities that lead to convective storms. Power densities comparable to 50,000 W/m^2 over some 2 km^2 have been produced in nature and can be used to gauge whether convective storms are indeed associated with such heat centers. Volcanoes with power densities of some 30,000 W/m^2 or greater have been known to spawn tornadoes, for example.[43] Thus, regional heating anomalies from a number of power parks would not only possess the potential to modify the global climate by influencing atmospheric long waves (which control regional climates), but might also have a more immediate and damaging mesoscale influence through the generation of severe convective storms or persistent cloudiness.[44] In fact, these mesoscale effects from concentrated energy generation would probably be felt long before global climatic changes from *thermal* energy releases were documented, unless, of course, the future patterns of energy use favor a distribution of power densities more uniform than the present and anticipated regional concentrations around the world's population and industrial centers.

Climatic Effects of Widespread Harvesting of Natural Energy Flows

Unnatural energy sources such as the burning of fossil fuels, nuclear power, and (someday) fusion power (the principle on which the sun and the hydrogen bomb work) all share the environmental side effect of introducing heat to the earth-atmosphere system that would not otherwise be present. However, natural energy flows (such as those from the sun, rain, and wind) can provide enough power to supply nearly all present energy consumption and a sizable portion of the expected future needs. Because harnessing natural energy may not, under some circumstances, require adding heat pollution to the global environment, it may be an attractive alternative (or at least supplement) to unnatural sources of energy.

Water Power: Too Little for Exponential Growth

Water power has been important in regions with large rivers and sufficient runoff from high elevations (e.g., in the American West and Scandinavia). Although it seems to have little potential to modify the global climate, regional or local weather conditions might easily be affected. The principal manifestation of such climatic effects could occur when water power is obtained by damming rivers. The great lakes so often created behind the dams cannot help but change the local temperature, evaporation rates, and perhaps cloudiness. Probably they are sources of at least some local and downwind influences, although it seems unlikely that these effects would be greater than regional in scale. However, if the currently powerful demands for clean energy continue to rise spectacularly, natural hydroelectric sources will soon be exhausted, unless major new technologies designed to augment the existing sources are quickly developed. If that happens, the most serious threat to global climatic patterns from water power may, paradoxically, arise from the human schemes to increase precipitation—rainmaking projects, for example. This has already been tried for the Colorado River basin. The risks associated with such deliberate climate modification plans, which can go well beyond cloud seeding, are discussed in detail later in this chapter and the next.

Solar Power: Not a Panacea, but a Promising Alternative

Solar power is often thought to be the most likely candidate for harvesting natural energy flows with minimum climatic impact. For example, the present energy needs of the entire United States could be supplied if only 1 percent of the country's continental land area were used to convert sunlight into electricity. The technological catch here is that capturing and storing the energy for use on demand is a difficult and expensive task, even with the best state-of-the-art methods. A number of possible storage techniques could be tried, including large flywheels, production of hydrogen as a fuel, and even such exotic methods of growing cellulose and converting it to alcohol. The latter approach to the collection and storage of solar energy operates through single-cell fermentation and enzymatic processes, and may prove economically feasible one day—provided some early tax help or research funds were made available.[45]

Nevertheless, some observers still say that the problems inherent in large-scale solar energy usage may never be overcome.[46] Yet, even if these obstacles are overcome (and only miniscule funding has been available for solar power relative to nuclear power research), harnessing solar power on a large scale can still cause climatic effects, particularly since the albedo of the very dark solar collectors will be much lower than natural surface albedos. The basic point emphasized throughout this chapter, which is inherent in each of the specific examples mentioned over the past few pages, is that the magnitude of any climatic effect of increased energy production will be directly porportional to the relative disruption of the natural energy flow patterns operating in the climatic system. Thus, if the future thermal power requirement is as large as half of 1 percent of all the solar power absorbed by the surface of the earth (see Table 5, page 170), then solar power farms covering only a small fraction of that global area would need to remove a very large fraction of the incoming solar energy. At the very least, regional climatic effects could be expected from such a substantial disruption of local natural energy flows. The heat rejected by inefficient conversion of solar energy to electricity in solar power parks would probably cause the same kinds of mesoscale effects referred to earlier in the section on the dissipation of power. (This is also a problem of geothermal power production.) Yet, if considerable care were used to minimize such alterations to natural heat flows, solar energy would become an immensely practical, long-term, renewable source of power, since no extra energy would be added to the earth-atmosphere system, except for that deriving from the difference in albedo between the surfaces of the earth and the collectors.

Let us examine this further. Harvey Brooks, Dean of Engineering and Applied Physics at Harvard University, pointed out[47] that the amounts of energy used indirectly in the construction and the conversion efficiency for solar energy collectors both suggest that the solar alternative will provide substantial heat additions to the environment. Nevertheless, he showed that energy could still be produced perhaps twice as efficiently by solar plants than by thermal ones.

Perhaps the chief environmental difficulties with extensive solar power use lie with the demographic changes that might easily accompany such energy concentrations. The fragile desert environment—where most solar power would be harvested because of the prevailing low amounts of cloudiness—would be hard-pressed to accommodate large population increases that often spring up around big industries. Thus, economic or other disincentives would be required to discourage these shifts. A better alternative might be to export the energy to the consumption areas, possibly in the form of hydrogen. This method of energy transport would have the added advantage of allowing storage in the supply chain from production to consumption,[48] creating reserves that would offer some degree of protection against the catastrophic effects of sudden malfunction in one of a few power parks.

Focusing on smaller-scale systems, the amount of solar energy incident on the roofs of dwellings in many parts of the midlatitudes is theoretically sufficient to supply the entire energy requirement of the houshold, plus some of the transportation energy requirements of the occupants.[49] The extra heat pumped into the urban climatic system would be limited to the difference between initial natural absorptivity (primarily that of the roofs) and the total absorptivity after installation of the collectors. Although this small-scale supplement is certainly not the answer to exponentially growing long-term global energy projections, it seems extremely foolish to ignore the potential utility of this clean and readily available energy source. Collectors are in general use in Israel, for example. Strong tax incentives to encourage the widespread use of solar house heaters would be a welcome step in the right direction. In fact, if mortgage sources did not discourage the building of atypical houses (ones with solar collectors prominently perched on the roofs, for example), many people would already be building them.

Wind Power: Another Useful Energy Supplement

Wind power is also an important supplementary energy source, but cannot possibly meet the global energy needs projected in Table 5 (page 170). Yet, some regions—such as the Continental Divide of the Rocky Mountains and the Great Plains in the United States—are amply supplied with near-surface winds and clearly warrant intense evaluation as sources of windmill power.[50] Again, the climatic effects will be proportional to the disruption of natural energy flows, but will probably be quite minimal. Even so, windmills would do more than merely trap wind energy that might otherwise be dissipated as frictional heat generated when the winds bump and scrape along the earth's surface. If windmills were built in substantial numbers over a sufficiently large area, the effective roughness of the earth's surface might well be altered and the near-surface wind profiles might then change significantly. Such a condition would have to be closely examined, since a slowing of the wind speed near the surface would be a reduction of the very energy that the windmills were intended to harness. Also, changes in the wind profiles could alter the climatically important rates of evaporation and heat transport away from the land.

Still another force must be considered: the gravity waves that transfer momentum up from the earth's surface into the fast-moving upper atmosphere and often act as a "drag" on the upper level winds in the vicinity of great mountain ranges such as the Rockies in North America.[51] A numerical experiment by William Bergen of NCAR has shown that increasing the height of the Continental Divide by ten m (with drag power equivalent to some absurdly large number of windmills) could increase the drag of the mountains on the upper air stream in that area by a few tenths of a percent. It is difficult to determine whether that would be sufficient to alter natural atmospheric long-wave patterns noticeably. The main point to remember is the one repeated throughout: that *no energy production system can be considered completely free of climatic side effects if it alters natural energy flow patterns.* Thus, all proposed schemes, using natural or unnatural resources, require careful scrutiny before their potential climatic effects can be estimated; and this examination must evaluate effects across the total system of energy sources. The many feedback mechanisms that operate in the climate system imply quite strongly that the behavior of the entire system may be different from the simple sum of the effects of individual components. (That is, the climate system is nonlinear.)

Carbon Dioxide and Dust: Indicators of "Gross National Pollution"

So far, estimates have been given of the order-of-magnitude response of the climatic system to local, regional, and global inputs of *thermal* energy. However, it is also well known that increases in the atmospheric concentrations of carbon dioxide (CO_2) and aerosols that accompany the combustion of fossil fuels (coal, oil, and gas) alter the radiative properties of the atmosphere, which in turn affects the heat balance and the climate, as discussed in some detail earlier. Building on an article written with my NCAR colleague William Kellogg,[52] I shall present some projected future concentrations of dust and CO_2 and estimate the order of magnitude of their climatic effects. Although this repeats some material presented earlier, it is useful here to provide a basis to compare the relative climatic influence of these by-products of fossil-fuel energy production to that of the direct heat that might be released in any of the energy-producing schemes now anticipated.

Since the beginning of the Industrial Revolution, the increasing use of fossil fuels has resulted in a steady buildup of the carbon dioxide content of the atmosphere. This gas is chemically quite stable in the atmosphere and, consequently, somewhat less than half of the added supply appears to have been dissolved into the oceans or absorbed by the plants of the biosphere, while the other half has simply remained in the atmosphere. It is estimated that the atmospheric content of carbon dioxide has risen from a pre-Industrial Revolution value of slightly under 290 parts per million (ppm) by volume to about 320 ppm and that by the end of the century it may rise to 385 [53] or even 400 [54] ppm. (The present energy crisis may well influence both of these estimates, since gas and oil will probably be replaced increasingly by coal as an energy source—unless advocates of nuclear power or the forces calling for reduced economic growth strongly influence current trends.) Lester Machta, head of the U.S. National Oceanic and Atmospheric Administration's Air Resources Laboratory, has made projections of the expected increase in CO_2 concentrations to the end of the century (Figure 27). The sometimes deceptive nature of exponential growth is evident. That is, CO_2 concentrations increased from a value of about 285 ppm in 1880 to about 320 ppm in 1970, slightly more than a 10 percent increase in 110 years. However, because CO_2 concentrations are growing exponentially (recall the parable of the lily pond), the next 10 percent increase will occur in about 20 years, and a 100 percent increase may be expected perhaps as soon as 50 years from now.[55]

Machta first estimated CO_2 increases in 1971 (Figure 27a), but then revised his estimate on the basis of new observational inputs in 1974 (Figure 27b). Many optimists like to belittle the prophets of doom whose early findings forbode catastrophes that disappear with further studies. But comparison of Machta's early projections to his later ones makes an important point: *Early estimates of possible future events are not biased toward maximizing those events, but are just as likely to be underestimates as overestimates.* Machta's 1971 early estimate has given way to more advanced knowledge that has *increased* the projected CO_2 levels, not reduced them. Although it is quite possible that a third estimate may be less alarming, the message that not all disturbing forecasts are necessarily exaggerations comes through clearly.

There are various estimates of the response of globally averaged surface temperatures to a doubling of CO_2 from about three hundred ppm to six hundred ppm by volume—a value projected to occur by about the years 2025 to 2040. State-of-the-art climate models unequivocally predict that such a doubling of CO_2 would raise the surface temperature of the earth. Although these predictions vary considerably, probably the best order-of-magnitude estimate that can be made *today* is for a surface warming by some 1.5 to 3°K globally and that the temperature increase in the polar regions might well be amplified severalfold.[56] But there is far less agreement over the magnitude and location of the warming than over the fact that CO_2 will warm. Projection of the CO_2 increase, granted the continuation of present trends to the year 2000, suggests, as said earlier, an increase in CO_2 concentrations of about 20 to 25 percent, a change corresponding to an approximately 1°K global surface temperature rise (plus the assumed amplification at the poles). Recall that it was shown that a change of 1°K in global surface temperature is as large a "natural" climatic change as any experienced on earth since the Little Ice Age, and the warming at the poles may be large enough to imply eventual changes in sea level.

While modern industrial societies have been spewing increasing amounts of carbon dioxide into the atmosphere, similarly large quantities of particles, or atmospheric aerosols, have also been belched aloft into the "sewer in the sky," as I once heard the atmosphere described at a meeting. As pointed out earlier, these aerosols arise directly from injection of dusts by coal-burning operations and other industrial sources, as well as from slash-and-burn agricultural practices;[57] and there is also evidence that they are created photochemically in the atmosphere from unburned hydrocarbon fuel vapors and sulfur dioxide under the influence of solar ultraviolet radiation. Measurements at a number of places across the earth have shown a steady rise in the aerosol content of the lower atmosphere over the past few decades, and sudden increases in the stratospheric aerosol content have followed the major volcanic eruptions, such as the eruption of Mount Agung in Bali in 1963 and of Volcán de Fuego in Guatemala in October, 1974.

But the long-term aerosol record *for the globe* is far from clear. You will recall that a particle floating in the lower atmosphere at mid-latitudes will have a mean lifetime of only several days, since the atmosphere cleanses itself regularly with rain. In the polar regions the lifetime of particles is probably longer, and in the rainy parts of the tropics, shorter. Thus, the reported increases in aerosol content are most noticeable near the sources—except for volcanic aerosols, which remain in the stratosphere, above the weather, for several years and are spread worldwide. Significantly, vast regions of the world, including most of the Southern Hemisphere, where industrialization is still very thinly dispersed, have apparently experienced little increase in anthropogenic (man-made) aerosols. Increases tend to be most pronounced downwind of the industrialized areas of the Northern Hemisphere and in tropical regions where the slash-and-burn practices are widely used to clear land. William Kellogg has concluded that much of the human component of the earth's aerosol burden can be closely matched to regions of high economic activity, as measured by the gross national product (GNP). Plotting the aerosol load in units of GNP led Kellogg to produce a world map of aerosol pollution, which he calls Gross National Pollution.[58]

Since many aerosols efficiently scatter and absorb solar radiation and, to a lesser extent, infrared radiation, the heat budget is influenced in regions where aerosol concentrations have increased. Over land with a moderately high surface reflectivity, typical tropospheric (lower atmosphere) aerosols tend to warm the atmospheric column in which they lie and at the same time decrease the solar radiation energy reaching the surface. Over the oceans, which have low reflectivity, aerosols tend to cool since relatively more sunlight is reflected back to space when aerosols are introduced over a dark surface.[59]

Aerosols not only affect the radiation balance, but also influence the formation of clouds and precipitation since certain kinds of particles can serve as ice nuclei (i.e., form ice crystals in clouds below 0°C) and condensation nuclei (form water droplets in warmer clouds). This nucleation effect may well, as suggested by some cloud physicists recently,[60] influence climate more than the clear sky radiation effect (since clouds are so important to the heat budget), but so far we cannot even assess whether this could warm or cool the climate, let alone its overall magnitude.

In any case, the effects of CO_2, and perhaps aerosols, on the earth's radiation balance would appear capable of disturbing the *global* climate much more and sooner than the potential thermal effects of energy use—at least for the next several decades, when fossil fuels will continue to be burned in quantity and the total human energy use will contribute no more than a tenth of a percent of absorbed solar power to the total power density at the global surface.

The Ozone Layer: It Protects Life on Earth

Other chemical constituents are released into the atmosphere in great quantities. Among the most important are the chlorofluorocarbons, CFCs (commonly known by the Du Pont trade name, Freon), which are used as refrigerants and as propellants in some aerosol spray cans. You will also recall the recent and almost simultaneous discovery by a number of scientists[61] that chlorine, one of the chemical elements in CFCs, was highly efficient in destroying the atmospheric gas ozone. Ordinarily, ozone (chemical symbol O_3) is produced high in the atmosphere (about twenty to sixty kilometers up in a region called the stratosphere) by the action of ultraviolet solar radiation on oxygen molecules; and the rate at which it is produced to the rate at which it is destroyed is dependent on the presence of other gases known as catalysts. Once produced, ozone absorbs much of the ultraviolet radiation from the sun, which can be biologically harmful on earth, since it causes sunburn and even skin cancer in humans and can affect the functioning of natural ecosystems in still largely unknown ways. (For example, ozone absorbs all the sun's radiation at a wavelength of 0.26 millionths of a meter, a wavelength that damages the DNA molecule that is the building block of life.[62])

Furthermore, ozone is important to the earth's climate because it absorbs some quantities of both solar and terrestrial infrared radiation, thereby affecting the energy balance of the earth-atmosphere system that determines the earth's temperature. Unfortunately, how changes in the ozone concentration might affect climate are far more difficult to determine than the effect of changes in CO_2 since changes in surface temperature from variations in ozone depend on such diverse factors as whether the total amount of ozone is increased or decreased, whether the height at which the maximum amount of ozone occurs is altered, or whether the latitudinal distribution of ozone is disturbed. James Coakley, an atmospheric physicist and NCAR colleague, made one of the early calculations of the effect of changes in ozone on the global surface temperature. He found that a uniform reduction in the total *amount* of atmospheric ozone would lead to a cooling of the earth's surface, but that a decrease in the altitude in the stratosphere where ozone has its maximum concentration can warm the surface. Similarly, an increase in total amount of ozone warms, but an increase in the altitude of maximum ozone concentration can cool the climate. Suppose we knew that some atmospheric pollutant, such as CFCs, could act to reduce the amount of ozone in the atmosphere. Then, before we could conclude that this would lead to a global cooling, we would still also have to know if the CFCs moved the altitude of maximum ozone concentration up or down. If the maximum moved up, this would enhance the cooling effect of a decrease in the ozone amount, but if the maximum moved down, that situation would oppose the cooling from the decrease in total ozone. Thus, it is conceivable, with a probability near 50 percent—depending on whether the maximum height changed with or against the change in total ozone amount—that a large change in ozone could significantly affect climate. But even the direction of any potential ozone-climate effect is difficult to determine.

Yet the situation is even more complicated than this. The pollutants that destroy ozone can themselves interfere with the transfer of both solar and infrared radiation. V. Ramanathan and L. Callis, scientists at NASA's Langley Research Center, developed a radiation computer model to estimate the possible climatic effects of adding pollutants to the stratosphere. For the addition of nitrogen oxides to the stratosphere (from a fleet of SSTs perhaps), they showed that although a uniform (in altitude) decrease in ozone concentration would cool the earth's surface, the nitrogen oxides absorb enough solar radiation to offset about half of this ozone effect.[63] On the other hand, Ramanathan's next calculations showed that CFCs (which also destroy ozone) are very effective absorbers of infrared radiation, and even at the relatively low concentrations at which they are found in the atmosphere, could lead to a greenhouse effect type of influence on the climate. This warming would be about 0.9°K if present trends in CFC use continue to the year 2000, he reports.[64] If this early forecast is proven reasonably accurate, it means that continued burning of fossil fuels and increasing use of CFCs could combine to lead to very significant climatic warming before the end of this century!

On the other hand, Ramanathan told me that preliminary calculations suggest that ozone would be increased if CO_2 were increased. This, he said, is because an increase in CO_2 would cause a decrease in *stratospheric* temperature, which in turn favors ozone production. This could partially offset any CFC effects, but the CFCs, CO_2, and increasing O_3 would all combine to provide an enhanced greenhouse effect. Clearly, more research is needed in this area. But the message remains: the complexity of the climate system is no grounds for complacency, and extreme caution is needed to minimize global-scale effluents.

A decrease in the total amount of ozone, regardless of the altitude of its maximum concentration and the subsequent climatic effect, will certainly increase the dosage of ultraviolet radiation at the surface and probably increase the incidence of skin cancer in humans,[65] particularly among Caucasians who are more damaged by ultraviolet radiation (Figure 28). Although most ozone is produced photochemically in the high tropical stratosphere, it is transported poleward by atmospheric motions and thus accumulates in high concentrations near the poles. The farther away from the equator one goes, the more ozone an ultraviolet ray from the sun must travel through. Thus, high latitudes filter out more ultraviolet rays and consequently decrease the risk of skin cancer compared to low latitudes. Thus, the incidence of skin cancer decreases significantly as one moves toward the poles.[66] There are two basic reasons postulated for this: (1) The incidence of skin cancer is proportional to the cumulative (over a lifetime) dose of ultraviolet radiation (which makes deliberate and repeated sunbathing rather unwise). For this reason, the disease is more common in light-skinned people in lower latitudes because the climate is warmer and people spend more time outdoors. (2) The closer one is to the equator (or the lower the latitude) the stronger is the ultraviolet dose. Thus, the person who spends one hour in the sun in Dallas will receive more ultraviolet radiation than one who spends the same amount of time in the sun in Minneapolis at the same time of year. Also, more ultraviolet radiation reaches the surface at higher elevations. Thus, the ultraviolet risk is greater on top of a mountain than at its foot.

These exposure factors are crudely taken into account in Figure 28, so the curves are not highly biased by the exposure factor, and are representative of the fact that less ozone overhead may likely lead to increased incidence of skin cancer—although there are still many arguments over this connection. Perhaps, in light of this, it is not surprising that light-skinned people, whose resistance to sunburn and perhaps skin cancer is much weaker than dark-skinned people, have over the millenia tended to be concentrated away from the tropical latitudes—in places such as Scandinavia—where their skins are better adapted to the lower levels of ultraviolet radiation at the surface. Maybe, the inability to spend as many hours outdoors searching for food was a serious disadvantage to people highly susceptible to sunburn. Of course, this ultraviolet radiation–anthropological "theory" cannot explain the relative absence of dark-skinned people from high latitudes since they would certainly not be damaged in the higher latitudes by sunburn the way light-skinned people would be hurt in the tropical areas by higher ultraviolet radiation levels.

But there are physical explanations that have been proposed to explain this phenomenon. Jacob Bronowski, author of the widely read book, *The Ascent of Man*, proposed an alternative radiation link to explain the geographical distribution of racial characteristics: "Man began with a dark skin; the sunlight makes vitamin D in his skin, and if he had been white in Africa, it would make too much. But in the north, man needs to let in all the sunlight there is to make enough vitamin D, and natural selection favored those with whiter skins." [67]

Perhaps, in reality, both of these climatic–anthropological connections are needed to help explain the distribution of racial types. I merely raise the question.

However, the important question here is how human activities can affect ozone. I have already said that chlorine in the stratosphere destroys ozone. So also do the chemical oxides of nitrogen, but not as efficiently as chlorine. Nevertheless, it is known that nitrogen oxides (chemically known as NO_x) can be injected directly into the stratosphere by the exhaust gases of high flying aircraft (the supersonic transport aircraft, the SST, for example), or may be cast up by a nuclear explosion; or may perhaps even percolate up slowly through the lower atmosphere from where they were produced at the earth's surface (as the result of biological processes, a chemical reaction with nitrogen fertilizers or from automobile exhausts, perhaps).

The Case of the Supersonic Transports (SSTs)

In 1971 the U.S. Congress was the scene of hot debates over conflicting testimonies from various scientists and economists. The controversy was about whether the United States should subsidize the Boeing Aircraft Company of Seattle, Washington, to build a prototype SST to compete with the English-French Concorde SST or the Soviet Tupolov-144 SST. Proponents ranged from "patriots" who claimed that it was our national duty to compete in SST construction, to business and labor sympathizers who argued that aircraft sales would mean profits, jobs, and foreign currency. Among the opponents to the SST appropriations were economists who argued that the plane was impractical and unable to pay for itself (some called it a flying Edsel); protesters of government subsidies for a private industry; and environmentalists concerned first with the sonic boom and then with the potential effects of SST engine exhausts in the stratosphere. (Only the last point is discussed here.)

The SST debate always amazed me since the supporters (the Nixon Administration, established labor, and the mostly conservative Congress-people and senators who often oppose big "centralized government") were arguing *for* government investments in a private industry, namely, Boeing. In contrast, the opponents (generally the more "liberal types" who approve of expensive social welfare programs) were repeating the capitalist "free market" line to the effect that private enterprise would have funded the plane, if it really were economically sound. However, the environmental opposition, though insufficient in itself to have defeated the SST in Congress, provided enough extra weight to tip over the teetering congressional debate to defeat the SST appropriations. (Extraneously, a number of analysts have since suggested that the widespread opposition to the SST emanating from many corners of the scientific community contributed heavily to the Nixon Administration's disaffection from science, which began with many prominent scientists' opposition to the Vietnam War. That disaffection eventually led to the president's abolition of the post of presidential science advisor and thus the exorcism of a direct scientific influence from the White House.)

Testimony before Congress over the potential climatic impact of a fleet of SSTs was heated, especially since little had been done by the advocates of the plane to estimate these effects—except to deny their importance. And little was known by the opposition about the photochemical processes that create or destroy ozone. (Ironically, one of the earliest warnings of potential stratospheric contamination from SSTs came from Halstead Harrison, a scientist who then was working for the Boeing Corporation in Seattle.[68] He has, subsequently, moved uptown and now is a professor in the Department of Atmospheric Sciences at the University of Washington in Seattle.) The American SSTs were to be larger and to fly higher than their European counterparts and would thus—their engines were bigger—have a much greater impact on the stratosphere. The opponents cited evidence to suggest that a fleet of five hundred Boeing SSTs could deplete the ozone layer by some tens of percent. Just before the SST vote in Congress, Harold Johnston of the University of California wrote a controversial paper that appeared in *Science* magazine,[69] warning that NO_x from the exhausts of a fleet of Boeing SSTs could substantially reduce ozone. Paul Crutzen, who was then a research associate at the University of Stockholm, had already published a paper pointing out the role of nitrous oxides in the ozone photochemical system.[70] Up to that time, some opponents had argued that the water vapor from SST exhausts would reduce ozone. Ironically, this early prediction now seems dubious: Water vapor injected into the stratosphere may well have the opposite effect of increasing ozone (because it ties up some of the nitrogen oxides, a fact not appreciated in 1971).

Johnston's paper was attacked by scientists who supported the SST on the grounds that its findings were based only on a simple model, and that many other chemical reactions that were poorly understood or entirely left out of Johnston's calculations could have changed his conclusions. The SST debate became so intense that even the character and integrity of some scientists were challenged by pro-SST congressional interrogators.

The SST Debate: More Political than Scientific. The political principle here is important, and that is why the history of the SST case is being discussed here in so much detail. Supporters of the aircraft argued that the uncertainty about its potential effects rendered state-of-the-art estimates inexact, and thus did not justify opposition to the project—especially since killing the SST funds would bring economic hardship to Boeing management and workers in Seattle, a city already economically depressed. Opponents, on the other hand, argued that enough was known to demonstrate that the order of magnitude of potential climatic effects from a fleet of SSTs was serious enough to imply the possibility of irreversible environmental damage, and that the burden of certain proof of the aircraft's harmlessness should rest on the supporters. Still others opposed the idea that a fleet should be built until its environmental effects were better understood, but supported subsidizing the prototype in the meanwhile. But some opposed that idea as impractical since so much would have been spent on the prototype that great pressure would be exerted to build a fleet.

An interesting point here is that most of the bitter scientific antagonists in the SST debate were probably in far greater agreement on what was known and unknown scientifically, and on the odds that state-of-the-art estimates would be correct, than they were over whether the evidence justified opposition to the planes. That is, the interpretation of the weight of the evidence that guided their opposition or support was based not only on the scientists' technical knowledge of the issues, but also on their *personal* philosophies—on whether or not they wanted the SSTs and on whether they thought the benefits of the project were worth the risks of ignoring the worst possibilities. This is not to suggest that most testimony was deliberately misleading, but rather that scientists, like most people, shade to some extent their perception of the merits of conflicting evidence with the shadow of their personal philosophy. Even if this bias is subliminal—which it is in most instances—that is no excuse. At the least, scientists have the special obligation to help their listeners discern where their scientific expertise crosses over into personal philosophies. In future cases where high degrees of uncertainty exist, division of scientific opinion on the merits of a project will be frequent, and undoubtedly attacks on the technical and even personal credibility of one witness or another will nearly always be grounded in political motives, not in the spirit of scientific inquiry. The issues facing future generations are too critical to permit the technical components to be obscured in attacks on the personal philosophies of experts, a tactic designed to discredit the witness. It is to be hoped that we can learn to recognize this ploy and separate a scientist's presentation of technical opinions from his or her political or other opinions.

Let me give one recent example of this problem of separating scientific and political opinion on issues with public policy overtones. A respected meteorologist who has engaged in computer modeling of the earth's climate often strongly opposes those who use his (or anybody else's) estimates of the potential climatic impacts of human activities, if these estimates are given as an input to the political process to affect policy. He recently stated his opinions on this matter quite clearly to a scientific panel that was examining potential climatic consequences of long-term energy use. During one of the discussions, he complained that too little is known of climatic theory to say very much with any certainty about the potential impacts of, say, projected CO_2 increases. On the other hand, some scientists there, including me, felt that current knowledge of CO_2 effects was sufficient to consider this problem a potential climatic barrier to long-term energy growth. Although most of us agreed fairly closely on much of what is wrong and missing in present theories of the climatic consequences of increasing CO_2 concentrations, we exchanged differing views over the question of whether existing knowledge has relevance to political decisions about the future growth of industrialization. These differences of opinion, then, were much less ones of technical substance than of personal philosophies on how to act in the face of uncertainty. This scientist often shows impatience over what he feels are the excessively simplified and premature predictions bandied about in public by some people. And I agree that there is a danger in emphatic public pronouncements based on highly tentative theories. Yet, I think that the dangers of waiting for certainty can often be greater than the risk of prematurely releasing disturbing theories to the public. At the very least, the appropriate public response to a tentative theory should be decided through the political process, not exclusively by the decision of a panel of scientists about whether to release a tentative theory. The role of the scientist should be to spell out the possibilities, not to make political decisions about what information to withhold.

As for the CO_2 question, I feel that we cannot afford to risk waiting too much longer for scientific certainty before acting—and that is primarily a personal view of how society should act in the face of this uncertainty. More scientific certainty about the potential effects of CO_2 increases on the climate would reduce the difficulty in making the correct political decisions.

Fortunately, at the panel meeting on climate mentioned earlier, there was little difficulty in differentiating between the participants' political opinions and their scientific ones. Thus, the controversy was healthy and it helped to clarify problems. Unfortunately, this is not always the case in public debates or congressional hearings, where controversy can be used deliberately by some to generate confusion, rather than to clarify differences and point out uncertainties in critical issues.

The Climatic Impact Assessment Program. Returning to the case of the SST effect on ozone, a three-year study called the Climatic Impact Assessment Program (CIAP) was commissioned by the U.S. Department of Transportation shortly after the defeat of the SST funds in 1971. I participated extensively in that CIAP program and remember well that the most hotly debated items among the scientists in the program were not so much the order-of-magnitude estimates of potential SST effects, but rather whether those early estimates should be released to the public via the final CIAP report before more scientific certainty could be established. I recall an incident in August, 1974, in England where a group of British scientists (whose country was already committed to the Concorde SST) were grilling the CIAP director, Alan Grobecker, over some numerical estimates that he was considering for publication in the CIAP report. The British group argued strenuously that the estimates were too tentative to publish, but as a participant at that confrontation I defended Grobecker's intention to publish them anyway. Both sides, of course, were really arguing politics, not science, even though none of us was particularly conscious of that at the time; the vocabulary of the debate was highly specialized. (Incidentally, one of the estimates I was arguing should be published has been supported by recent findings, and one has been contradicted.) That one of these estimates was wrong (actually incomplete) is scarcely surprising, since CIAP had asked the scientists in the project to make their estimates as quickly as possible in order to suit a political decision deadline (the report was due in early 1975), not a scientific schedule cognizant of the need for sufficient time to achieve considerable technical certainty.

Congress wanted a "technology assessment" on the SST in three years. We thus had little choice but to estimate as best we could. Or we could have refused, as many did, to participate in CIAP because of the deadline. Those of us in favor of participating in the program and then making our estimates public before reaching a satisfactory scientific consensus had a favorite argument: "Would you prefer to estimate the SST impact, or have Senator Goldwater guess it for you?" As it turned out, some of the early estimates of the 1971 debate were changed very little by new findings, while others looked wrong. Three years of intensive CIAP studies failed to disprove Johnston's early warning that a fleet of 500 Boeing SSTs would reduce ozone by tens of percent, but these years of study vindicated to some extent the supporters of the Concorde, since their smaller engines and lower cruising altitude imply that a few hundred of them would probably be of only minor consequence to the stratospheric ozone shield, as the work of Paul Crutzen,[71] an atmospheric chemist at NCAR and NOAA, shows (Figure 29).

Nevertheless, after the CIAP Report of Findings appeared in early 1975, several editorial opinion-makers across the United States decried the prophets of doom and overzealous environmentalists who were blocking progress, as exemplified by those who were wrong in 1971 about the SSTs being environmentally dangerous. But these promoters of progress forgot, overlooked, or didn't know that a number of so-called environmental activists were, it turns out, remarkably astute in their early estimates of the ozone-destroying potential of a fleet of Boeing SSTs—the plane that the U.S. Congress was considering funding in 1971.

One CIAP participant, Thomas Donahue, president of the Solar Planetary Relations Section of the American Geophysical Union (AGU), helped to arrange under AGU auspices a technical review of early CIAP documents. After the final CIAP Report of Findings was published [72] in December, 1974, Donahue sent an angry letter to *Science* magazine. He bristled over the lack of sufficiently prominent statements in the report to the effect that the 500 Boeing SSTs voted down in 1971, in part from the early calculations of several scientists, would have had serious environmental consequences. "No such clearly stated specific conclusions are found in the Report of Findings, although they can be deduced by someone familiar with the history of the study or a perceptive person who goes to the trouble of reading fine print, footnotes of the tables, or who pursues some of the statements in the report to their logical ends . . . Those who raised the alarm have been effectively discredited and stand accused of providing damaging counsel to this country. I hope," Donahue concluded, "that this letter will repair a little bit of the damage that has been done." [73]

Grobecker replied that no such discredit was intended in the CIAP report and readily admitted that "the alarm in 1971 created by fears of such an effect [i.e., NO_x emissions] was an important consideration in causing cancellation of U.S. plans to build an SST fleet and has had a useful result in stimulating stratospheric and other researches necessary for such an avoidance." [74]

Donahue replied that he was pleased to read Grobecker's comments and expressed hope that the "director's letter goes a long way" toward explaining what the CIAP report really said.

Unfortunately, most citizens and opinion-makers will never read the Donahue-Grobecker exchange in *Science* magazine, but will remember the early stories about overzealous environmental activists. This is typical of how the public is insufficiently informed on such complex technical issues as the SST debate. Even though some early SST–environmental estimates were wrong or simplistic because of the meager data available in 1971, I strongly believe that it was the duty of the estimators to warn of the dangerous plausibilities they foresaw *and* of the uncertainties of these potential effects; they also had the right as citizens to demand that more certainty be established *before* going full speed ahead with a project that would be very hard to abandon once the go-ahead was given. Congress wisely agreed. The analogy to nuclear power, especially current plans to build a large number of plutonium-producing breeder reactors, is instructive, as we will discuss later on.

When "Inert" Isn't Always Inert: The Case of the Aerosol Spray Cans

Returning to the ozone reduction problem, the bizarre case of the aerosol spray cans requires immediate evaluation. Some spray cans contain certain substances (deodorants, insect repellants, paint, cleaners, hair spray, and many other things) propelled by a number of chemicals, most notably an ostensibly harmless family of "inert" gases, chlorofluorocarbons (CFCs). As also mentioned earlier, a number of research groups originally concerned with the photochemistry of ozone had, over the past few years, independently discovered the phenomenal ability of some forms of chlorine present in the high atmosphere to destroy ozone. But, since there was no obvious significant source of stratospheric chlorine, let alone any hint that atmospheric chlorine would increase in concentration, little attention was paid to this knowledge.

However, in 1974 Mario Molina and F. S. Rowland concluded that CFCs—which are relatively insoluble in water and apparently do not interact with the oceans or degrade biologically—could be easily broken down by the action of ultraviolet radiation of the kind present in the high stratosphere; and one of the components resulting from the breakdown of CFC is chlorine. Following the Molina–Rowland discovery, several groups that were engaged in constructing models of the photochemistry of the stratosphere (through funding largely motivated by the SST problem) calculated the effects on ozone of increases in chlorine. The groups were those of Ralph Cicerone and Richard Stolarski at the University of Michigan, Michael McElroy and Steven Wofsy at Harvard University, Paul Crutzen, and Julius Chang at Lawrence Livermore Laboratory, among others.

Rowland was quoted as describing his sudden realization of the CFC–ozone link: "There was no moment of 'Eureka!' really. I just came home one night and told my wife 'the work is going very well, but it looks like the end of the world.' "[75]

The media have been showing considerable interest in the potential effects of the exponentially growing use of CFCs, and some states have already banned their use. The production of CFCs (Figure 30) shows a trend of dramatic growth,[76] whose continuation has become questionable. Let us examine this controversy.

What, then, might continuation of growth in the use of CFCs mean for the ozone shield? Calculations of Crutzen[77] and his colleague

Ivar Isaksen are typical of those made with a simple one-dimensional model. This model indicates that a 10 percent annual increase in CFC production and use continuing to 1998 could reduce the ozone shield by about 11 percent (Figure 31). Natural fluctuations of ozone concentrations from year to year can be 5 to 10 percent; thus, current theories of ozone depletion cannot easily be verified experimentally in the next few years, because any change in the ozone layer due to CFCs would be masked by the natural variability. Unfortunately, researchers have found that by the time a change would be large enough to verify the theory, it would be ten more years until the maximum reduction of ozone would be felt.

Crutzen and Isaksen simulated termination of CFC use as of 1998, and their most interesting and worrisome finding is that the ozone concentration goes right on decreasing for ten more years after CFC production is cut off (Figure 31). This is predicted because the CFC released into the lower atmosphere seems to have no known sinks to remove it, and thus a decade or so will expire before it spreads and reaches sufficiently high into the atmosphere to be broken down by ultraviolet radiation. Early calculations suggest that perhaps a century would lapse before the ozone concentration returned to "normal."

Estimates of the biological effects of ozone reduction are, like estimates of its potential climatic effects, surrounded by controversies and uncertainties. Nevertheless, like climatic effects, order-of-magnitude estimates of the anticipated ozone decreases can be given: Thomas Maugh and Allen Hammond, research news writers for *Science* magazine, reported that "a 5 percent ozone depletion might cause anywhere from 20,000 to 60,000 additional cases [of skin cancer] per year in the United States alone. Fortunately, 90 to 95 percent of such cancers are curable";[78] that is, some annoying treatments can remove most of the diseased areas. But even if all people afflicted asked for medical treatment, could there still be some 1,000 to 6,000 cases of incurable cancer caused in the United States alone from the 5 percent decrease? No one is certain whether increases in ultraviolet radiation also cause incurable forms of skin cancer.

Even less certain, but potentially more serious, is the effect that increased ultraviolet radiation might have on natural ecosystems. Whether cell damage from increased dosage would be repaired in time to prevent serious changes in biological productivity is an open question. Experiments were carried out under CIAP to irradiate various species with increased ultraviolet. Also, a National Academy of Sciences report on this problem was cautious about the extent to which present knowledge could say anything definitive about the ecological effects of ozone decreases, but it did offer a strong warning about the danger of our ignorance: "Because of the relationships between species in ecosystems, damage to one species might jeopardize an entire ecosystem. Hence, the potential effects of any elevation of the present [ultraviolet radiation] reaching the earth's surface should be taken most seriously and studied thoroughly." [79] This report of a panel of scientists did not suggest how to determine when enough studying had been done to recommend political action.

When might a 5 percent decrease in ozone occur? Calculations suggest that such a depletion would occur by 1990 (Figure 31). But this depletion could occur even if worldwide CFC use stopped in 1980, because the 1990 ozone reduction effect arises from CFC released at the surface a decade earlier. A slow recovery of ozone concentrations back to normal seems to follow decades after the CFC cutoff. The political problem is clear: How long do we study before we act?

Ban the Can?: How Much Proof Is Enough? We must ask whether something might be seriously wrong with the estimates of Crutzen's group, or the Harvard group, or the Michigan group (since all are in general agreement). Basically, the critics of their CFC–ozone theories are attacking the calculations: first, because they are based on a less than perfect understanding of the photochemical processes occurring in the stratosphere, which govern ozone production; and, second, and perhaps more important, because they are based on one-dimensional models while the real atmosphere is three-dimensional. For example, an industrial group called the Chemical Specialties Manufacturers Association (CSMA) reported to their colleagues their views of the ozone controversy in a newsletter: "Molina and Rowland themselves admit their findings are based on assumptions, not facts. The same is true of the findings of [the other groups]," CSMA states. The report goes on to quote Igor Sobelev of Kaiser Aluminum and Chemical Company, who doubts the CFC–ozone theory even though measurements made by "James Lovelock of the

University of Reading, England, show that the amount of fluorocarbons present in the lower atmosphere closely approximates the amount of fluorocarbons manufactured." Sobelev argued "there is still a possibility that a tropospheric sink exists."[80] However, a recent report to *Science* magazine by a team of scientists from the U.S. Naval Research Laboratory concluded that indeed "the evidence provides no indication that significant sinks exist"[81] for the destruction of CFC in the lower atmosphere. In addition, industry-defender Sobelev argues that the three-dimensional nature of the real world "could mean that different chemical reactions will predominate. And the possibility exists," Sobelev insists, "that one of these reactions could result in the deactivation of reactive chlorine."[82] However, Sobelev neglects to mention that the sword of uncertainty cuts in both directions. That is, his unknown chemical reactions are just as likely to make the situation worse as better.

Finally, the CSMA document concludes that in light of still missing pieces of knowledge in the CFC–ozone puzzle "the scientists in the field are not in agreement over the importance of the Molina–Rowland theory, but they do agree there is much work to be done before any reasonable action can be instituted."[83]

Notice the political implications of the word "before." Probably Sobelev and Rowland would agree far more easily on the technical uncertainties of the CFC–ozone problem than over the question of whether regulatory action on CFC emissions should await more scientific certainty. Unfortunately, if the Molina–Rowland theories are right, such certainty may not be achieved much before a few thousand additional skin cancer cases are recorded or some ecosystems are irreversibly damaged. Furthermore, the CSMA assertion that scientists in the field agree that more work is needed before regulatory actions are justified is distortion for the apparent political purpose of buying time for the multibillion dollar CFC industry (although substitutes may already be available and thus little economic hardship may befall the industry).[84]

Two prominent scientists in the field told me in conversation that they might not strongly oppose another year or two more study before the imposition of governmental regulations, but both expressed worry that if they are right, or if they are too low in their estimates—which is just as likely as their having overestimated the ozone depletion—the price of giving the CFC manufacturers more time to establish

"By George! I think you've done it!"

Drawing by Dana Fradon; © 1974 *The New Yorker Magazine, Inc.*

such certainty may be very high indeed for thousands of people and unknown numbers of plants and animals. On the other hand, two other respected scientists in the field said they favor immediate legislative regulation.

It is extremely difficult to make value judgments, weighing some unknown thousands of potential skin cancer cases or speculative amounts of damage to ecosystems against the hardships of possible unemployment and economic loss for a fair number of industrial employees and stockholders (again assuming no substitutes for CFCs are immediately available). It should be noted, however, that the CFC industry, unlike the SST manufacturers, has had no advanced warning that their product might be environmentally dangerous, and thus the recent findings have given them little time to absorb the shock. No one is collapsing in the street from small amounts of ozone depletion, nor will they be in the foreseeable future. The effects, if any, of small changes on the ozone layer will be felt, unlike the economic impact, only years later in the skin cancer statistics; larger depletion, however, might have more immediate and serious effects on the ecosystems, which supply us with basic functions such as pest control, as well as on the climate.

As a result of all this uncertainty, regulation of an influential industry—CFC manufacturers in this case—is controversial since the knowledge of the harmful effects of the CFCs is based only on a plausibility case against their continued use. This presents an especially unhappy political dilemma for a government confronting the economic disarray of the middle 1970s. Deciding when precisely to stop studying and start regulating merely to insure against a probability, albeit a possible irreversible one in terms of biological beings, is at best a touchy political question involving substantive matters as well as the difficult problem of proper timing. "We must be careful to cry wolf at the right time," [85] admits Michael McElroy of Harvard. But we must also remember that every extra hour of grace granted the CFC industry before regulations are imposed may be bought at the expense of innocent people—including many on the other side of the earth who have no say in our decision of whether or not to ban the can.

"What the Hell Else Slipped By?"

Two other issues need mention in connection with the question of gases released at the earth's surface that may eventually work their way upward into the stratosphere and adversely affect the concentration of ozone. One is nitrous oxide, N_2O, commonly known as laughing gas. A number of years ago, Paul Crutzen worried that N_2O, a by-product of chemical reactions in the soil or oceans, might be increasing as a result of the rapid growth in the use of nitrogen fertilizer, the mainstay of Green Revolution agricultural productivity.[86] If so, and if N_2O destroys ozone, then this realization could be very disconcerting because the increasing use of fertilizer is absolutely essential in any planning effort designed to enhance world per capita nutrition standards. If population growth continues, and it shows regrettably little inclination to level off soon, the new mouths must be fed somehow. And so emerges another growth dilemma: How are we to increase fertilizer use to improve the quality of life and sustain population growth when that very growth in fertilizer use may lead inevitably to increases in the concentration of N_2O? This might lead to decreases in the ozone concentration and the attendant risks of damaging biological life by enhanced dosages of ultraviolet radiation. We are once again confronting an extremely difficult value judgment that is punctuated by immense uncertainty, involves substantial trade-offs, and constitutes yet another possible climatic limit to exponential growth.

More recently, Crutzen re-examined the N_2O question and then he concluded that "it is likely that there exists a large, but still unknown, sink mechanism for N_2O in the atmosphere." He also said that at present rates of nitrogen fertilizer use "it does not seem likely that man can influence strongly the biological production and atmospheric abundance of N_2O." However, since vastly increased amounts of nitrogen fertilization are being considered as part of world strategies, he is cautious and warns that "if the global production of nitrous oxide were to increase by 20 percent, then a decrease in total ozone by about 4 percent is estimated. It is therefore important to monitor the concentrations of atmospheric N_2O and learn more about its production and destruction mechanisms." [87]

In June, 1975, I attended a workshop on the stratosphere at NCAR where Michael McElroy, the atmospheric chemist from Harvard, was discussing this same subject. He did not challenge Crutzen's calculation on the effect of N_2O increase on ozone (happily, since Crutzen was in the audience), but he strongly disagreed with the paper (of another author)[88] that Crutzen used to estimate the natural to artificial (e.g., from fertilizer) sources of N_2O. Thus, McElroy contended that the potential seriousness of a fertilizer–ozone crisis could be even greater than previously thought, because the relative fraction of atmospheric N_2O traceable to agricultural practices may be many times greater than assumed. For example, using the figures for nitrogen fixation rates from R. Hardy and V. Havelka,[89] scientists with the E. I. Du Pont de Nemours Company, we can estimate the relative fraction of fixed nitrogen that is attributable to fertilizer manufacture. Nitrogen fixation is the process by which atmospheric nitrogen gas, N_2, is converted to a form usable by plants. Dentrification is the process that undoes fixation, that is, biologically fixed (usable) nitrogen is reconverted to either nitrogen gas or nitrous oxide, N_2O.[90] Up until now these processes have largely been carried out naturally by microscopic bacteria or algae in soils or waters, but the amount of nitrogen fixation by the industrial process that makes nitrogen fertilizer is becoming significant. The Hardy and Havelka figures, which are admittedly crude, estimate that 175 million metric tons of nitrogen are fixed biologically and that about half of this is related to agriculture. Abiological processes, such as lightning or combustion, produce about another 45 million tons annually. In 1974 about 40 million metric tons of nitrogen were fixed in the manufacture of nitrogen fertilizer. If these estimates are used, then present fertilizer manufacture already accounts for 15 percent of all fixed nitrogen (i.e., 40 divided by the sum 40 + 45 + 175). But, to meet food needs over the next twenty-five years, projections of nitrogen fertilizer use have been suggested as high as ten times present use. "In the absence of alternative technologies for producing fixed nitrogen," Hardy and Havelka suggest, "about 200 million metric tons of fertilizer nitrogen will be required annually by the end of this quarter century"—a 500 percent increase over present usage. Let's use their estimate.

Assuming that 200 million tons are produced in the year 2000, we must ask: What percent of increase of fixed nitrogen over today's levels does this represent? A simple scaling argument can give the answer: a 62 percent increase (i.e., 200 minus 40, divided by the sum 40 + 45 + 175).

Would this 62 percent increase in fixed nitrogen lead to a 62 percent increase in denitrification and thus a 62 percent increase in N_2O production? If so, how long would it take to convert the extra fixed nitrogen to N_2 or N_2O? Would a 62 percent increase in N_2O lead to a 12.4 percent decrease in ozone (merely scale from Crutzen's estimate of a 4 percent ozone decrease from a 20 percent increase in N_2O to obtain a 12.4 percent ozone decrease from a 62 percent N_2O increase)?

These questions were argued vigorously at the NCAR workshop, and few definitive answers could be given because so little is known of the natural nitrogen cycle. Some workshop participants thought the Hardy and Havelka estimates for fixed nitrogen amounts in the biosphere could be off by as much as a factor of 10. If these values for natural fixation are really underestimates, then the problem becomes far less serious. But if they are overestimates by a factor of 10, then the projected use of 200 million tons of nitrogen fertilizer in 2000 A.D. would imply a 156 percent increase in fixed nitrogen. Scaling this to Crutzen's calculation suggests a 31 percent decrease in ozone. And, if fertilizer production in the year 2000 were 400 million metric tons, the effect would nearly double (assuming linear scaling again).

The numbers given here are *not* predictions. They are given primarily to illustrate that the potential magnitude of the nitrogen fertilizer effect could be large indeed. The question is still far from settled, but plainly demands immediate resolution, since vast increases in agricultural activities and fertilization are the key to food self-sufficiency hopes in the developing world. In my view this case is so serious that it deserves perhaps three CIAP studies, if they could lessen the uncertainties soon.

The general message to be derived from this case is a familiar one: The potential risks from the unexpected side effects of human activity grow in proportion to the scale of the activity and can begin to become quite worrisome as the activity escalates to the point where global disturbances are possible (e.g., with CO_2 or CFC release). Although I fully recognize that arguing for the restriction of human endeavors once they become "large" makes one vulnerable to the accusation of being blindly "anti-progress," I can't help but recall the poignant question McElroy was reported to have asked soon after the discovery that "harmless" and "inert" gases, CFCs, could pose such a serious environmental hazard: "What the hell else slipped by?" [91] I share that feeling, especially since our pollutants are becoming increasingly global in scale; and we are releasing thousands of synthetic compounds with unknown properties, with which most organisms have had little evolutionary experience.

Before we leave the chlorine–ozone problem, the question of carbon tetrachloride (CCl_4) in the atmosphere merits some discussion. CCl_4 is a volatile chemical often used as a powerful solvent, yet its abundance in the atmosphere is, ostensibly, far greater than can be explained by industrial use. At the same time that CFC concentrations have been increasing—to the point that a sizable fraction of the ozone-destroying chlorine atoms in the air that are expected to increase come from that source—CCl_4 has been discovered to exist in abundance comparable to the CFCs. That discovery raises many questions: Where did so much CCl_4 come from, how is it removed from atmosphere, can it penetrate the stratosphere and destroy ozone, and will it increase with time? These questions are all unanswered, but exceedingly important.

There has also been speculation that the chlorine compound chloroform is a byproduct of chemical reactions in rivers and estuaries, where the source of chlorine would be the reagents employed in sewage treatment or water purification processes.[92] If this hunch is right and chloroform can reach the stratosphere and destroy ozone, still another difficult value judgment involving major trade-offs, and thereby a further climatic limit-to-growth dilemma, could appear. Since sewage treatment and water purification are crucially important technologies if health standards are to be improved in the less developed world (and maintained in the developed world), chlorine use may well grow dramatically in the next few years. In this instance, too, further population growth will only exacerbate the situation, increasing the already unmet need for large quantities of pure water. However, if the expansion in the use of chlorine causes increases in the atmospheric abundance of chloroform and ultimately a reduction in ozone, then a dilemma arises: Are we to choose ozone reduction, with its attendant increase of damaging ultraviolet radiation, or impure water and untreated sewage, with the attendant general degradation of health standards? Whether the chlorination–chloroform–ozone speculation proves serious remains to be seen since at present it is an argument based on little more than pure speculation. (Ironically, an alternative to using chlorine for water treatment would be to use ozone, which can be produced in a highly energy-dependent industrial process. And who could argue that such a process would cause depletion of the ozone layer! However, ozone in the *lower* atmosphere is extremely reactive and toxic and is a dangerous pollutant.)

But the fertilizer–N_2O–ozone connection is supported at least by some order-of-magnitude calculations, albeit based on rather incomplete knowledge of the nitrogen cycle. And, worse yet, fertilizer increase is a must if a growing world population is to be fed over the next few decades, and there is no known substitute. We have already grown dependent on this technological fix, whether we like it or not. Of course, more knowledge is needed quickly to see whether there is any substance to these speculations, but nonetheless the growth consequence emerges once again: The larger the technological fix and the greater our dependence on it, the greater the risk that its potential environmental side effects will be global and serious; there is also an increased risk that we will be faced with the harrowing prospect of choosing between the removal of a technological prop on which the maintenance of our growing population depends or suffering the impact of global environmental damage.

One more ironic aspect of the chlorine–ozone controversy should be mentioned. The paper in this book is nearly white, most likely because it has been bleached with chlorine in a process that can produce chloroform.[93] Therefore, it is conceivable that the more copies of this book we print, the more ozone it will destroy! Of course, I must rationalize: If people become alerted to the problems discussed in these pages, then the potential ozone damage is worth it.

More seriously, I indulge in this speculation merely to point out that we are only beginning to understand the potential impact our global effluents may have on the environment; a great deal must be learned quickly about the workings of the atmosphere so that we can estimate in time the extent to which our activities influence natural processes.

If the reader does not yet feel that he or she has read the ultimate in science fiction-like doomsday scenarios for the future, I will now recount an event that should bowl over even the most jaded. In 1973 I was invited to be a member of a panel in a National Academy of Sciences study, which was requested by the Arms Control and Disarmament Agency (ACDA). The panel's task, I was told, would be to draft a report on the potential climatic consequences of a large-scale nuclear war. I was incredulous! This seemed to be the modern classical example of "locking the barn door after the horse has escaped." I recall being thoroughly stunned by the realization that, in this often ridiculous world we live in, some people actually need to be clubbed with the knowledge that a nuclear war can cause severe global environmental damage before they become convinced that there can be no winners in such a conflict. Perhaps the need for this report was the ultimate proof of Dostoevsky's musing that "the world stands on absurdities."

At any rate, I went to the meeting and amidst all the loose chatter of a "10,000 megaton war" versus a mere "5,000 megaton war" I discovered that such wars could almost instantaneously reduce the ozone shield by some 50 to 75 percent. Only by assuming the very much unproved contention that the photochemical processes associated with ozone production are reversible after such a large jolt, could it be calculated that the ozone shield would restore itself in some five to fifteen years, depending on whose computer model results one believed.

I also learned that the quantities of dust likely to be kicked into the stratosphere from such blasts might be equivalent to those from a large volcanic eruption. Given these inputs, I estimated the possible climatic consequences with the same sorts of caveats climate modelers always use about order-of-magnitude possibilities for climate modification. Perhaps, I speculated, a few tenths of a degree to a few degrees Celsius surface temperature cooling would follow the ozone reduction and the stratospheric dust cloud. No, I couldn't rule out the possibility that this sudden ozone reduction might trigger an irreversible climatic change—such as a new ice age—but I gave that little probability since the estimated changes in stratospheric composition were not anticipated to persist more than a decade or so.

One last Strangelove-like recollection. "Social assumptions" were to be avoided as far as possible in the report. For example, guesses of the effects of the destruction of the fertilizer industry on the livelihoods of the noncombatant nations were to be resisted. Since one purpose of the report was to estimate the damage a nuclear holocaust might bring to noncombatant nations, the hesitation to include such social assumptions as this struck me as a good example of the old platitude about "throwing out the baby with the bath water." Nevertheless, I made a social assumption when writing the following lines for the report on the possible climatic effects on a nuclear war.

> In trying to assess the potential climatic consequences of a nuclear war, it must be realized that the climatic influence of human activities (e.g., industrialization, farming, or deforestation) could be drastically altered. Although it is still not universally agreed among climate specialists, the influence of human activities on climate may already have become significant, and almost surely will become more significant if present growth trends continue. Therefore, if a significant fraction of the continental areas in the Northern Hemisphere now under the plow were to be abandoned after a nuclear war, the surface albedo (or water-bearing characteristics) of several percent of the earth's surface would be changed by an amount equivalent to the difference between the albedo of plowed fields and the albedo of the vegetation that would replace them (presumably several percent darker than farmed areas). While prewar climate shifts from the conversion of natural vegetation areas to farmland may already have occurred as a result of agricultural practices, these shifts would have evolved slowly in comparison to the speed at which postwar natural vegetation would replace prewar farmland. Therefore, a rapid climatic shift could be perceived by the postwar surivivors whose food crops would have already been adapted preferentially to prewar climates.

Similar arguments could be advanced in the case of atmospheric aerosols (dust). That is, if the atmospheric dust loading from prewar human activities had, in fact, slowly caused a climatic shift, the sudden elimination of the source of anthropogenic dust could cause a rapid postwar climatic response.

The major point for climate purposes about social assumptions is that at present levels of knowledge, we cannot assume that potential climatic shifts resulting from direct atmospheric effects of a large-scale nuclear war will prove any larger than the possible climatic changes resulting from differing levels of human activities in a postnuclear war period. In fact, the disruption of general patterns of industrial and agricultural practices over large areas might well produce longer-lasting effects than the transient ones caused by the nuclear impulses themselves.[94]

This portion of my material was hardly used in the NAS–ACDA report [95] because, I was told, it relied too heavily on a social assumption. (One of the privileges of writing a book such as this is the opportunity to include in it material rejected elsewhere.) Despite the toning down of the crucial importance of such social assumptions, the report still goes a long way toward accomplishing its purpose: proving to the world that no one "wins" a nuclear war.

Deliberate Climatic Modification Schemes

Why Tamper with the Climate?

The preceding discussion of the inadvertent consequences of a military activity (nuclear war) leads to one final point in the discussion of some examples of inadvertent climatic modification, a point that sounds paradoxical: namely, that there are potential inadvertent consequences from advertent climate modification schemes. Such unintended side effects are likely to occur because we are still relatively ignorant of the workings of the climatic system, and could easily misjudge or ignore the impact on the system of any deliberate changes we might seek to induce. Such advertent changes might include a massive and continuous effort at weather modification with conventional cloud-seeding techniques, the attempt to melt the Arctic Sea ice, the injection of stratospheric dust to offset the inadvertent climatic warming from CO_2 increases, or the operation of a particular climatic modification scheme to gain tactical military or economic advantages. These cases of deliberate climate modification schemes have at least one thing in common: They all can backfire and cause unexpected side effects that could not be anticipated, given the imperfect knowledge of climatic theory available at the time they might be attempted. The potential for interstate conflicts to arise from the implementation of such scenarios is high, as shown in the next chapter; but here I'll merely discuss several examples of advertent climatic modification schemes that have been, or might soon be, proposed because the technology is already at hand or believed to be near. Why would anyone want to modify the climate? A short list of reasons can easily be given:

(1) To offset an inadvertent climatic change from exponential growth.
(2) To relieve drought.
(3) To relieve or prevent flooding.
(4) To increase fresh water supplies.
(5) To offset a gloomy long-range weather forecast.
(6) To stabilize the climate.
(7) To improve food production.
(8) To gain military or economic advantage.

Setting aside for the moment the question of whether it makes sense to alter the climate, let us pick up on a number of points I reviewed together with William Kellogg [96] and re-examine some of the schemes that have been suggested for modifying climate on a hemispheric or global scale, schemes that have so far been considered to be on the fringe of science fiction. The range of possibilities widens rapidly if one imagines the financial resources of the major world powers available to carry them out. These schemes are summarized in Figure 32, and some of them will be described briefly in the next few pages.

Melting the Arctic Sea Ice

As mentioned earlier, one perennial suggestion—none of these should be considered as firm proposals—is to eliminate the Arctic Ocean ice pack. This layer of drifting ice that covers most of the Arctic Ocean varies in average thickness from less than 2 m in summer to about 3 m in late winter,[97] and if it were removed, the characteristics of the northern polar regions would likely be dramatically different. An open ocean could result in much more moderate and quite possibly more snowy winters around the Arctic basin, with January temperatures some 10 to 15°C warmer than now.[98] Nobody knows whether this could backfire and start another glaciation of northern Canada and Europe because of increased snowfall, but such an outcome is a definite possibility. Furthermore, a change as large as elimination of the Arctic Sea ice would almost surely cause important climatic changes in places far from the Arctic basin. Of course, the temperature rise in the Arctic resulting from either the currently increasing carbon dioxide concentrations or the growing global thermal pollution that have been described earlier might be quite sufficient to melt the ice without any extra effort. This has been pointed out [99] by Mikhail Budyko, a Russian scientist. There are, however, a number of ways to help the process along.

Spreading black particles such as soot by aircraft is one of them. Gary Maykut and Norbert Untersteiner, ice dynamicists at the University of Washington,[100] say a 20 percent decrease in the reflectivity of a large area of the ice would cause it to disappear in a period of about three years, simply because it would absorb great quantities of the solar heat energy that is normally returned to space. Another suggestion is to dam the Bering Strait and pump water from the Arctic Ocean into the Pacific, thereby drawing warm Atlantic water in from the other side and raising the surface water temperature enough to melt the ice pack.[101] A third way might be to detonate "clean" thermonuclear devices in the Arctic Ocean to fragment the ice and stir up saltier, warmer water from below. Diverting northward-flowing rivers that add fresh water to the Arctic Ocean would speed the process, since the present surface layer of low salinity (and thus of lower density) a few tens of meters deep is partially replenished by these rivers; if it were eliminated, the pack ice would grow less rapidly in winter.[102] This latter effect might easily be achieved inadvertently, simply by diverting those rivers that normally feed fresh water into the Arctic to the grain lands in the more southern regions of Canada and the U.S.S.R.

Science Fiction Indeed!

Although all of these projects may sound unrealizable at present, I am not at all talking about the impossible or writing a science-fiction novel. For example, a short unobtrusive story appeared on page 8 of the February 16, 1975, *New York Times:* The headline read "Atom Device Used for Soviet Canal," and the subtitle was "Russians Describe '71 Test, First Nuclear Explosion for Big Water Project." The purpose of the project, the article explained, would be to use some 250 nuclear charges to excavate a "70 mile canal, which would link the north-flowing Pechora River with the south-flowing Kolva River" in order to replenish the water level of the Caspian Sea, which has been plagued by a drop of over eight feet in the past thirty-five years. "The test, which produced a crater 2300 feet long, 1100 feet wide at its widest point and 35 to 50 feet deep involved three charges of 15 kilotons each, buried at a depth of 400 feet about 500 feet apart." The article quoted a Soviet report to the effect that the test "had demonstrated the idea's feasibility." No mention of the potential effects of the removal of fresh water from the Arctic Sea was given, although it did report that the nuclear blasts were conducted only five miles from civilization. Only "light damage in the form of cracks in ovens and in wall plaster was observed in eight buildings of the nearest town," the Soviet report said, but no mention of radioactive contamination was apparent.

Science fiction indeed!

If the Arctic Sea ice pack were removed, the magnitude of the equator-to-pole temperature gradient that influences midlatitude storminess would be changed, and quite possibly significant climatic changes would be produced in other areas of the globe, though on a scale that is now uncertain.

Let's continue with a few more examples of deliberate climatic control schemes.

It has also been suggested that a massive extension of present cloud-seeding techniques could be used to modify precipitation patterns and release latent heat on a regional or hemispheric scale. By attempting the regular "steering" of hurricanes (or typhoons, as they are called in the western Pacific), the climate of hurricane-prone regions might be considerably altered, or at least regularized. One way to do that might be to pump cold water from the depths of the ocean to the surface directly in the paths of the hurricanes, since they are known to draw much of their energy from warm surface waters.

Cloud-seeding experiments have already been carried out by NOAA's Project Stormfury in the tropical Atlantic Ocean to test the feasibility of lessening the force of the maximum winds in hurricanes. A number of theoretical computer experiments have suggested that this technique may be successful,[103] and Project Stormfury cloud-seeding airplanes conducted a number of experiments to test these theories.[104] Hurricane Esther was seeded in September, 1961, Beulah in August, 1963, Debbie in August, 1969, and Ginger in September, 1971. The results are still tentative, but seem to suggest that wind-speed reductions are quite possible. Cecil Gentry, director of the National Hurricane Research Laboratory, wrote that "a reduction in the maximum winds should result in less damage to trees and shrubs, so that the effects of modifying the storm should be beneficial in this regard."[105] He is aware of the danger of potential side effects from seeding, but argues that the seeding is "not expected to destroy hurricanes, change their tracks, or have much effect on the rainfall."[106]

But there are dissenters. Swedish meteorologist Hilding Sundquist constructed a computer model[107] of hurricane development suggesting that seeding would be followed by effects opposite those predicted with the U.S. model. Furthermore, it could be asked how we know that seeding will not affect storm tracks, since we don't yet adequately understand what governs the natural course of a hurricane. The perception that a seeded hurricane was knocked off track is perhaps more credible to a hurricane victim than an educated guess of a hurricane modifier about the impact of a hurricane modification experiment. There is a danger here because of the conflict that could arise if one state thought that the modification efforts of another state had brought it damaging weather.

Thus, these may be risky schemes unless there is reasonably certain knowledge of the full consequences of their implementation. And a proposal to tamper systematically with hurricanes headed for the southern United States, for example, would raise great concern south of the border, since the gulf coast of Mexico benefits considerably from the rainfall brought by these same hurricanes.[108] Similarly, the Japanese are very upset by American and Philippine plans to seed typhoons in the Pacific. Japan's agriculture and water supply count on the rainfall from the typhoons, and the Japanese fear the inadvertent consequences of experiments of this kind, such as increased damage from storm surges. Fortunately, the scientists involved in the project are fully aware of the risks of some nations' perception that a hurricane modification experiment "caused" some damaging change to the storm. Unfortunately, not all modification projects or proposals are supported by such responsible people, as we will see shortly.

Let's consider another way to tamper with the climate. Recall that certain atmospheric aerosol particles (e.g., stratospheric dust) have a tendency to cool the earth on the average, and that when injected into the stratosphere they remain there for several years, creating a more prolonged cooling effect at the surface. If, then, we are concerned about the climatic effects of a generally rising temperature because of increased atmospheric concentrations of carbon dioxide and thermal pollution, for instance, why should we not simply inject into the stratosphere just enough of another kind of pollutant to counteract the warming trend? [109] Perhaps a fleet of supersonic transports could help by creating a kind of "stratospheric smog" around the optimum height. In a more fanciful vein, why not scatter even larger particles (or mirrors) in orbit around the earth, where they may remain in place even longer?

One could go on with such suggestions—some to cool and some to warm vast regions of the earth, some to change the patterns of rainfall, some to protect from damaging storms, and so forth. The execution of such schemes might improve the current climate for some, or offset a predicted deterioration of climate for others, whether that deterioration were induced by natural causes or by human activities. The next chapter will consider whether we should use any of our newly acquired powers for climate control or stabilization, but first I want to elaborate upon a few of the sinister possibilities inherent in tampering with the climate. It is imperative that these possibilities be brought into the open now, to spur efforts among nations toward a timely agreement against their unwise or unilateral use before the Pandora's box of irreversible climatic disasters is fully flung open.

Make Rain, Not War

Rainmaking provides the simplest example of the dangers accompanying advertent weather modification. Despite the claims of cloud seeders that weather modification really works and the counterclaims of its detractors that careful statistical evidence is lacking, there is an important point for our purpose: A *real event* has already demonstrated the existence of a mentality predisposed to employing risky environmental modification schemes for potential military advantage. This operation was carried out despite the military's uncertainty from the outset that it would work, and despite the positive knowledge of its potentially dangerous side effects (e.g., flooding). I am referring to the United States' use of cloud-seeding in the 1960s in an effort to transform the Ho Chi Minh Trail in Vietnam to a sea of mud. The proponents of the scheme claimed that it was morally preferable to dropping bombs, but I am not at all sure that such rationalization is justifiable. Even though this instance of limited operational use with questionable success may render the moral question moot, it is fair to ask what might happen if the mentality that led to this operation were extrapolated to larger modification projects, as the technology to carry out grander and more dangerous schemes becomes available?

For example, we know now that chlorine or bromine injected into the stratosphere can destroy ozone and permit increased amounts of solar ultraviolet radiation to reach the earth's surface. Why not, then, attack a mortal enemy with every trick in the arsenal? One can imagine sending a fleet of stratospheric bombers laden with a hundred tons of bromine over an enemy's territory and then releasing a "bromine bomb." If one timed the injection to coincide with a period of weak stratospheric circulation known as a sudden warming, the potentially harmful effects on the lives and crops below might last for weeks. And, even though the aggressor nation would eventually suffer increases in its own dosage of ultraviolet radiation, after a few weeks the ozone "hole" created over a distant enemy would have been considerably diluted by mixing with ozone adjacent to the hole and, thus, the reduction of ozone to the aggressor would be relatively light. Or, better yet, the attacking party might inject an antidote chemical to counteract the bromine effect on ozone several days or weeks later and thus minimize further the ozone decrease over its own territory. Frightening, but simple; and only a mild example of what modern ingenuity will increasingly be able to conjure as knowledge of the workings of the atmosphere increases.

To Learn or Not to Learn?

Before the chapter ends, I must confront squarely the argument that my fears about abuses of the atmosphere as our knowledge grows are unwarranted. The simple solution to these dangers, it will be argued in some quarters, is to end research on the atmosphere—learn no more, and the knowledge cannot be misused. Some have suggested that this be done in connection with certain types of biological and genetic research, where a discovery of a new microbe with no natural or laboratory enemies could end the human race. And there have been a few scientists who have declared a moratorium on such work.[110] Such an argument has been made in the case of nuclear technology on many occasions.[111] But this approach is ineffective if applied unilaterally. Every nation that conducts such research would have to cease doing so, or else those who did stop would risk severe disadvantage, possibly to the point of encouraging the worst inclinations of the others. Perhaps my objection here is nothing more than a contemporary version of the old problem of military escalation: Our new arms stimulate our adversaries to build better versions of their own, which in turn stimulates us to build still more powerful weapons, and so on. But a *unilateral* disarmament only encourages potential adversaries to use their superior weapons or knowledge. Disarmament specialists argue these points frequently.

For the case of the atmosphere, a far better approach than placing an embargo, unilateral or otherwise, on new research would be to open all new knowledge about the climate system to public scrutiny and to work to obtain multilateral agreements among all nations, allowing large-scale use of atmospheric technology only under international control. This approach is especially appealing in the case of climate modification because no one seems quite willing yet, or so I would hope, to implement large-scale sinister (or otherwise) climate modification schemes. Recall that in the area of nuclear development France and India, among others, refused to sign a treaty limiting their operations or experiments, because their own capabilities in the area of nuclear devices were not as advanced as some of the nations who had signed the treaty. This example illustrates that if we wait until the technology nearly exists, it is probably too late to control its further development and its use.

Finally, there are compelling intellectual and humanitarian reasons to continue research on the climate and to learn as much as we can as quickly as possible. These reasons have been enumerated on several occasions throughout this book, but, to recapitulate, they include the need to understand and predict the natural fluctuations in climate that have wreaked havoc with our crops in recent years; the need to understand and confront the growing evidence that climatic limits to growth exist; and the urgent need to develop more precise knowledge of these limits to help world planners avert the kind of tragic mistake that might follow too rapid expansion of largely irreversible agricultural, industrial, or demographic developmental patterns. The price of continued ignorance of the workings of the climatic system may have already become unacceptably high (to the Sahelians, for example); but we must set up machinery without delay to ensure that newly acquired knowledge is always in the public domain, thus improving the chances that new findings will be channeled only into constructive efforts to reduce human misery. Such actions will help assure a future secure from yet another horrible abuse of human inventive genius.

7

The Politics of Climate

After a severe rain and hail storm, a Caro, Michigan, farmer suing professional rainmaker Irving D. Krick for $37,500 in damages to crops explained, "We're saying you shouldn't mess around with the environment unless you know what consequences your messing around can cause."

—MICHAEL REINBOLD
Michigan farmer (1974)

If you tell me the technology [for weather or climate modification] is here, I tell you the need for formalized organization and regulation is apparent.

—HOWARD TAUBENFELD
Addressing North American Interstate Weather Modification Council on the legal aspects of weather modification (1975)

Serious problems of public policy relating to weather modification are coming into national focus. Policies developed in the next few years will likely have widespread impacts on the national welfare and, perhaps, on international relations. Clearly, the public as well as the scientific community and the executive and legislative agencies of government must be well informed on these issues.

—ROBERT G. FLEAGLE
Chairman, Department of Atmospheric Sciences, University of Washington, Seattle (1973)

Environment Versus Economics: A Public Policy Value Judgment

Industrialized societies are often plagued by instances of local degradation of air quality related, of course, to atmospheric pollution. In principle, the degree of such pollution can be kept tolerably low by the development of a variety of regulatory laws and agencies. In the United States, for example, the regulations set or enforced by the federal Environmental Protection Agency and state and local agencies are often used to improve air quality or to prevent it from worsening. But the proponents of pollution controls, who are much more interested in the preservation of environmental quality, the health of the populace, and the aesthetic circumstances of our lives than in economic development for its own sake, are often in sharp conflict with those whose chief concerns lie in the economic sector.

On a small scale these differences in outlook were apparent when I lived in New York City. Some landlords bluntly informed their tenants that they would have to raise rents in order to pay for trash collection, or for the installation of expensive pollution control devices on their smoke-belching incinerators—sobering threats for even the most vociferous environmentalists (especially the less affluent ones). Modern urban living that is clean is also very costly, we have learned. And, not unsurprisingly, the "twilight game" was often played: Quick puffs of thick smoke rose above the tops of various neighborhood apartment buildings, where landlords (and perhaps even some tenants) hoped that in the dim light no one would notice their less "expensive" manner of getting rid of garbage, and provided yet another source of urban pollution. It is even more sobering to imagine the burning that went on after dark!

The previous chapter showed that urban centers often have different climates from the outlying areas and perhaps may even alter the atmospheric conditions a few hundred miles downwind of the city center. The warm air near the surface of cities or industrial sites is transported upward, forming clouds that can cause precipitation; under some circumstances the extra particles in the urban atmosphere nucleate a sufficient number of cloud droplets to enhance precipitation, but in other cases can actually reduce rainfall by "overseeding." However, the important questions in this chapter are concerned not so much with how the climate may be modified, but rather how both intentional and accidental changes in climate might be controlled; who is responsible for such changes and how such responsibility can be assessed; and who should pay to stop the climate modifiers, assuming that stopping them is the verdict attained through the political process.

What Have They Done to the Rain Lately?

Acid rain and snow are easily measurable examples of unplanned climate modification caused by human activity. Here, the widespread combustion of fossil fuels (coal, gas, and oil), energy sources that often contain sulfur compounds, injects large quantities of sulfur dioxide (SO_2) into the atmosphere. Then a chemical reaction involving the SO_2, water vapor, and sunlight forms sulfuric acid droplets, which are later washed down to earth by precipitation. Similar reactions can be documented for nitrogen oxides, produced mainly from automobile exhausts.[1]

Such acid precipitation has had noticeable effects in Scandinavia and the northeastern United States. Sulfuric acid and other pollutants are produced locally in Scandinavia, but considerable quantities are also carried there from industrial centers in Central Europe and the British Isles. The acid-rain problem is especially serious in Scandinavia because the soil there is already highly acidic. B. Ottar, a Norwegian scientist wrote:

> The acid precipitation has therefore caused a critical acidification of rivers and lakes, and in southern Norway the salmon are disappearing in several rivers and several thousand trout lakes have become biologically "dead waters." Studies also indicate that there is reason to fear a future reduction of forest growth due to excessive leaching of calcium and other exchangeable cations from the soil.[2]

Acid precipitation is also thought to damage buildings, pieces of art, and even women's stockings.

In the northeastern United States, acid rain and snow may be responsible for a reduction in the rate of forest growth over the past twenty years.[3] Acid precipitation leaches nutrients from plants by reacting with chemicals in the plant tissue. Similar leaching occurs in the soil.

One of the most potentially serious aspects of the acid rain is the possibility of increasing the amount of nitrous oxide produced in the soil through the process of denitrification, as explained in chapter 6.

If the food needs of the growing world population are to be met, then perhaps a 500 percent or greater increase in the use of nitrogen fertilizer can be expected by the year 2000;[4] and that increase in nitrogen fixation will ultimately result in increased denitrification.

The important point of all this in regard to acid rain, however, is that denitrification, the conversion process by which fixed nitrogen is transformed back into nitrogen gas, N_2, or nitrous oxide, N_2O, depends on the acidity of the soil. Particularly perplexing, as Michael McElroy has pointed out,[5] is the prospect that increasing the acidity of the soil will increase the fraction of N_2O produced relative to N_2 by denitrification. Although N_2O is relatively inert in the lower atmosphere, as pointed out earlier, it can be decomposed by ultraviolet light into ozone-destroying components in the stratosphere. Thus, if the amount of fixed nitrogen were increased, by cultivating more nitrogen-fixing plants[6] or through the use of more nitrogen fertilizer, and if the acidification of the soils from acid rain or heavy irrigation were able to raise the atmospheric concentration of nitrous oxide, then these by-products of regional-scale human activities could have the potential to deplete the ozone layer. Obviously, the wide range of uncertainty about present estimates of the magnitude of this nitrous oxide/fertilizer/acid rain/ozone predicament is an example of how our continued ignorance of natural environmental processes may result in intolerable alternatives—ozone destruction or severe restrictions on industrial emissions and agricultural productivity. It is of monumental importance that these processes be more completely understood, at least as long as most civilizations continue to choose technologies that often produce effluents on a global scale as their means to procreation and prosperity. And this urgency is reinforced because there is no apparent substitute for nitrogen fertilizer today.

Interestingly, in 1970, before the question of a nitrous oxide–ozone connection was of concern, a biochemist, C. C. Delwiche, noted in an article in *Scientific American* that the ecological aspects of too much nitrogen could be serious. He commented that "the ingenuity that has been used to feed a growing world population will have to be matched quickly by an effort to keep the nitrogen cycle in reasonable balance."[7] That need has certainly not diminished today.

Acid precipitation may also have contributed to the acidification of some lakes and rivers in Scandinavia and Canada, as well as to considerable fish mortality and damage to buildings, pieces of art, and even women's stockings; nor is it pleasant to the eyes and lungs. However, correcting conditions that cause or worsen such undesirable atmospheric effects as acid precipitation is expensive and difficult, because the solutions may well require severe controls on industry or transportation. Worse yet, the pollution that affects an area can easily originate hundreds of miles away in a region under a different political jurisdiction. Furthermore, it may be especially difficult to trace definitively the sources of acid pollution, and, in fact, a recent paper questioned whether acid rain is caused by human activities at all.[8]

But, assuming it is at least partially caused by human activities, what can be done to alleviate it? To lessen the local intensity of such pollution a number of remedies can be used: the movement of industry to more remote locations,[9] the development and use of alternate technologies, and even the abolition of certain industries in extreme cases where abatement procedures are ineffective.

The more developed nations, however, are irreversibly committed to industry as the chief source of their employment and, concomitantly, the principal means to maintain their relatively high standards of living. And most developing countries are striving equally hard to achieve high degrees of industrialization so that their own material aspirations might be ultimately realized (although there is a growing core of dissenters in those countries who are now calling for an abandonment of the Western industrial model).[10]

The most efficient and practical manner to develop and carry out industrial production has been to centralize it, often in industrial parks near urban centers where there are an ample work force, effective transportation systems, and other advantages. Ironically, it is becoming increasingly evident that much of today's industrial production and its polluting by-products are lowering the very quality of living that they were so instrumental in raising since the beginning of the industrial age. (This observation involves a difficult value judgment between environmental quality and economic standards.)

Although dealing rationally with the complex array of economic environmental problems presents many obstacles, it is more feasible within a large country such as the United States, Canada, or the Soviet Union, since most of the pollution generated in these societies is likely to affect areas primarily within their own political boundaries. But in Europe, for example, because the countries are much smaller in area, it is much more likely that the harmful effluents of one country will be deposited in another, as with acid rain. When the deleterious effects of industrial or social activities in one political jurisdiction are carried downwind into another that possesses no administrative control over the source, the only solutions are to complain, but let the situation continue; to turn to interstate legal mechanisms, which are often ineffective; to appeal to good neighborliness, which is equally likely to prove ineffective; or finally to resort to more unpleasant or even hostile actions.

It is far from easy to extract agreements from industrialists to reduce polluting emissions even within a single political jurisdiction. For example, the Environmental Protection Agency in the United States won a long struggle with the U.S. auto industry for implementation of stringent emission standards by 1975. But because of the onset of economic troubles and ensuing unemployment in the auto industry in 1974–75, considerable pressure has been exerted to delay, ease, or even lift the restrictions, despite the health and aesthetic problems that automobile exhausts have created for many major American cities. Now, the U.S. auto workers union has backed the managements' call for relaxing controls. The problem is further complicated by the lack of proven technologies that can significantly reduce the emissions, and the confusion created by technical bickering over this uncertainty is being cited by some as a good reason to delay all regulation until more is known, which is probably what will happen.

The Case of the Colorado River

Where one region's polluting activities cross political boundaries into another, the difficulties confronted in working out a mutually acceptable solution are multiplied. Cases involving disputed claims on rivers, for instance, are found in many corners of the world. For example, the Colorado River has been the basis of perennial disputes between generally friendly neighbors, the United States and Mexico. The Colorado, whose waters come from the rain and snow of the western slope of the Rocky Mountains, is dammed and tapped in many places. The river is drained and diverted along a thousand miles of its course for uses as diverse as irrigating crops, keeping lawns and golf courses green, and filling the swimming pools in Los Angeles. That city, despite its proximity to the Pacific Ocean, has an annual rainfall less than that found in the "dryland" farming regions of the western Great Plains where much of the world's wheat is grown primarily without irrigation (which explains the label, dryland farming). Because there is such intense competition for the limited resources of the Colorado River, the highest bidder could get the most water. And there are many bidders. The river is counted on to supply water for the farmers' irrigation needs; for keeping much of the populace of distant Arizona or California alive; for local industrial, recreational, and residential needs along its course; and perhaps in the near future for extensive use in developing oil shale and coal production. So great has the demand become for water from the Colorado River that a weather modification program to increase snowpack in the Rocky Mountains has been tried.

With all the drains on the waters of the Colorado in the United States, the greatly diminished and much re-used flow that enters Mexico has become perilously salty and filled with minerals in recent years, often to the point that it is unfit for irrigation use in this food-needy developing nation with a rapidly growing population. Strong complaints from Mexico have finally led to an agreement in which the United States has consented to construct water desalination plants at the Mexico–Arizona border to repurify the water. But desalination requires heating the water, a vastly energy-dependent process, and the United States' willingness to carry out large-scale desalination in the future remains to be seen (since the agreement with Mexico predated the energy crisis).

Before leaving the example of the Colorado River, it seems appropriate to observe that here, too, is a case in which society's dangerous vulnerability to technological fixes is all too evident, and the need to apply Genesis Strategy principles in this (or any) large-scale technological solution to development is clear. That is, when the Colorado River was first dammed and diverted and its waters transported hither and yon, the absolute size of the project was dependent on the use of estimates of two variable factors: first, the average amount of water the river could provide (based on data compiled some forty years ago); and, second, the likely water needs, given the levels of population, affluence, industry, and agriculture at that time. As the years passed, the demand for water rose steadily. But the amount of water has been uncertain, and so additional projects for increasing the useful supply have been undertaken.

One project designed to dam great quantities of water for controlled use was the controversial Glen Canyon Dam, which flooded the magnificent canyon and formed a huge reservoir called Lake Powell. Aside from spoiling a scene of rare beauty, the dam appears to have other drawbacks. Gilbert White, director of the National Hazards Project at the University of Colorado, reported in a TV program[11] that the porous stone walls of the steep canyon that are now the sides of Lake Powell may well be soaking up tens of percent of the precious Colorado water trapped behind the dam in that huge reservoir. However, he told me that it is unknown whether these spongelike walls will eventually saturate or will instead continue to soak up water indefinitely into the future. The Glen Canyon project, which may eventually prove an enormous folly, is another example of how large-scale technological projects are rarely (if ever) foolproof. Furthermore, we are forced by this example to confront a growing danger: that many people are becoming dependent on grandiose, but shaky technologies, dams and diversion projects, among others, and that many will suffer enormously should such technologies ever fail. To be sure, without the technology of long-distance water diversion, the populace of Southern California could not exist where it does today,[12] nor would the region have grown to such proportions in the first place. But metropolitan Los Angeles *does* exist, and its citizens and works quite simply must have many times more water than they required originally to survive. Technology allowed Los Angeles to develop, and now it must keep the metropolis alive.

As Los Angeles's water needs grow, the walls of Glen Canyon could be soaking up unknown amounts of precious water like a gigantic sponge, dramatically illustrating in the process that serious omissions or misjudgments are quite possible in the best-intentioned designs of technologists. The increasing need for water to support growing populations in dry areas led to the construction of a dam—a technological fix—whose unforeseen effects may well prove to be the wasting of vast quantities of the very water it was intended to harness.

What are we then to do, now that Mexico is demanding its share of clean water from the Colorado River; that energy industries, claiming to help the United States develop domestic sources in the West, are beginning to seek an allotment; that agricultural interests are demanding water for irrigation; and that Southern California is willing to pay top dollar to prevent itself from withering away? The harried Bureau of Reclamation is turning to still another technological fix, one fraught with uncertainty and controversy: namely, an experiment with extensive cloud-seeding in an effort to augment snowpack in the Rocky Mountains, thereby increasing water flows in the Colorado basin.

"Colorado cannot be allowed to become an energy colony," said Governor Richard Lamm of Colorado in 1975. "We cannot spend our water and our resources generating Los Angeles's electricity. We've got to come to some fair compromise."

To Fix a Fix with a Fix: Cloud-Seeding in the Colorado Basin

Since the winter of 1970–71, the Bureau of Reclamation has been conducting a cloud-seeding program in the San Juan Mountains of southwestern Colorado. The Colorado River Basin Pilot Project is designed to evaluate the effectiveness of seeding winter clouds to increase snowfall in the high mountain areas of the Rocky Mountain West. Some of these clouds are caused by updrafts created by air forced upward as it hits the mountains. In geographic terminology, the mountains are an orographic barrier, and the clouds produced from such forced uplift over the mountains are called orographic clouds. Such clouds are considered to be good candidates for successful precipitation augmentation operations through cloud-seeding.

The results of the pilot project will be considered in the development of a large-scale program called Winter Orographic Snowpack Augmentation (WOSA). Such a project could, of course, present serious environmental, political, social, and economic problems, most of which are discussed by Leo W. Weisbecker in *Snowpack, Cloud-Seeding and the Colorado River*.[13] He provides in that booklet the highlights of a much longer technical report in which the Stanford Research Institute did a technological assessment of WOSA in the upper Colorado River basin for the National Science Foundation.

Some of the most obvious hazards of snow augmentation are the avalanches and floods that might occur as unintended side effects. Avalanches pose the greatest threat to the mountain highway driver, often traveling to or from a ski resort, as well as to hunters, cross-country skiers, and snowmobilers. But the effects of spring runoff floods have even greater potential to wreak devastation.

Still another possible, and much more important, effect of WOSA would be its general impact on the environment. Variation in either the temperature or the amount of soil moisture encourages growth in some plant species and discourages it in others; changes in vegetation, in turn, influence wildlife populations.[14] Since WOSA would cause a long-term average increase in snowfall (according to its proponents), the result is likely to be a gradual movement of certain mountain and highland vegetation—including trees—to lower elevations. Mountain sheep and goats that customarily forage in high wind-swept areas during the winter could have more difficulty in finding snowless areas in which to graze.

At the same time, the increased level of moisture would prolong the growing season for forage plants, on which deer and elk as well as cattle and sheep depend in the spring, summer, and fall. And, if snowfall begins earlier in the fall or continues later in the spring, the animals will need to forage still longer on the winter ranges at lower elevations. Eventually, because winter ranges are usually much smaller than summer ones, the competition for forage in the winter areas will become very great among the big game, and between game and livestock. For livestock, which can easily be moved to "greener pastures" or fed a supplemental diet, the problem is relatively minor, but game populations are greatly affected by how much time they must spend foraging on sparse winter ranges.

The whole ecological system will eventually become involved if the basic food input, vegetation, is disturbed. Populations of various species of herbiverous mammals, birds, and insects cannot help but reflect changes in the abundance of the primary food source. In turn, the predators on these vegetarian species will also be affected. Eventually, nature's scavengers, those creatures that feed on carrion or are responsible for the decomposition of organic matter, will also respond to the changes. Even the character and formations of the soil will in time be altered by a change in the kind of vegetation growing on it. Whether or not these potential changes are for better or for worse is hard to judge and depends on many factors, including one's personal viewpoint. But, if the changes to the environment are rapid, then adaptation of organisms whose lifetimes are long could be difficult since they would have little time for an evolutionary adjustment to the rapidly altered environment.

Other forms of life would also be affected by augmented snowpack. Snow on ice-covered mountain lakes blocks out the sunlight that aquatic plants need to photosynthesize (transform carbon dioxide to oxygen). Both fish and decaying organic matter already present in the water consume what little oxygen these plants can produce in the winter, and if there is a severe insufficiency of oxygen, the native fish, insects, and other aquatic animals present will suffer. Increased snow cover would make such winter kill more widespread and frequent.

The increased snowfall would also cause more snowmelt runoff into mountain streams. The additional waters would cut and scour the stream beds more deeply. Material needed by aquatic insects for shelter or by fish for egg nests would be swept away. Furthermore. with increased snowfall, the period of high meltwater flow would be lengthened, and the natural habitats of some forms of aquatic life would tend to be carried downstream, where conditions might be much less favorable for their survival.

Augmented snowfall might also lower the average temperature of high mountain headwater streams and lakes. Since changes of only a few degrees can have a profound effect on aquatic life—its growth, breeding, feeding habits, and so on—we should, at the very least, attempt to assure ourselves that such changes are tolerable, and reversible as well.

In addition, generally wetter conditions would continue for longer periods in the mountains if augmentation were successful. Such conditions would encourage the growth of certain insect populations, including those of disease-spreading ticks, mosquitos, and flies.

The Stanford Research Institute report further indicates that a full-scale snow augmentation program in the upper Colorado River basin would require use of about eight thousand pounds of silver iodide for seeding operations during each winter season. The environmental effects of such large doses of silver iodide are still unknown, although a report by biologists Charles Cooper and William Jolly [15] some years back considered the likelihood of serious effects small—at least for the quantities of silver iodide anticipated in experimental programs. "Some materials that are present in water in low concentrations can build up to dangerous levels in moving up through the food chain," Weisbecker says. Mercury is one example of a chemical that has grown to dangerous levels in swordfish and tuna, which, in turn, are eaten by people.

In short, Weisbecker notes, "We will have the capability to modify the weather on a regular basis over widespread areas long before we could have a detailed understanding of its possible environmental and ecological effects." As with the chlorofluoromethane–ozone controversy, we face difficult value judgments, but in this case have even less certain information to go on.

Environmental effects aside, there are economic ramifications as well—both positive and negative—that would come with augmented snowfall, according to the Stanford Research Institute report. In general, there would be additional water for hydroelectric power and for agricultural, municipal, and industrial needs. In the Colorado mineral belt, mining would certainly be curtailed by longer periods of snowcover, and perhaps, too, by increased flooding. On the other hand, ski areas would benefit with a longer winter. But, there might be a curtailment of the seasons for hunting, fishing, and other summer recreation, seasons which are dependent on the length of time between the disappearance of the snowpack in the spring and the beginning of snow accumulation in the fall. Despite a shorter summer pasture season in the high mountains, cattle and sheep raisers could benefit from more snowfall because the pastures in the lowlands would become better and larger.

And, primarily, if the seeding were effective, there would be an important increase in runoff. This, of course, is the purpose of the entire project and the main economic benefit that would need to be weighed against potential economic or ecological losses. William Douglas of the Bureau of Reclamation in Denver was reported to suggest that cloud-seeding could produce a 10 to 15 percent increase in yearly snowfall, and that the cost of this extra water could be perhaps as little as one-tenth the present price of fifteen dollars per acre foot.[16] Despite the drawbacks, this could be the crucial aspect of the experiment. In summary, there would be economic winners and losers, but their identification would be a complex task, assuming that snowpack enlargement is possible in the first place.

Before we leave the case of the Colorado River, let us examine yet another aspect of this case that vividly demonstrates why large-scale technological projects designed to improve the human condition—if they absolutely must be attempted—should be planned in a manner that guarantees flexibility, reversibility, and a large margin of safety. When the Colorado River Compact was drawn up some forty years ago by the Bureau of Reclamation,[17] the size of the project was based in part on the amount of river runoff that had been measured in the Colorado River basin for a number of the preceding years. It is now recognized that by sheer chance the precipitation (and thus the river's runoff) in the basin was unusually high for many years before the compact was implemented.

By analyzing tree-ring widths for a sample of trees collected in the Colorado River basin, dendrochronologists at the University of Arizona have been able to reconstruct the general pattern of precipitation in that region over the past few hundred years.[18] As in any time series, these data show considerable variability: changes in amount of rainfall from year to year and several instances of rainfall above or below the average for several consecutive years. Perhaps the most striking feature of these tree-ring records is their revelation that a sequence of many years of above-normal rainfall occurred precisely during the period used by hydrological planners to define "normal" rainfall for the river basin.

Drier years followed in due course, and because of this it has become more difficult to deliver the quantity of water of a certain salinity that was promised originally to the various users when the compact was drawn up. Given the present extensive use of the river water, its salinity would have gradually increased with less rain, even if the total amount of water diverted had been left unchanged.

The dangers of living close to the margin in cases of water supply were pointed out by a National Academy of Sciences report in 1968: "Money invested in water development may be jeopardized if present estimates of flow are in error by 10 to 20 percent, a range that is, in fact, rather small for estimates of natural phenomena. Moreover, the problems of the Colorado will be encountered elsewhere as use builds towards the total available water."[19] Through thoughtful planning, though, we can avoid some of the proven problems of water diversion.

Thus, one dam led to more dams, and now increased natural precipitation or a very expensive desalination process are needed merely to restore the earlier water levels. Meanwhile, the number of people dependent on the water from the Colorado has risen drastically, as have standards of living. A weather modification via snowfall augmentation is being considered to "solve" the problem in the long run: that is, to fix a technological fix with yet another fix.

Climate Control: The Ultimate Technological Fix

When rivers are shared by hostile nations, solutions to water allocation problems are seldom negotiated; they tend to be found, if at all, in conflict. In the case of the atmosphere, however, the causes of climatic change and the effects of human activity on climate are much more difficult to sort out than determining who may be using excessive quantities of water from, or adding minerals to, a river.

We have seen that human activities in support of industrial, hygienic, agricultural, demographic, and even military objectives possess the potential to modify climate inadvertently, and that their local and regional effects can already be demonstrated. But the global climatic effects of human activities cannot *yet* be proved larger than those natural forces responsible for worldwide climatic variability over the past few thousand years (such as those causing the Little Ice Age, for example).

However, if present land use, energy consumption, aerosol spray-can use, and industrial development patterns simply continue (and do not even expand dramatically with the improving lot of the less developed countries), then it is not unlikely that the global climatic effects of human activity may be felt above the "noise" levels of natural climatic variability sometime after the year 2000. In addition, regional climatic anomalies, severe urban heat islands, or increases in the frequency of severe local weather (tornadoes, for example) might well be felt even sooner,[20] and some of these unintended climatic changes could be both global and irreversible. But abandonment of the causative industrial, demographic, and agricultural patterns of our societies would be economically or politically devastating if not accomplished smoothly over a generation or so. Thus, strong pressures are very likely to develop in favor of maintaining present growth rates and developing means of *deliberate climate control* to offset inadvertent changes from our activities. (A number of such climate control schemes have already been proposed.)

We know also that world food supply and demand are now in extremely close balance and may remain so for several years at least; therefore, even small fluctuations in the climate, such as those of 1972 or 1974, will be capable of creating food shortages and price spirals in well-fed countries and touching off famines in poor ones. Might not climate modifications to alleviate drought, prevent floods, or improve agricultural productivity be proposed as a consequence? Should weather modification on a local or regional scale be considered merely another tool—along with modern irrigation, chemical fertilizers, pesticides, and mechanization—in the technological arsenal of a modern agricultural nation?

Now that some hope of minimal funding exists for the climatological sciences, perhaps our knowledge of how the climatic system actually works will improve markedly in the next few years, allowing rapid development of our predictive skills. Suppose then, with our newfound ability, we were able to predict several bad climatic years for crops somewhere. Might not those nations for which such forecasts spell disaster propose climate stabilization schemes? And even if we remain unable to predict or understand the system very well for the foreseeable future, might not some nations attempt climate modification operations out of desperation, when confronted by several consecutive years of bad weather that endangers their food supplies? Or might they not at least attempt some large geophysical countermeasure? For example, the Soviet Union, having recently suffered its second serious crop failure in three years, might decide after all to divert its northward-flowing rivers southward to its grain lands without sufficient regard for the possible effects that removing this major fresh water source could have on the stability of the Arctic Sea ice. Similarly, the example of American actions of cloud-seeding the Ho Chi Minh Trail in Vietnam provides evidence that some nations will be likely to experiment with weather modification schemes for hostile military purposes, if they believe the technology is in hand. Thus, given the established propensities of human nature, it seems fair to assume that those with the mentality that favored small-scale weather warfare would be willing to try large-scale climatic modification strategies for political, economic, or military advantage, if the technology were (or were *perceived* to be) available. And I should emphasize here that such possibilities are not at all science fiction; as we have seen, many have already occurred on a small-scale basis.

Despite our ignorance of the complete inner workings of the climatic system, we do know enough already to realize that every part of the climatic system cannot be "kicked" very hard in one place without "bulging" elsewhere. The atmosphere, oceans, land, and ice, which collectively make up the climatic system, are coupled, internationally shared resources whose workings do not respect national sovereignties. Thus, any manipulations—intentional or unintentional—of that system have the potential to cause or, perhaps even worse, to be perceived as causing, climatic changes elsewhere on earth. This reality or perception implies that manipulations affecting, or perceived to be affecting, the climate have enormous potential for either creating directly or providing a pretext for international conflicts. A few examples or scenarios of potential conflict situations follow.

Operational Weather Modifiers: Military and Civilian

Cloud-seeding to increase rainfall is already an operational program of a state agency in South Dakota. And the cloud-seeding of hailstorms to increase the number of cloud droplets, thereby decreasing the average size of hail particles, is already common practice in the Soviet Union and the People's Republic of China (and is being done experimentally, at least, within the United States and Switzerland). "We now have a 114-member flight–hailstorm group," says a Mr. Chen, who is a spokesman for a Chinese weather modification team. "Whenever hailstorms gather, we fire the guns under unified command to disperse them before the hailstorm comes into shape." According to Chen, the hailstorm campaign has revealed the falseness of Confucius's teaching that "everything is decided by heaven . . . a fallacy that the decadent exploiting classes used to deceive people." [21]

Climate modification for "defense" purposes, already used by the U.S. Air Force in Southeast Asia, is being seriously studied by the U.S. Department of Defense. Henry Hidalgo of the Institute of Defense Analyses (an organization that often does contract studies for the Defense Department) began a study of advertent climate modification with this statement: "The fundamental question of advertent climate modification has to do with the theoretical evaluation of the *feasibility* of controlled modification of the earth's climate, in the early 1980s, by a world superpower for either political or military purposes."[22] Incidently, Hidalgo's study was clear in its warnings about the dangers of climate modification, given our present ignorance of the climatic system.

On March 20, 1974, several high-ranking officials of the Department of Defense told members of the Senate Foreign Relations Committee about the seven-year $21.6 million program of cloud-seeding to induce rain over trails in Laos, North Vietnam, South Vietnam, and Cambodia. The briefing became the first public admission of weather modification used as a weapon of war. The text of the briefing, released by Senator Claiborne Pell of Rhode Island, is described in an article by Deborah Shapley, which appeared in *Science* on June 7, 1974.

Most of the presentation was made by Lieutenant Colonel Ed Soyster of the Joint Chiefs of Staff; other Department of Defense spokesmen were Dennis J. Doolin, Deputy (Pacific Affairs), and Major General Ray Furlong, Deputy Assistant to the Secretary of Defense (Legislative Affairs). Soyster explained, "The program was to increase rainfall sufficiently in carefully selected target areas to further soften the road surfaces, cause landslides along roadways, and to wash out river crossings. These events normally . . . occur anyway during the height of the rainy season. By seeding it was intended to extend the period of occurrence beyond the normal rainy season and to supplement the natural rainfall as required to maintain the resultant poor traffic conditions."

Pell asked Doolin whether the rainmaking that went on in Southeast Asia was related to the heavy floods that occurred at that time in North Vietnam. Doolin denied the possible connection: "At the time of the heavy flooding in North Vietnam there were no rainmaking operations conducted. . . . The flooding in North Vietnam, as you will recall, generated widespread civilian suffering and that was never the intention nor the result of this program."

It would be quite difficult, I believe, to determine whether or not flooding in seeded-cloud areas was the result—at least in part—of weather modification or whether such flooding was generated by totally natural processes. Apparently, Soyster provided no data on how much rain falls "normally" in the area that was seeded. Rather, his argument on the success of the program was based on the decrease in enemy movements during times of active cloud-seeding. However, Doolin did agree with Pell's assessment of the effectiveness of the effort: "An elephant labored and a mouse came forth." Doolin then added, ". . . it looks to me like when you are getting twenty-one inches in a given area, and we add two inches, if I was on the bottom, I do not think I would know the difference between twenty-one and twenty-three." (Runoff that causes floods certainly would be affected by the difference between twenty-one and twenty-three inches of rain, however.)

Officials of this kind—that is, technically narrow and relatively uninformed—can be dangerous when they have the power to direct secret weather-modifying operations. They are, I fear, well scattered among the military, government, and political systems of many nations. Even more regrettably, a few scientists, who certainly ought to know better, can always be found to favor the clandestine operational use of such a technology before its many ramifications have been adequately considered. Ralph Nader has called this "professional insanity." [23]

Not all military aspects of weather modification need be sinister, however. At a recent conference I attended on weather modification, one of the speakers, Pierre St. Amand of the Naval Weapons Center, was trying to convince the audience that cloud-seeding really worked. St. Amand was a promoter of cloud-seeding in Vietnam and has been opposing current efforts to obtain a ban on military use of weather modification. He argued that we did not necessarily need careful statistical experiments to demonstrate its success. "The navy uses it to clean the airplane," he explained. "Just fly over the ocean and seed the biggest old cloud you can find, and then wash the salt off the aircraft by flying under the cloud!" [24] Perhaps the American taxpayers would prefer more statistical proof?

The Weather Fix: "When in Doubt, Try It Out!"

The Sahel provides yet another example of how well-meaning but short-sighted officials (in this case primarily from Africa, France, and the United States) are all too prone to try isolated technological fixes to solve complex problems. Since the Sahel is dry, they reasoned, the more water that can be pumped into the region, the better off the Sahelians will be. Apparently, the officials did not foresee the vicious spiral of animal and human overpopulation, of overgrazing, and the resulting disastrous imbalance of a delicate ecological system caused by the infusion of such "unnatural" amounts of water. Or perhaps those consequences were not considered important to the Sahelians or to the Westerners who supplied the technology. Michael Glantz, a political scientist who is heading a multidisciplinary study of the Sahelian drought,[25] collected no less than half a dozen schemes that have been proposed to modify the climate of Africa, schemes fraught with uncertainties that are seldom brought up in the arguments made by their proponents.

Climatologists have learned that differences of more than a few tenths of a degree Celsius in the mean global temperature can cause climatic changes on earth significant to some societies, especially if such changes are coupled to the likelihood of climatic variability. For example, as one moves poleward through the midlatitudes the average growing season decreases by some ten days for each one degree Celsius cooling,[26] proving quite clearly that *small changes in long-term averages—even if these changes are far smaller in magnitude than shorter-term variations—can be very significant.* Similarly, an increase or decrease of a few inches of rain can be the proverbial last straw and cause flooding in a wet area or drought in a dry one. In a semi-arid region, for example, a few inches of rain can make up a considerable proportion of the yearly rainfall.[27]

Even some U.S. government agency personnel exhibit the attitude that technology can save humanity with such fixes as climate modification. Referring to the midwestern drought of July, 1974, a Bureau of Reclamation official claimed that "if the techniques developed in the West over the last eighteen years had been used in time, rainfall might have been brought to drought-stricken fields in Kansas, Iowa, and Nebraska, where lack of moisture severely damaged corn, wheat, and soybean crops this year." [28] Peter Kuhn, an atmospheric scientist with the National Oceanic and Atmospheric Administration, and his colleagues have studied the possibility that thin cloud cover can inhibit severe storms over hot spots—areas such as plowed, moderately dry fields or cities, which absorb and retain more solar energy than the vegetated areas surrounding them and thus generate the kinds of thermal instabilities that precede severe weather. "If scientists could create artificial cloud cover over hot-spot areas, it might one day be feasible to use this technique to inhibit severe weather activity in certain thunderstorm-prone areas of the United States," Kuhn says.[29] Thus, the possibility of large-scale climate and weather modification attempts is not just the product of the hyperactive imagination of a modern-day Jules Verne. Such schemes are already being tested and implemented on a small scale. As long as these attempts are experimental—that is, aimed at uncovering the prospects and problems of weather modification—they pose little threat and could prove beneficial. It is when they become operational programs *before* their full consequences are understood that they can be dangerous. All too often these operations are undertaken without much public knowledge or concern about the potential negative effects.

In January, 1975, representatives of fifteen states from the United States, three Canadian provinces, and Mexico met to organize the North American Interstate Weather Modification Council. Most of the speakers extolled the virtues of their art of weather modification and discussed how existing techniques could be employed to alleviate drought and thus reduce food shortages and hunger. One of the council's resolutions was that "federal agencies actively request and utilize the capabilities and inputs of state and local interests, acting alone or in cooperation with other states, in planning, developing, and implementing emergency weather modification programs during drought situations." [30] The group seemed very reluctant to accept the concept of federal controls on weather modification; they seemed to consider such activities by governments as premature and unwarranted interference in private enterprise.

At that session I cautioned that large-scale weather modification operations have the potential to create conflicts because too little is known to rule out the possibility that unusual weather events occurring in or downwind of a seeded area are connected with seeding operations in the clouds above it. Howard Taubenfeld, professor of law at Southern Methodist University and a pioneer in the study of the legal implications of weather modification, pointed out to the delegates that it is not consistent to claim, on one hand, that weather modification really works and, on the other, to argue that centralized regulation is premature. If it really works, he warned, people are going to demand some local measure of control over their clouds and over the operations of the weather modifiers, whether they are in private enterprise or not.

Taubenfeld's point was echoed by Eugene Haas of Colorado University's Institute of Behavioral Sciences, who has been sampling public opinion on weather modification for a number of years. Haas points out that in the Southwest, along the Pacific Coast, and in other areas without sufficient local water supplies, population is continuing its rapid and essentially unchecked growth (mostly by migration). At the same time per capita water consumption is growing, and it is widely assumed that the only viable solution to the water shortage problem is to search for more water supplies (as opposed to conserving water or limiting migration). "Thus the proponents of weather modification have a strong ally in the demographic forces already underway," [31] Haas says. In discussing various perspectives on weather modification, he explains, "It is a social fact that most human beings individually and collectively in groups and organizations seem to operate primarily on the basis of perceived self-interest. . . . Both the proponents and opponents of weather modification reflect this self-interest orientation." [32]

Since one of the principal goals of an elected official is to be

Drawing by Dana Fradon; © 1953 *The New Yorker Magazine, Inc.*

reelected, it follows then that weather modification, like more mundane issues, tends to be viewed as a possible means to reelection. Haas says:

> This can be seen in the fact that Federal Legislators whose constituents in general or home-based powerful interest groups are perceived to favor weather modification have been supporters of federally financed weather modification efforts. Such projects have the potential for a two-pronged payoff for the elected official—the federal expenditures bring additional income into the state or district, and the project itself offers evidence that the politician is supporting a scientific effort to solve a significant problem for his constituents.
>
> . . . . Within limits the President also has something to gain and little to lose from backing weather modification. When the Governor of a state, pointing to the provisions of the 1970 Disaster Relief Act, which authorizes the expenditure of federal funds to lessen the impact of an impending disaster such as drought, requests federal assistance through the Federal Disaster Assistance Administration, it is to the President's advantage to approve the request. Never mind if the majority of atmospheric scientists in the country believe that such emergency drought relief cloud-seeding efforts are premature. Most voters and powerful politicians are not scientists. If it is believed that an emergency rainmaking effort will win more votes than it will lose, the effort will be authorized even when there are very severe budget restraints. Reelection is the name of the game. But, again, it should be remembered that the use of such decision criteria is not unique to weather modification nor are the criteria used by members of only one political party. It is part and parcel of the political process.[33] [The "political process" as we now use it, I should add.]

At the Denver weather modification conference, Haas said that although most people he interviewed would probably not want to block attempts at weather modification, there is overwhelming sentiment that permission to modify the weather should be granted at the local level. Although it was apparent to me that most of the people at the conference were unhappy with the various reservations that Haas, Taubenfeld, and I expressed, the weather modification organization should be commended for inviting speakers whose opinions about weather modification differ considerably from their own. Indeed, as I pointed out to the conferees, a little restraint now may even prevent the occurrence of conflict that would hasten governmental control of weather modifiers.

As a case in point, consider the disastrous flash flood in Rapid City, South Dakota, that occurred on June 9, 1972. On that day, only a few hours before an intense downpour, the South Dakota School of Mines and Technology, under contract with the Department of the Interior, carried on a cloud-seeding operation known as Project Skywater near Rapid City.[34] It is not surprising that a class-action suit seeking hundreds of millions of dollars in damages is being brought against the federal government, on the grounds that the seeding operation was inherently dangerous and that the government was negligent to contract it out. Meteorologists have argued at length about whether the cloud-seeding soon before the torrential downpour led to the flood.[35] Even now, I suspect that all but the most partisan antagonists retain reasonable doubts about the effect of the seeding program's contribution, simply because it remains nearly impossible to establish conclusive cause and effect linkages (let alone magnitudes) in any single weather or climate modification experiment. In the Rapid City case, though, there is an accepted legal authority, the courts, to adjudicate the disputes, and its ultimate ruling can be enforced. But suppose one of the bitterly opposed nations in the Middle East, for example, were conducting weather modification experiments, and its neighbor downwind became aggrieved because it suffered, or imagined it suffered, as a result of the experiments. Would the matter be more likely to end in the World Court [36]—or in some form of military action?

Since cause and effect are so extremely difficult to unravel with weather and climate issues, and since no formally assembled body of impartial experts has been established, genuine blame would be difficult to assess in any particular case. What is worse, perhaps, is that experts around the world would probably align themselves with the combatants on political, rather than scientific, grounds if a climate-related dispute flared. Because progress in climate research necessarily depends on cooperative working relations among the scientists of rival powers, the potential damage to scientific progress, let alone the other possible consequences, is frightening.

Perhaps we should consider creating a panel of "impartial" international experts to adjudicate (or at least mediate) such disputes before one explodes. A number of internationally selected scientists, economists, lawyers, and political and social scientists, for example, could be assembled into a formalized body; together, they would act as watchdog and mediator for potential disputes related to atmospheric cause-and-effect uncertainties such as acid rain or the inadvertent consequences of deliberate weather modification. If the body met regularly and gained the respect of the international community, its existence could go a long way toward lessening the danger of conflict between nations practicing weather modification (or suspected to be causing inadvertent climatic changes) and nations feeling damaged by such activities.

If a dispute were to occur before the panel's delegates were chosen and functioning, it could be too late to avoid conflict. Moreover, with polarization and partisanship being accepted factors in world diplomacy, it would be very difficult to establish such a panel if conflict over weather modification had already erupted. How would power be assigned to such a panel? How would its constituency be determined? These questions are as familiar and nearly unanswerable as those that accompany any effort to share power and responsibility multinationally.

Oil in the Arctic: "Exxon-eration" Versus Caution

The Arctic Sea region has been mentioned a great deal in this book because it is suspected to be a fragile environment, both ecologically and climatologically. For example, a Canadian study by Larry Bell, of Water Associates Ltd., on the ecological effects of oil spills in the Arctic warned of that region's special fragility: "The whole arctic marine environment is in a very stable thermal state. Anything that upsets that balance could have a drastic effect. For example, the animals evolved to that stability." [37]

The oil companies have been considering tanker trips across the Arctic Sea in summer to transport Alaskan oil. Several years ago in a controversial article in *Science*, William Campbell of the U.S. Geological Survey and S. Martin of the University of Washington warned that the danger of oil spills in the Arctic was a particularly acute problem since dark-colored oil could mix with the sea ice, altering its thermal balance (much as spreading black soot on the ice, as earlier discussed, would do).[38] Such a spill could lead to the loss of part of the ice mass; but in any case the danger of oil spills through ice-covered water and the suspected peculiar fragility of the Arctic environment suggested to Campbell and Martin that the region was no place for oil industries, at least until the meager knowledge of the Arctic's climatic balances and workings is improved.

In late 1974 a number of researchers at the large oil firm Exxon Company wrote a comment to *Science* on Campbell's and Martin's original paper. The Exxon workers' rebuttal stressed that some experiments in a few regions in icy waters suggested that oil spills were only a local problem, and that spills "cannot significantly change the heat balance of the entire Arctic Ocean." [39] Although they admitted that "the Arctic environment will require special precautions to minimize the risk of an accidental spill and special techniques for cleanup in ice-infested waters," they still insisted that "concern about the possibility of a significant alteration of the heat balance of the Arctic Ocean from a major oil spill appears to be unwarranted." [40] In other words, the oil companies should be careful, but proceed anyway—for the oil must get through.

Campbell and Martin replied that one cannot conclude from a few limited oil spill experiments under controlled conditions what "the dispersion of a medium-to-large oil spill in pack ice" might be. Thus, they continue to believe "that until more relevant studies are carried out, a cause for concern still exists." [41] Perhaps this technical debate, which is merely another version of the classic argument between short-term economic interests and long-term environmental ones, may be summed up as: "Exxon-eration" or Caution?

Regardless of the ultimate resolution to the technical questions, the principal difficulty resembles the question of controls on the chlorofluorocarbons industry. The industry people claim that there is insufficient proof that their activities are harmful enough to justify economically painful regulations, whereas the more environmentally oriented people argue that the polluter should bear the burden of proving that the effects are negligible—especially in cases where irreversible or global effects are possible. The political problem here is how to weigh the conflicting or missing evidence and how long to wait for more information before acting to hedge against the worst possibilities. Normally, political actions follow crises, but, unfortunately, some crises in climatological and ecological systems may be discovered only after they have become irreversible and have spread across political boundaries. In the case of the Arctic, Bell's philosophy was in agreement with that of Campbell and Martin. Bell warned, "The general problems of prevention, containment, and control of oil spills in the Arctic are going to require the development of a whole new technology. Contemporary technology of open-water capability will not be effective. This must be approached with a whole new outlook and it should be done before extensive operations are allowed in the Arctic."[42] Here again, I can't escape notice of the word "before." On one hand, Bell, the environmentalist, has said that the environment should be considered "before" industry is allowed to exploit the Arctic, while, on the other hand, Sobelev, the protagonist of industry, argued that evidence against aerosol spray propellants be more firmly established "before" we ban the can.[43] Both scientists have made judgments that have a predominantly political flavor, and these opinions must not be confused with their scientific qualifications.

An Age-Old Plan for Survival: The Genesis Strategy

A strong case for tapping the oil sources in Alaska is made by the combination of the problems associated with energy shortages and the economic dislocations and loss of employment that would result from postponement or cancellation of the project. To avoid such unfavorable circumstances, many would trade off the uncertain risk of damaging the Arctic environment at some future time for the certain and immediate economic benefits of extracting oil from Alaska. This is a crucial point. An industry should not, on the one hand, argue at an early date that suspected environmental impacts from their proposed activities are too uncertain to be allowed to halt their progress, and then, at a later date, claim that they cannot absorb restrictions without unacceptable economic dislocation. The Alaska pipeline, for example, has been built with full awareness that it may prove to be environmentally dangerous; thus, economic hardship from a premature commitment to construction should not be grounds for a reprieve to the project should it prove too risky. The SST manufacturers and the nuclear power plant builders are other examples of industries that have been warned in advance of a potential risk from their production. But the chlorofluorocarbon manufacturers, on the other hand, were not aware long in advance of the potential effects on the ozone layer from their products; perhaps society should look upon this industry's plight with far more sympathy than on that of an industry that was aware of possible trouble from the outset.

In my view it would be far preferable to avoid (or at least minimize) such dilemmas in the first place by adopting the Genesis Strategy of building greater flexibility into the technologies that support our society. If we provide ourselves with sufficient technological alternatives, we could eliminate our extreme dependency on only a few types of energy, a dependency that forces us to consider tapping potentially marginal resources in fragile ecosystems. Proper planning for controlled growth is our best alternative, at least for the long term. We will come back to this point later, but first, another example of the potential climatic consequences of rapid technological change.

If those poor countries that use primarily traditional slash-and-burn agricultural practices were to abandon their customs suddenly and adopt mechanized clearing methods, there would likely be a measurable decrease in the amount of atmospheric aerosols in the tropics. Such a change might at first appear merely to be a healthy return to the natural state, and therefore a favorable development. But we must remember that slash-and-burn agriculture has been practiced for centuries, and therefore any slowly evolving climatic changes it may have caused have already been incorporated into the overall climatic norm. However, if rapid industrialization and mechanization of agriculture were to infiltrate the already cultivated parts of the tropics, the high aerosol levels that have slowly developed in the atmosphere as a result of slash-and-burn practices could be quickly eliminated, thus altering climatic patterns in as little as a generation. Not that such changes would necessarily be adverse in the *long run*; but rapid changes are almost always problematical for food crops in the short run, since most crop plants are chosen for their adaptation to the immediately preexisting climatic conditions.

"No-Fault Climate Disaster Insurance"

Coming back to the central question of deliberate climate control, let us now suppose a climate disaster were foreseen, and then forecast. Would not at least some countries, those most threatened perhaps, propose climate stabilization measures, assuming that they could agree among themselves how, precisely, the climate should be stabilized and who would implement the stabilization scheme? And in view of the potential for economic or military advantage, who among the controllers would deal with errors or side effects that might affect a third party—that is, if a cause-and-effect chain could be established beyond a reasonable doubt?

Rather than restrict the discussion to an enumeration of a variety of potential disasters, I will accept what seems the collateral obligation of the critic: to offer solutions where possible. One possible solution is a proposal that William Kellogg and I made in 1974 in an article for *Science*[44] on climate stabilization. We readily admitted that we had raised many more questions of potential climate-related conflicts that we were remotely capable of answering, a situation that has scarcely changed since then. One "modest proposal" we did offer, however, was adoption of global "no-fault climate disaster insurance." This plan suggests that if a large segment of the world truly imagines that the benefits of any proposed climate modification or stabilization scheme outweigh the risks, the members of that group should express willingness to compensate those (and possibly even a few of their own number) who lose a more favored climate, as defined by past statistics. Such compensation should be made without much debate whether the victims were negatively affected by the results of the scheme itself or by the natural course of the climate. After all, experts could (and undoubtedly would) argue well on both sides of cause-and-effect questions without resolving all reasonable doubts in the public's mind. One could argue the case for either side in clear conscience, at least until we learn much more than we now know of causes and effects in the weather mechanism, and thereby become better able to separate fact from speculation.

Returning to the "modest proposal," a less ambitious trial of the original insurance plan to cover ongoing weather modification efforts could be made within a single country. This may be an appropriate experiment even now. In the Rapid City case, for example, if the majority of the people of South Dakota were to agree that weather modification operations were likely to do more good than harm for the greatest number of people in the state, then a state-wide insurance premium might be levied and a no-fault weather modification insurance policy then be issued to every citizen likely to be affected by the operation. Of course, it might be argued that the natural variability of the weather assures there will always be substantial risk that a damaging storm will occur despite a weather modifying operation. This risk may well be great enough to raise the premiums for such weather insurance beyond a level that the majority would find acceptable, and thereby curtail potentially valuable projects. But until cause and effect can be traced with more certainty, it seemed to us that compensation, as well as benefit, must be spread more equitably among all potentially affected people.

In offering this proposition, we intended that it apply only to operational weather or climate modification projects, not to small-scale research experiments. The latter are crucial to the acquisition of the kinds of understanding that will ultimately lead to knowledge of climatic cause and effect, and will help us to decide whether climate modification should be an option to increase food production, for example. But even in these limited projects the cooperation of those affected locally is essential to an experiment's success.

For example, when NCAR began a hail suppression experiment in northeastern Colorado in 1972, two staff members, John Kennedy, then with the hail project, and Henry Lansford, NCAR's information officer, talked extensively with the farmers and ranchers in the area to enlist their cooperation—or at least work toward their understanding of the project. They also set up a citizens' committee and maintained local involvement in the project.[45]

In 1974, after I had given a lecture on climate modification at Colorado State University, a rancher told me that if it weren't for the citizens' committee and the frankness and general availability of Kennedy or Lansford, some of the NCAR hail suppression aircraft would quite likely have been moving targets for suspicious ranchers with high-powered rifles. He was probably right; a number of violent incidents have occurred in connection with cloud-seeding operations in the San Luis Valley in Colorado, an area where little effective effort to obtain citizen involvement has been undertaken.

Control the Climate Controllers, Not the Climate

Inadvertent consequences unfavorable to some, or to all, might easily result from any of the modification schemes mentioned thus far. Moreover, even before the technology becomes (or appears to become) sufficiently sophisticated to justify operations, the pressures on governments and private individuals to utilize it will become increasingly severe. But what if we become truly able to anticipate the consequences of human activities on climate? What then? (This question, too, was touched on in the *Science* article Kellogg and I wrote.) Suppose we had the ability to predict the effects of a perturbation to the system (such as an increase in the concentration of CO_2) or the ability to forecast seasonal anomalies months in advance (and there is some hope that this could be done in the decades ahead). And suppose we granted the existence of some semiperfect operational scheme to stabilize the climate. We would still confront the agonizing decision about whose climate should be preserved, whose improved, and whose sacrificed. Take, for example, the differing attitudes of Japan and the Philippines toward typhoons, cited previously. Perhaps agreement could be reached if it were simply a matter of stabilizing the present climate or preserving the climatic *status quo*. But, regrettably, we have no international mechanism, institution, or treaty to decide what might constitute an overall improvement, let alone to tackle the question of who would be responsible if a scheme produced, or were perceived to produce, unexpected results in someone else's territory.

It is likely that many more schemes will be proposed for control of the climate than for control of the climate controllers. Yet, whether or not purposeful climate control is ever needed, attempted, or achieved, the problems of inadvertent climate modification, climate prediction, and feeding a growing world population suggest, at a minimum, the special timeliness of studying potential climate-related crises and conflicts. But studying is only a first step. A meeting called to write a climate modification treaty must follow closely as a second step.[46] In any case, the understanding and anticipation of either natural, inadvertent, or even purposeful climate change and its consequences for society must continue to be a major interdisciplinary research goal—and all new results must be available to the world community for study and criticism. How else can the political and social risks of underdevelopment, characterized by a widening gap between rich and poor, be effectively weighed against the climatic and environmental dangers of development and growth, which are intended to close that gap? In this case, ignorance is not necessarily bliss. The most troublesome aspect of this dilemma may be that we probably have little time left to reduce ignorance, or to encourage bliss.

Part III

Climate Change and the World Predicament

The global economy must achieve a balance between food production and population growth and must restore the capacity to meet food emergencies. A condition in which one billion people suffer from malnutrition is consistent with no concept of justice.

—HENRY A. KISSINGER
U.S. Secretary of State,
Address to the U.N. General Assembly,
April 15, 1974

When the well is dry, we learn the worth of water.

—BENJAMIN FRANKLIN

Our country does very well at solving rather straightforward scientific–technical problems, when we place high priority on the goal. Nothing illustrates this better than our space program . . . We do poorly, however, when real innovation, especially social innovation, is a precondition for achieving gains.

—WALTER ORR ROBERTS
A View of Century 21 (1969)

So far I have concentrated on presenting an analysis of climate-related issues that forbodes potential danger. The most imminent and far-reaching of these dangers is the possibility of a food–climate crisis that would burden the well-to-do countries with unprecedented hikes in food prices, but could mean famine and political instability for many parts of the nonindustrialized world.

But analyses, no matter how rational, are simply insufficient by themselves to prevent such disasters unless they are also accompanied by concrete proposals for action, practical steps addressed to the general public in clear language.

Most of those who write about the world predicament do attempt to offer solutions in addition to their analyses. The next two chapters present a number of such proposals, a concrete basis for political actions that could be implemented now if there were a will to do so. Many of these will be similar in some ways to the suggestions of others, and those are generally referenced. Many others may not be entirely new, and apologies are offered to those who are not properly cited; in fact, I generally cite recent literature in which more classical works are referenced. However, attribution is far less important than public attention; the questions raised by the debate over the world predicament require immediate and widespread public discussion. That is best achieved through exposure to a wide-ranging debate.

I am aware that many of the dramatic reforms proposed here might be strongly challenged by certain interests and that battles over their worth must be fought in the political arena, a relatively unfamiliar terrain for a physical scientist. Nevertheless, the dangers associated with climatic change are merely a part of the entire world *problèmatique*, and to view them in isolation from the rest of the world predicament would be to repeat the mistakes of many narrowly specialized observers who have examined the prospects for the future *only through the tunnel of their expertise*. Therefore, what follows deals not only with climate, but also with agriculture, population, energy, economics, morality, politics, and other subjects. Viewed together, these begin to illustrate the urgency and enormity of the difficulties and opportunities for corrective actions that loom as the twenty-first century approaches.

8

The Politics of Food

It is inadmissable that those who have control of the wealth and resources of mankind should try to solve the problems of hunger by forbidding the poor to be born.

—Pope Paul VI

Commenting on the World
Food Conference,
Rome, November, 1974

He no play-a the game, he no make-a the rules.

—Earl Butz

U.S. Secretary of Agriculture,
at the World Food Conference,
Rome, November, 1974

Whoever could make two ears of corn, or two blades of grass, to grow upon a spot where only one grew before would deserve better of mankind, and do more essential service to his country, than the whole race of politicians put together.

—Jonathan Swift

Gulliver's Travels,
Voyage to Brobdingnag,
Chapter VII,
Part II

Many Americans who grew up in the forties and fifties remember being scolded by their parents when they left food on their plates, "Finish your meal! There are people starving in China." Some of us would wonder how the leftovers could be gotten to the hungry Asians, anyway. Obviously, our waste would not be collected and distributed, and so in clear conscience many of us ignored our parents' pleas and ate the amount that suited us.

Today, there is a similar problem: The food needed to feed the world can be found—but not necessarily in all the places that it is needed. These days, we hear little about starving masses in China. Rather, the hunger belt seems to have shifted to other parts of Asia, India, and Africa.

The Hungry Look to the United States

How extensive is this misery, and what are the prospects for alleviating or eliminating it? Don Paarlberg, chief economist of the often optimistic U.S. Department of Agriculture, has said, "These are exciting times. Ours is the first generation that could hope to wipe the specter of Malthus from this globe." [1] However, wiping away the horror of starvation as a check to overpopulation often suggests a scenario of a highly developed, technological world with an almost universally high standard of living. Thus, as the world grows smaller through an increasing network of telecommunications and grows even larger in population, many of the poorer undeveloped countries are looking increasingly to the United States and other rich countries for salvation. "We [Americans] will be expected to quite literally sweep back deserts, move mountains, and make the heavens rain," [2] says William Mullen, a feature reporter for the *Chicago Tribune,* who spent months traveling through hunger-stricken parts of Asia and Africa during the famine-punctuated summer of 1974. He and Ovie Carter, a photographer for the *Tribune,* later won a Pulitzer Prize for a series of articles on what they saw.

Mullen also tells of visiting hunger- and drought-stricken farms in India with photographer Carter. At one point, a middle-aged man furiously pedaled up to their car, exhausted from the extreme heat of the drought-stricken Indian interior. They waited five minutes until he recovered sufficiently from his ride of several miles to speak. The man was editor of the weekly paper in a small town nearby. "I heard you were in Singimarie, and my village is even worse. It's getting worse everywhere. We can't help ourselves any more. You are Americans, you are the richest people in the world. America can do anything. You must help us, or we will all be dead." [3] He then pedaled back into the heat.

In India, Dr. M. S. Swaminathan, director general of the India Council for Agricultural Research, told Mullen that "President Kennedy said we could wipe out hunger within our lifetime, given the political will to do so. I think he was right, and I think he proved his point when you put a man on the moon. He wanted America to put a man there within ten years. There was political will to do so, and it was done." [4]

The World Food Conference: Long on Blame, Short on Negotiation

In sharp contrast to the Indian journalist's desperate plea for aid from the West are some harsh words about the richer countries. At an Indian conference on population, Prime Minister Indira Gandhi said, "The current [world] economic predicament has arisen largely from the powerful nations' inability to manage the international economic system justly or efficiently. The threat of their life styles has led some persons to blame the growth of population in countries like India. . . . The most relevant and revealing fact is that a tiny minority in the affluent countries is using up food, petrol, and other essential commodities out of all proportion to their needs." [5]

The World Food Conference, held in Rome in November, 1974, provided a natural setting for the airing of such international resentments and prejudices about excess in both population and affluence. The next few pages will excerpt a number of instances where controversies erupted in Rome. These are not intended as a comprehensive coverage of the conference, but rather as illustrative of a number of conflicting political views of the famine issue and the fact that delegates were often quicker to blame each other than to work out compromise solutions.

In his address to the conference, President Luis Echeverria Alvarez of Mexico blamed the affluent countries for both the food crisis and inflation. He said:

> The rise in prices, the confrontation of markets, and the crisis of raw materials cannot be attributed to the third world. In the final analysis, this situation has been determined and brought about by

> the inability of the great industrial nations to submit their production model to a system of international solidarity and shared development and interdependence based on equity and justice.[6]

Ironically, Echeverria later "made a bid for leadership among the conference gourmandisers in a caviar reception on a Rome roof garden," [7] according to *The New York Times*. Perhaps the crowning touch of irony was the caviar itself, a protein-rich delicacy that serves as a classical symbol of wealth and indulgence.

Whether or not such a reception was in good "taste" for a world food conference was questioned by *PAN*, a feisty tabloid circulated at the conference by a group of uninvited journalists (mostly British). One of *PAN*'s more memorable stunts was to set up a weight scale outside the conference doors and dare the passing delegates to be weighed and to pay a "fat tax," a three-dollar donation to the famine relief fund for every excess pound they registered.

Earl Butz Versus the "Fuzzy-Thinking Do-Gooders"

A month before the conference, Jean Mayer, a renowned nutritionist and columnist at Harvard University, wrote that a 10 percent yearly decrease in meat consumption in the rich nations would yield enough grain to feed 60 million starving people.[8] In addition, he said, the decrease would be good for rich people's consciences, their health, and their pocketbooks. Mayer has also suggested that Americans observe two meatless days per week (replacing meat with vegetables, fish, cheese, or eggs—all of which use no grain or much less grain for production), and that they stop drinking grain-based liquor (hard liquor or beer) or reduce consumption to one grain-based drink per day. Mayer recommends wine, which is produced from grapes, as a viable alternative. Similar sentiments have been expressed by Lester Brown, formerly of the Overseas Development Council and one of the United States' leading food experts, and others, like Senator Hubert Humphrey.

Norman Borlaug, Nobel Prize-winning developer of a "miracle" wheat strain, suggested that Henry Kissinger and a dozen other key diplomats be locked in a room for ten days without food. Borlaug quipped that only by such extreme measures would the delegates truly empathize with the world's hungry and take action for immediate relief, rather than restrict all their attentions to planning far into the future.

On the other hand, Earl Butz had an answer for the Mayers, Browns, and Humphreys of the world:

> I wonder if some of those ill-informed, fuzzy-thinking do-gooders who suggest we eat one less hamburger per week to release more food stuffs for the world are really serious. If so, they could make the first onslaught on this noble goal by reducing our cat and dog populations by 50 percent, which likewise would suddenly release more grain for the world.[9]

Butz repeated his opinion that with enough profit incentive for farms and industry, farm production can continue to expand. Those who understand this point, he said, are not impressed by the critics' "persistent cries of doom." However, he later retracted his remark on pet food: "The world is not on the brink of chaos and famine, and neither are America's pets," Butz said. "We are certainly not to the point where we need to get rid of our household pets or remove meat from our diets." [10] Of course, his critics did not suggest those extremes either; they merely were arguing for small changes in habits, not wholesale alterations to diets. Although most people and pets in the United States were well fed at the end of 1974, American food reserves were dwindling, much of the world was perilously close to famine, and the world food supply was especially dependent on weather because of low reserves. Nevertheless, Secretary Butz argued that to give the proper incentives to farmers the U.S. government should still keep out of agriculture and hold back on commitments to give grain away at Rome. On the other hand, a delegation of U.S. senators at the conference argued that the United States should commit additional food aid immediately to help raise the 10 million metric tons of grain that the conferees had determined would be needed in a number of critical food-deficient regions to combat famine in 1975.

To this suggestion, the leaders of the U.S. delegation to the conference replied that the purpose of the conference was to solve long-term food problems. Edwin M. Martin, deputy chairman of the U.S. delegation, told a *New York Times* reporter, "The conference was not called to get food to people tomorrow but to lay out a plan of action to prevent the crisis we have now from recurring." [11] This position was echoed by the delegation head, Secretary of Agriculture

Earl Butz, who said, "The immediate problem of world hunger is not going to be solved by this conference." [12] Something of a partisan bickering then developed between Butz and a number of Democratic senators who called for a vastly increased contribution of food aid to alleviate the hunger problem. Senator Dick Clark of Iowa said, "It would be unfortunate to say that we are so interested in future generations that we do not worry about the half-billion who may die now." [13] Butz accused the senators of looking for partisan political gain and of placing the United States in a defensive position, stating: "If we go in with a bag full of goodies we are going to come out in bad shape. Our capacity to deliver is limited this year." [14] The food conference followed shortly after the disappointing U.S. harvest in the fall of 1974.

Secretary Butz also contended that the Arab oil embargo was more damaging to the Third World than to the United States because it had led to an increase in the price of fertilizer. Fertilizer, a petroleum by-product, is especially precious in highly populated regions, and when scarce is often unaffordable by those who need it most desperately. The developing countries should look not only to the United States for aid, but also to the oil-producing Arabs, Butz said. The Democratic senators did not oppose Butz on that point; in fact, Senator Henry Jackson said, "We feed the world. I think we have to seriously consider how far we're going to let them [the Arabs] injure the poor in our country and the poor in other countries of the world," [15] by raising the price of petroleum and thereby driving up fertilizer prices (and profits for the fertilizer industry).

The "Pits Would Go Crazy": The Debate over Food Reserves, Free Markets, and the Chicago "Grain Trust"

In the U.S. government a basic debate about the American role in the world food situation has been reported to be raging between the State Department, which wanted nearly to double the food aid program in 1974 to about $1.8 billion, and the Office of Management and Budget and the Council of Economic Advisors, both of which opposed any increase in aid. The President's Council for International Economic Policy, the Treasury Department, and the Agriculture Department [16] took a more moderate line favoring small, conditional increases. A high-ranking U.S. delegate to the World Food Conference reported:

> If we announce what we would like to give, the traders in the Chicago grain pits would go crazy. Because the prices would rise, the money we could spend would buy less, and the developing countries that must import their food would have to pay more. It would just be counterproductive.[17]

He never suggested re-examination of the grain market structure, however. Food reserves stabilize prices, and Secretary Butz (who has often been accused of suggesting policy that benefits agribusiness and large farmers)[18] has been effective in keeping the government from building up grain stocks. And stocks have dropped precipitously since 1972, sending the cost of some commodities soaring—or at least wildly fluctuating. Butz commented:

> This is the first time in 20 years that the government has not been in the commodity business. We now own no cotton, no corn, no wheat, no grain sorghum, no soybeans. . . . Some people are very upset by this, the people who depended on the United States government to carry their inventories for them, at taxpayers expense.[19]

But with the U.S. government out of the stockpiling business, some contend, it is possible that private traders and foreign governments now owning all the grain stocks will permit speculation and profiteering to occur in the market, with even socialist governments taking on the worst image of the capitalists. Commenting on such possibilities, Lester Brown said:

> No question about it. I think that is why Earl [Butz] would like to see the government not be in the business. Because he comes from an agribusiness background, and the more uncertainty there is, the more prices of grains fluctuate, the more money the traders who know what's happening can make.[20]

Regardless of which policy is best, Brown pinpointed the real contrast between the sacrifices of the rich and those of the poor; he said that higher energy costs and wheat and rice prices that had tripled in a year "make it possible to look at grain futures published on the financial pages of newspapers and correlate them with the death rate in developing countries," particularly in Asia, Africa, and South America.[21] And, once again in 1975, the Soviets, plagued by unfavorable weather, were back in the marketplace, buying up tens of millions of metric tons of North American grain—and prices can only go up when this happens. The world market in basic foodstuffs has been determined by price rationing, which means, according to one Chicago grain exporter: "If foreign buyers have the funds, they get the stuff. If they don't, they go hungry. Obviously, this situation can't go on forever." [22]

To sum up, Secretary Butz feels there is little wrong with American agriculture that could not be cured with a free market and less government interference. He has spent much of the last few years preaching that gospel—arguing that exports of food should be increased, that food prices are not yet sufficiently high to provide adequate income to producers, and that agriculture is, first and foremost, a business and not a way of life. He has little sympathy for inefficient "family farms," where they still "knock potato bugs off potatoes with a wooden paddle." [23]

On the other side, in the fall of 1974 the National Farmers Union published a report saying that the farm policies carried out in the first five years of the Nixon Administration were directly responsible for the world food shortage and threatening famines. The union claimed that in those five years "nearly $4 billion more was paid to farmers to prevent food production than it would have cost for the Government to buy the same amount of food from farmers." [24] And, in late 1974, Harry Walters, World Bank economist, commented, "If 1972 had been a better year and 1974 had been a better year, we would have had grain running out of our ears. We wouldn't be here talking about a world food crisis." [25] But they were bad years, bringing unfavorable weather at an especially inopportune time, when the U.S. government was liquidating its food stockpiles. In any case, on one side of the debate there are economic and political factions that fear surpluses more than deficits because surpluses lead to reserves, which stabilize prices and lessen profit incentives of some farmers and grain traders. Thus, they are willing to accept the risks of shortages, price rises, and perhaps even famine in bad years like 1972 and 1974 because incentives are necessary, they believe, to maintain high productivity—something both sides agree is desirable. On the other side of this political argument are elements concerned with the economic consequences of price fluctuations for consumers and small producers, and the potential famine possibilities that might follow a continuation of the policy of no governmental food stocks.

In short, the lesson one might learn from this apparent Babel of conflicting opinion is that the world food problem is compounded by a staggeringly heterogeneous array of demographic, political, social, economic, transportation, agronomic, ecological, and climatic factors; and any attempt to understand, let alone devise, a long-term food security strategy is bound to fail, unless a multidisciplinary or "systems" approach is devised that considers the interaction of seemingly unconnected factors. Now let us take a closer look at some of these constraints on world food production, and the problems of increasing food production and reducing the pressure of growing populations on food supplies.

Meeting World Food Needs: Can It Be Done in Time?

The Malthusian debate over which will increase faster, population or food production, is just as hotly fought today as it was two centuries ago. The arguments and controversy at the World Food Conference in 1974 proved that. But what are the prospects for increasing production? That question has evoked much optimistic prophesizing from all corners of the world. The chief of the Soviet delegation at the conference, Nikolai N. Rodionov, said that he saw no reason for pessimism and that Soviet specialists had concluded that the world had enough land to increase its population tenfold—to some 40 billion people [26] (despite the tens of millions of tons of grain the Soviets have been importing yearly now, when the world's population is a "mere" 4 billion).

Such optimism is often reflected by government officials and by some proponents of the Green Revolution. Roger Revelle, director of Harvard University's Center for Population Studies, says that if one-tenth of the potential gross cropped area (that is, the land that can be farmed productively) were used to grow nonfood products, such as cotton or tobacco, and if technology and other purchased production inputs (such as fertilizers, irrigation water, farm tools, and machinery) equivalent to those used in Iowa corn farming were applied to the other nine-tenths, "a diet based on 4,000 to 5,000 kilocalories of edible plant material could be provided for between

38 and 48 billion people, between ten and thirteen times the present population of the earth." [27]

Optimism similar to Revelle's has been expressed by many others. One of the local newspapers in Boulder reflects the ease with which such soothing reassurances are accepted; in an editorial the paper quoted a "food expert" to the effect that starvation in the world is not necessary, given all the excess available land to grow food and that starvation is the fault of the starving. Therefore, they contended that the United States should follow the "natural law of survival of the fittest" and not consider it to be our responsibility to help food-insufficient countries out of starvation. "The first requisite of survival is to feed ourselves," the editors counseled. The letter I sent in reply to the paper (which I never saw in print) appears in the footnotes.[28]

But, Revelle's optimistic projections are for the very long term, and this crucial detail can be easily overlooked. Furthermore, his projections are made under the assumption that all of the requisites can be provided: fertilizer, pesticides, water, seeds, transportation systems, good weather and, not least, stable ecosystems. The World Food Conference, on the other hand, made clear the immediacy of the food crisis in the 1970s. What then is the prospect for the interim?

To answer that question the U.S. Department of Agriculture did a study on an increase in food production versus an increase in population to 1985. Its conclusion was that under the best conditions world food production will vastly exceed the needs of even a rapidly increasing world population, and under the worst of conditions food production will be sufficient to just barely provide all with an adequate diet.[29] According to USDA projections, then, the world can do no worse than muddle through until 1985, for there will be sufficient land, technology, good weather, environmental stability, and food to stay ahead of population growth. But is this a realistic projection?

Do my local paper, Don Paarlberg, and the USDA give adequate attention to the obstacles confronting their optimistic outlook? Roger Revelle, despite his apparent optimism, does recommend considerable caution.

He emphasizes that much of the earth's potentially arable land is of poor quality and that a tremendous capital investment—between \$500 and \$1,000 per hectare (2.7 acres)—would be needed to extend significantly the areas that are already cultivated. An even greater obstacle, he says, is "the uneven distribution of potentially arable land with respect to the distribution of population." [30] He concludes that without the rapid development of intensive irrigation agriculture, "population growth will almost certainly keep up with increasing food supplies and the final outcome would be a more desperate situation than the one that now exists," [31] (i.e., Boulding's Utterly Dismal Theorem).

Let us examine in some detail how much cultivable land is actually available.

It is evident that the only hope for Asia's self-sufficiency is a vast increase in yield per hectare (Figure 33b). On the other hand, the efforts of Africa and South America to achieve self-sufficiency require a double-pronged approach: not only increasing yields per hectare, but also increasing the total number of cultivated hectares. In North America, because farming technology has already reached an extraordinarily high level, increasing yields per hectare can only be accomplished through a vastly more energy-intensive agriculture or a clearing of lands less than ideal for most crops—unless new ecologically sound "miracle" technologies are developed rapidly. But from a global perspective, it would be a far more efficient use of available resources to employ existing technologies to increase yields in less developed countries. For example, ecologist Barry Commoner, a pioneer of the environmental movement, has said that "the efficiency with which fertilizer nitrogen is converted to grain—the resource productivity—is about 50 percent higher in Asia than it is in Illinois." [32] In other words, the level of fertlizer use in Illinois is already so high that adding more fertilizer there would not increase production as much as it would in India, where fertilizer is used only sparingly because of its relatively high cost.

Indiana to India Shift

In chapter 4, it was shown how fertilizer use in the United States corresponds closely to increased yield per hectare. The question is: Will increased application of fertilizer continue to produce proportionally large increases in crop yield? The answer at this stage appears to be a weak "yes," but it is apparent that there is a law of diminishing returns for highly energy-intensive agriculture. The degree of enhancement in crop yield to be expected with any increased level of fertilizer use has been calculated by Decker (Figure 34). Where there are low levels of fertilizer use, a considerably larger proportional increase in yield per hectare results from each unit increment of fertilizer than in areas with high levels of fertilizer application. This saturation effect, indicative of any law of diminishing returns, is evidenced by the flattening out of the curve in Decker's figure. Thus, from the perspective of maximizing the efficiency of world food production, the optimum policy would be to apply fertilizer to those fields that would gain the largest increment in productivity per unit fertilizer input.

Following up on Commoner's remarks, perhaps this principle of global efficiency could be called the Indiana to India shift of beneficial technology. The obvious difficulty here is procuring political acceptance of a policy advocating the principle of *global* efficiency rather than national interests, a political problem that was illuminated at the World Food Conference when the developing countries accused the affluent countries of overconsumption. Yet with the increased prices of petroleum and fertilizer and with present world political systems oriented toward short-term national interests, it is unlikely that the Indiana farmer who can afford the price of fertilizer and knows how to use it is going to sit by quietly while his tax dollars pay for sending fertilizer to farmers on the Indian subcontinent who may not be able to pay for it—or even know how to use it wisely.

To return to the question of increasing the productivity of North American agriculture, agronomist Louis Thompson [33] has shown yields per acre obtainable in an experimental agricultural station in Iowa and the average yields per acre of the Iowa farmers over the past few years (Figure 35). He reports that the Iowa research yields have increased little recently relative to the average yield of the farmers; this suggests to Thompson that unless some unforeseen breakthrough in agricultural technology appears soon, the farmers will realize maximum yields from their lands in just a few years. Furthermore, the large amount of energy necessary to realize this extra yield from the highly technological farming of Iowa confirms once again that the law of diminishing returns has not yet been repealed by the sheer force of human need.

In an article in *Science*, John and Carol Steinhart of the University of Wisconsin have traced the energy component in the U.S. food system.[34] Their study demonstrates how farm output has risen with increased energy use and how, in recent times, there has been a leveling-off of that very food output despite still increasing energy inputs (Figure 36). Thirty to forty years ago only a few calories of energy input to the food system could buy a calorie of food output, they show, whereas in 1975 it took more than ten calories of energy input to get the same result (Figure 37). From a global point of view, the same message comes through: The most *efficient* means for increasing world food production is the Indiana to India shift.

In reality, however, food yield increases from the agricultural systems of poorer developing countries may not be as easily achieved as Figures 34 to 37 suggest. Merely providing energy input to farmers in these lands carries no automatic guarantee that the increase in food output will be nearly as high as it could be, or as it has been in the United States when such energy was "applied." For high yields require not only technological inputs, such as fertilizer and pesticides, but also careful management of these technologies relative to the local climatic, ecological, and soil conditions. Because a good deal of arable land remains unused or underutilized in the developing world, the major technological optimism for realistic increases in world food production in the immediate future may not be unjustified. However, the primary improvements in agriculture depend on the cultivation of new lands and the careful husbanding of existing agricultural areas in the developing nations (which are largely tropical).

The Green Revolution in Tropical Agriculture: Panacea or Disaster?

But what does it take to realize such increases in tropical production?

In 1965 a great wave of optimism arose over the future of the Green Revolution. After many years of work a number of scientists—Norman Borlaug, who received the Nobel Peace Prize, is probably the best known—finally achieved success in breeding high-yielding varieties of corn, wheat, and rice that could be adopted by the agricultural systems in the developing world. Special institutes were set up to improve and implement the Green Revolution methods, and a great improvement in food productivity was achieved.

Yet, while the Green Revolution was being touted, the well-traveled agricultural specialists William and Paul Paddock argued in their book *Famine 1975* that it would not prevent the onset of "times of famines," which they predicted in 1967 would begin around 1975:

> . . . synthetic foods will not turn aside man's hunger a decade from now, nor fish from the sea, nor irrigation with water from desalinization, nor culture by hydroponics, nor fertilizer, nor the cultivation of new land in the vast, untapped jungles or arid wastelands, nor land reforms, nor socialistic controls, nor capitalistic initiatives. Nor will research produce in time new seeds, new techniques, new plants for the farmers to sow and reap. All of these combined can be the salvation of the twenty-first century. Any one of these could achieve a sudden leap forward in progress before the end of this century. Yet none, alone or in combination with each other, will have major effect on the food crises of the 1970s.[35]

The fact that famine became a real threat in the mid-1970s is not a full vindication of the Paddocks' position, for the Green Revolution techniques were able to increase dramatically the wheat yields in Mexico and India, for example, and rice yields in Asia; and food security and reserves were at reasonably high levels by 1972. But, despite the progress of the Green Revolution, the weather troubles of 1972 and 1974 and the agricultural policy of nations with food surpluses (primarily the United States and Canada) wiped out the optimism of many Green Revolution boosters and recalled the Paddocks' original contention that "overnight miracles . . . do not happen in agriculture."[36]

Robert Chandler, who has directed the International Rice Research Institute in the Philippines, commented in defense of Green Revolution methods:

> Those who take a more gloomy view of the future should consider the relatively short time during which the new technology has been available and the fact that it is still actively being improved. . . . Rice yields are so low now that there is much room for improvement. I believe it is entirely possible to double average rice yields in South and Southeast Asia within the next two decades.[37]

But this will take large quantities of fertilizer, among other things. The principal problems with fertilizer are not only its expense and scarcity (or its potential to pollute waterways or create by-products that could destroy the ozone shield), but also convincing farmers to use it—particularly subsistence farmers in the developing countries. In *Famine 1975* the Paddocks also attacked fertilizer as a "panacea," arguing that farmers will not use it unless they know that they can make money doing so, if their government even permits them to make money. They also pointed out that to be economically profitable, the fertilizer must be correct for the soil, crop, and climate with which it is used.[38]

Another important element of the Green Revolution is irrigation: Water may well be a fundamental limiting factor. And there is concern about the long-term effects of uncontrolled well-digging on the total supply of ground water.[39]

In a more recent book, Green Revolution critics William and Elizabeth Paddock argue that irrigation has been the "lifeblood of the new cereals" and that "on nonirrigated land the new varieties do no better than the standard ones. . . . Ford Foundation's Lowell S. Hardin says that 'if one looks at a map the land where this new technology, this Green Revolution, applies is a postage stamp on the face of the earth.' "[40]

Rice, which accounts for 40 percent of Asia's total food-energy supply, depends heavily on irrigation and is grown almost entirely on irrigated or rain-fed paddy fields. Together, China and India have nearly half the earth's irrigated land, and, not surprisingly, rice is the mainstay of the diets in those countries. The advantages of irrigation for rice and other crops are manifold: It contributes directly to higher yields, expands the possibility for profitable use of other yield-raising inputs (particularly fertilizer), and is essential for increasing multiple cropping in many areas of the world.

"Like many of man's other interventions in natural cycles, his reshaping of the hydrological cycle has had unwanted side effects," cautions Lester Brown while describing a well-known difficulty of technological farming. Diversion of river water onto land can raise the underground water table. As the irrigation water percolates downward, it accumulates and can gradually rise to within a few inches of the surface, inhibiting the growth of plant roots by waterlogging and making the surface soil salty (perhaps increasing the ratio of N_2O to N_2 produced by denitrification, as explained in chapter 7).

Brown points out that such a situation occurred in Pakistan after decades of irrigating with water from the Indus River. Pakistan was losing twenty thousand acres of fertile cropland annually by 1960, and its population was growing 2.5 percent yearly. At Pakistan's request, an interdisciplinary team of researchers went there in 1961 to study the seriousness of the problem. After careful analysis they recommended a system of tubewells to lower the water table by tapping the water underground. Implementation of this solution brought large quantities of fresh water to the surface, which, as it percolated back into the ground, also washed the salt downward. The plan was a considerable success. "By 1966 the amount of abandoned land being brought back into production began to exceed that being lost. The continuing progress made since then, bringing the problem under control, constitutes one of the most exciting successes in the short history of international technical assistance." [41]

In sum, much of the success of the Green Revolution depends on the breeding of new high-yielding plant varieties, varieties whose seeds often require carefully controlled but massive doses of both irrigation water and fertilizer to live up to their high potential. As a result, the Green Revolution methods can directly help only those who can afford them or who know how to employ them. As D. B. Singh, vice chancellor of the G. B. Pant University of Agriculture and Technology in Pantnagar, India, said:

> Nobody has been able to invent a cow which will milk without being fed. How can we have bumper crops if rains fail, irrigation doesn't come, electricity is not available and fertilizer is scarce and available at abnormally high prices? [42]

Once, when I mentioned my impression that Green Revolution strains, despite their high yields under optimal environmental conditions, could produce lower yields than conventional strains unless adequate supplies of fertilizer and irrigation water were available, I encountered strong opposition from a number of agriculturalists and a plant physiologist. Miracle crops are not necessarily more sensitive to environmental inputs than conventional varieties, they said; we were at a conference on food–climate interactions in Sterling Forest, New York, in December, 1974. I was informed that any knowledgeable plant breeder can usually pick a strain that yields more than conventional varieties *even without irrigation and fertilizer*. So why, then, I was asked by these Green Revolution supporters, do so many people attack the Green Revolution?

I argued that an important distinction had to be made between absolute yield and yield variability. Let us consider a hypothetical case, employing the best current methods, where the yield of a miracle crop is substantially higher than those of the conventional varieties—but only within a narrow range of environmental factors such as temperature, irrigation, or fertilizer. We can see that the yields of these environmentally sensitive miracle varieties can be even lower when the environmental factors are not favorable, such as in a stress environment (Figure 38a). But plant breeding, I was told, has progressed to the point that yields need not be lower in most instances. Rather, by careful selection of miracle crops, some insensitive varieties can have higher *absolute yields* than conventional species over a wide variety of environmental factors—including stress conditions (Figure 38b). Pest resistance, however, is not among the environmental conditions envisioned in this schematic figure.

Given that some miracle crops do better under fertilizer or water stress than some conventional varieties, why have I not yet abandoned the contention that miracle strains, even the insensitive ones, are still more environmentally sensitive than conventional ones? The reason is simply this: Both miracle and conventional strains exhibit yield variability when the environmental factors change, but miracle crops vary more. For instance, if the normal environment (shown by the dashed vertical line in Figure 38b) were suddenly changed to a stress environment (shown by the dotted vertical line) because of drought, irrigation failure, or a fertilizer shortage, for example, the crucial question would be, Which variety would have a larger drop in expected yield—conventional or miracle? Obviously (see Figure 38b), the *variability* of yields with differing environmental factors is greater for the miracle strains than for the conventional ones. However, these insensitive miracle varieties are still more productive overall than conventional ones since they have higher absolute yields even in environmental stress conditions. The danger, however, of such insensitive miracle crops, assuming that they really do exist, is that their higher variability of yield may well produce much larger *fluctuations* in total food production than is found with conventional crops. And if food reserves are already low, as we know they are, then the world population becomes dangerously vulnerable to the high variability of these high-yielding strains. The lesson here is not to abandon these high-yielding miracle crops, but rather to be prepared for their potentially high variability by maintaining adequate food reserves as a cushion—a cushion that can be more easily built up in good years by use of these miracle strains.

A look at world history recounts an important example of the danger confronted by relatively large human populations depending on a few high-yielding varieties, and shows that the Green Revolution started long before the mid-twentieth century. We can go back to the Irish potato famine in 1846, for example. Because of the introduction and cultivation of a miracle crop, the potato, both the Irish population and the *temporary* carrying capacity of the land grew rapidly during the late eighteenth and early nineteenth centuries. Then, in the summer of 1846, weather conditions favorable for the initiation of a potato blight hit the country, and some 1.5 to 2 million of the 8 million Irishmen starved to death. In the following decades, another 2 million emigrated, mainly to the United States, and 4 million remained, most living in poverty. Even though part of the problem in Ireland was related to the distribution of food, which was controlled by the English landlords, the sudden collapse accompanying the potato famine proves that potential disasters from undiversified agriculture are more than speculative theories. "When such a thing as a Green Revolution occurs," say William and Elizabeth Paddock, "its name will be Disaster if it arrives ahead of a Population Control Revolution." [43]

Diversity in Tropical Agriculture

The Genesis Strategy I have been advocating is based on the theory that there can be great danger when diversity is eliminated, especially in an ecological system. In the Turrialba, Costa Rica, meeting in 1973, attended by a variety of North American and Latin American scientists concerned with agriculture and its impact on tropical ecosystems, the most notable feature was that the Latin American scientists shared with their North American colleagues a feeling of great caution about development in the tropics, a caution that was expressed in the conference report, entitled *Fragile Ecosystems*.[44] Those Latin American scientists, who also understood the urgent needs for agricultural and industrial development in their home countries, nevertheless approved this statement:

> Much of the economic development throughout the American tropics has been achieved only at a significant environmental cost. If there is any lesson to be learned from conservation ecology, it is that economic development and progress can only be sustained within a framework of regional environmental diversity and stability, i.e., short-term economic gain at the cost of regional environmental stability is self-defeating in the long run. At present there is a clear lack of an adequate conceptual base upon which to base sound, sustained development plans in the American tropics that would optimize economic benefits and environmental stability.[45]

This frankly expressed fear of our present ignorance of these matters should be taken very seriously, for it comes not only from North Americans observing the developing countries, but also from indigenous scientists who regularly see the general poverty and deprivation of their fellow countrymen. Although there are many good references on various aspects of tropical ecosystems,[46] I strongly recommend that those who look to the tropics as the easy solution to world food problems at least read *Fragile Ecosystems* before making optimistic statements about the millions of arable acres that are itching for the plow.

This is not to suggest that the tropics have generally poor potential as farmland, but rather to warn that they must be developed with extreme caution if they are to be productive and be preserved. Shifting cultivation, for example, when practiced successfully by subsistence farmers in tropical areas of low population densities, can minimize ecological impact by short cropping periods and long fallow periods, that is, farming the land relatively infrequently. However, as population becomes denser in the tropics and land becomes scarcer, the fields will be allowed to lie fallow for shorter periods; in addition, the frequency with which fires are used to clear the land will probably increase, adding to the deterioration of the environment. There are other disadvantages to further cultivation of the tropics: Much of the unused land that is available for cultivation lacks the appropriate soil structure, is not very fertile, and is without means of irrigation or drainage. Thus, the authors of *Fragile Ecosystems* conclude that "a major potential lies in the more intensive use of the land now under cultivation," [47] rather than in merely clearing new lands.

One means of achieving more intensive use of land is through multiple cropping; its advantages are manifold. Multiple cropping consists basically of planting different crops in the same field, often at overlapping times. The International Rice Research Institute has been able to obtain extremely high yields with this method.[48] In one case they planted rice as the initial crop in early June, following with sweet potatoes about ten to twenty days before the rice was harvested. This sequence could be followed by planting other crops, depending on the local soil and climate conditions. Multiple cropping in the tropics can cause yields per acre to be higher there than in temperate regions since there is no winter recess, and it can also provide a balanced diet for the local population. Fertilizer leaching (washing out) is reduced and attacks by insects and other pests may be abated by interruption of their normal life cycles. Finally, multiple cropping also maximizes either the use or conservation of irrigation and sun energy alike.

Some parts of the tropics, then, do indeed have excellent productive potential, but realization of that potential requires an understanding of the region's fragility and special problems. There have already been instances of development where ecological considerations were ignored; for example, the construction of reservoirs and irrigation systems in the tropics has often tended to spread water-borne human and domestic-animal diseases. "In addition, many of the transportation systems, resettlement programs, and nutritional changes associated with such projects have adversely affected the pattern of disease," say the authors of *Fragile Ecosystems*.[49] Several case studies of tropical reservoir projects have pointed to the relationship between water-borne disease and riverine projects. "These studies reemphasize the vital necessity for all tropical river-basin projects involving water stabilization to require intensive ecological studies early in the planning process," [50] they add.

One of the most stubborn of these water-borne diseases is schistosomiasis. It is transmitted by snails that carry the parasite, infecting humans through penetration of the skin or through drinking and causing recurrent fever and diarrhea. The tiny worms (schistosomes) reproduce in the human liver. Their eggs are excreted in human waste and often end up in irrigation canals, sewers, and ditches, where they latch onto the snails. As mentioned earlier, tremendous spreading of schistosomiasis has occurred recently in poorer regions because of increased densities in rural human populations, poor sanitation, and stabilization of irrigation and reservoir impoundments. Control of this debilitating disease will clearly require an adequate understanding of its ecology. Thus, research must be done to fill in the gaps in knowledge, and it should emphasize regional systems, especially where the potential exists for people to spread the disease in areas that have not yet been infested. The Amazon basin, for example, is being settled by migrants from northeast Brazil, a region infected by schistosomiasis. Unfortunately, the best soils for agriculture in the Amazon basin are found in areas of low acidity, where snails, like crops, fare best. Thus, colonization and population growth may be heaviest precisely where the possibility of disease is greatest. The Turrialba conference was emphatic in its conclusion that despite the need to exploit the productive potential of the tropics, severe damage to the ecosystem would eventually eliminate any such possibilities.

Without diversification of agriculture, then, a region can suffer huge losses of crops and human lives because of one type of insect or disease. Again, the Irish potato blight of 1846 is a good example, and the rice blight epidemic that led to the starvation of a million Indians in the 1940s is another. The Food and Agriculture Organization (FAO) of the the United Nations reports that in 1959, during the last major locust plague in Africa, the insects ate, in the period of a month, a year's supply of grain for 1 million people.[51] A more recent example occurred in Zanzibar in 1974, with cloves, which had become one of the country's principal export crops; the clove trees were infected with a disease that turned them gray and shriveled the clove buds. Thousands of trees were burned in order to stop the infection from spreading. However, as one article reported:

> The clove disease has had one good effect. It has forced Zanzibar to diversify its agriculture. There is a drive on for chili production, more valuable copra is being produced, tobacco growing is being intensified and many farmers are going over to cinnamon and ginger growing.[52]

Earlier it was mentioned that the costs of clearing new lands for agriculture could average as high as $1,000 per acre; at current world production averages, in which one acre feeds one person, an expenditure of about $80 billion per year would be needed just to keep up with present population growth rates and eating standards. Fortunately, farmers can feed considerably more than one person per acre with adequate amounts of fertilizers, pesticide, irrigation, and selected seed stocks, and the $1,000 figure may be too high; Revelle suggests it is two or three times lower, which is close to Paul and Anne Ehrlich's median estimate [53] of $28 billion per year. But, in any case, the anticipated costs of increasing food production ahead of population growth are in the tens of billions of dollars per year. Even this investment, according to Norman Borlaug, would merely "buy time" in the face of the current "monstrous population growth." [54]

How might the developed world help to increase food production in the developing countries without damaging their fragile ecosystems? Even if the ecological difficulties could be overcome, the capital costs of expanding agriculture in these lands are immense. Thus, by providing economic, technological, and scientific help, the wealthier countries would make an important contribution.

The Demographic Transition and the Transfer of Technology

Third World leaders such as President Echeverria of Mexico and Prime Minister Gandhi of India have argued that population control in poor countries can only be accomplished when nations achieve higher standards of living. If that is the case, then food production will have to increase at a considerably higher rate than population (and be transferred to the people, not merely concentrated in the upper echelons). The costs of such rapid development and sustained productivity will be considerably higher than the tens of billions of dollars needed to keep food production abreast of population growth; they are likely to run into hundreds of billions of dollars annually. Swedish sociologist Gunnar Myrdal (who correctly predicted the American race riots of the 1960s more than a decade before they happened) has recently argued that the developing world can move toward zero population growth only with rapidly rising standards of living brought about by a massive and immediate "transfer of technology" from the richest nations:

> What should be transferred is a technology that engages the latest results of scientific research, adjusted to the highest possible utilization of the labor force . . . . It is evident that the transfer of technology must proceed on a much larger scale if increase in the production of material necessities is to overtake the increase in the population of the underdeveloped countries. The need for aid in the larger part of the underdeveloped world must therefore be recognized as substantial. Moreover, this need cannot be thought of as a short-term exigency; it will be long-lasting.[55]

But Myrdal also points out that a one-sided system, with aid coming from the developed countries, will not be sufficient in itself to raise living standards in the developing countries. Rather, the less developed nations must give much higher priority to agricultural self-sufficiency than they have, and the developed countries must adapt their technology, which has been traditionally energy intensive, to "the task of raising yields in agriculture through highly labor-intensive technologies." An important element, Myrdal argues, is "the demand for a well-planned land and tenancy reform. Such reform has been on the agenda in all underdeveloped countries; with only a few exceptions it has been botched."[56] Many others have argued similarly.

As for the help that developing countries are receiving from their own universities and research institutions, Myrdal has harsh words: ". . . many of them are bent on producing results that will impress us rather than solve their own practical problems."[57] He argues that an important part of our aid should be a reorientation of our own research efforts toward their problems. In essence, Myrdal can be counted among many who have asked the developed world to make contributions to the developing world; at the same time, he insists that the underdeveloped countries slow down their preoccupation with industrial growth and reorient their own priorities toward the immediate practical problems of agricultural self-sufficiency with labor-intensive technologies. Both developed and underdeveloped regions will not easily agree to such mutual efforts, and the implementation of what I will call a Global Survival Compromise may be as slowly accepted as was Rousseau's idea of the social contract, the compromise that is the philosophical foundation of modern democracy made by the rulers of Europe two centuries ago. Unhappily, today's exponentially growing world may have no more than a generation or two to adjust to this new idea.

And Roger Revelle, despite his often quoted optimism about the prospects for world food production, also agrees that a massive, expensive, and immediate transfer of technology must take place to prevent collapse. He says, "The new technology must, however, be highly labor-intensive. Otherwise it will swell the stream of migrants to the city slums." [58]

An interesting dilemma follows from the suggestion that we create labor-intensive technologies in the developing countries: Technology has been developed primarily to increase the relative effectiveness of each laborer and to reduce the drudgery of repetitive, hard work. Thus, adoption of labor-intensive technologies (many workers spreading fertilizer by hand, e.g.) would imply a lower effectiveness for each worker compared to the output of one worker using machine-intensive technologies (one driver on a tractor spreading fertilizer). In addition, labor-intensive technologies may also require difficult and repetitive tasks when compared to machine technologies. But labor-intensive practices are not "make work" or featherbedding schemes, but rather attempts to replace capital, which is scarce in most developing countries, with labor, which is abundant. Yet, until developing countries can establish industrial economies that can gainfully employ those workers who would be displaced by machines, the implementation of labor-oriented technologies, even if they are below optimum efficiency, seems clearly the best interim strategy for those countries now unable to feed or employ a large fraction of their citizens. This is a perfect example of "economics as if people mattered," [59] a coinage of Ernest F. Schumacher, who addresses the need for new thinking about large technologies in his book *Small Is Beautiful.* Thus, along with higher levels of agriculture and technology there is a need for a high level of social and economic development. Revelle's assumption that the less developed countries will achieve such levels is the basis for his optimism. He says that such achievements would "likely lead to a cessation of population growth long before the maximum carrying capacity of the level is reached." [60] (That is, they would accomplish the demographic transition that has been approached by most industrial nations whose population growth rates began to decline only after high standards of living were widely attained.) This is the sticky part of the argument because two of Revelle's assumptions are not at all universally accepted. First, as discussed earlier, many argue that the carrying capacity has already been or will soon be reached. And, in any case, they say that it will be reached long before the tremendous momentum of population growth can be stopped, since it takes a few generations to achieve zero population growth even after fertility rates reach replacement levels (i.e., if each couple merely reproduced itself).[61] The other hotly debated assumption is that a demographic transition can be attained by improving the material standards of living of poor people or by any other demonstrable noncoercive method.[62]

A number of population experts met with a variety of generalists and specialists to discuss such issues in the summer of 1973 at the Aspen Institute for Humanistic Studies. A report of their discussion by Thomas Wilson of the Institute reflects recent thinking on a number of population problems, including the question of how to achieve demographic transition:

> The curtailment of population growth is seen partly as a matter of family planning services, partly a matter of income, partly a matter of literacy and education and, by no means last, full access to basic social services for the poorest parts of the society. It is also a matter of child labor practices and school leaving age, of employment opportunities for women and their age at marriage and of yet other factors, whose weights vary within and among societies and which can, to some extent, be influenced by systems of incentives and disincentives tailored to the conditions of each society.[63]

And, if that long quote induces suspicions that we don't really know exactly what causes population rates to change and that no general statement can be made as to whether the demographic transition will work, then the message has come through.

Although we do know that birth rates have been lowered in many of the wealthier countries, including the United States, we cannot be sure that the same would happen in different cultures if *only* their prosperity were increased. Unfortunately, there is little time left to find out what really can bring about a demographic transition. A closer look at some of these critical issues follows.

Developing Agriculture in the Less Developed Countries

I have been guardedly optimistic that food production can be increased substantially in the developing world, barring sudden ecological or climatic catastrophes or major technological breakthroughs. However, it has also been said that this optimism must be infused with the reality of the hardships and uncertainties that are inextricably associated with such progress. Large capital expenditures will be necessary, and considerable social and cultural changes may also occur. In addition, the bureaucratic inefficiency and corruption that often seems to engulf government officials must be lessened considerably before real progress toward self-sufficiency and eventual demographic transition can be achieved. (Indeed, corruption and inefficiency are not found only in developing nations, but can be especially deadly there because they have little margin for waste.) Several specific case studies drawn from primary source material—mainly accounts of journalists who spent considerable time in a number of developing countries—are presented here to give an inkling of the enormity of the social and cultural problems that need to be dealt with in assessing the likelihood of future disasters.

The Case of India. Perhaps it is appropriate to begin with India since it is most often cited as a hopeless famine case. Indian officials, notably those from the ruling Congress party, were unhappy with the idea of taking food from the United States during the late 1960s; they vowed in their political campaigns to eliminate the humiliation of begging food from America and promised Indian self-sufficiency in food. By 1972, adoption of Green Revolution methods and favorable weather in India had actually led to a modest economic surplus in wheat and prompted a proclamation by a euphoric Congress party that India had achieved self-sufficiency in food production. The fact that the Indian monsoon had been extremely favorable during the five-year period since the inception of Green Revolution techniques was not mentioned. Later, India's "self-sufficiency" was smashed by the reality of monsoon failures in 1972 and again in 1974 and by the world inflation stoked by India's Third World "friends," the oil-producing and exporting countries.

The Indian government then blamed the overindulgent affluence and the excessive consumption of the rich countries for India's famine problems. On the other hand, attacks against the agricultural policies of the government by the Indian press have increased (at least up to the summer of 1975, after which the government invoked emergency powers, including censorship). *The Indian Express* said in early summer of 1974: "Famine conditions, widespread destitution and starvation deaths are being reported from different parts of the country. It is, of course, a set official policy not to admit starvation deaths. But that cannot hide the ugly reality." [64] An important distinction between starvation deaths and starvation-related deaths should be made clear. Most humans can live on diets far below minimum standards of nutrition, and will die of starvation only if this substandard condition is prolonged or worsened. On the other hand, a malnourished person, particularly a child, can easily die of a disease (measles, for example) that is normally not fatal. Innumerable such starvation-related deaths occur and are rarely considered as starvation deaths by local governments.[65]

In response to the 1974 famine, West Bengal Chief Minister Sidhartha Shankar Rey said, "We shall deal with this situation on a war footing." But V. B. Kulkarni, a columnist in *The Indian Express* voiced a differing view: "In actual fact, agriculture has so far received priority only in rhetoric." [66] Apparently, Indian officials will not admit the country's problems in public because that would be an admission of a failure of the avowed promises of self-sufficiency.

Thus, when India conducted an expensive nuclear blast in 1974, it was only natural for journalists such as Kulkarni and donor nations throughout the world to become perplexed and alarmed at India's apparent priorities. Prime Minister Gandhi, however, then retorted angrily to the outcries and questions:

> The achievements of our atomic scientists are misrepresented and we are lectured on the need to have the right priorities. The most relevant and revealing fact is that a tiny minority in the affluent countries is using up food, petrol, and other essential commodities out of all proportion to their needs.[67]

What Ms. Gandhi failed to mention was that the deadly spiral of wasted resources, more easily coped with by the developed nations, could only be exacerbated by India's initiation into the nuclear club. When a country with a certain economic, locational, or political status acquires nuclear technology, other countries of similar status all too often feel compelled to acquire that technology. After India's "peaceful" nuclear blast in 1974, Pakistan's Prime Minister Ali Bhutto told *Time* magazine, "We will eat leaves and grass, even go hungry, we will have to get one of our own. We have no alternative." [68] One need have no doubt that some Western nation will sell them that nuclear capacity.

Despite India's significant gains in food production and promise for further success, some serious social and political problems need to be overcome to attain long-term food self-sufficiency. For example, at the height of the famine crisis in July of 1974, wheat farmers in Punjab, India's most productive wheat state, were deliberately withholding their supplies to drive up prices and increase their profits. The food minister, Chidambara Subramanian, appealed to the farmers to sell their surplus in order to avoid mass hunger in India. No outside nation, he said, could at that point feed "this elephant of teeming millions." [69]

At the same time, despite Ms. Gandhi's plea that losing courage has never helped—"The people in India have never been crushed under any burden." [70]—energy was so scarce that fertilizer plants in India were operating at 60 percent capacity and electricity for irrigation pumps was often unavailable. The bold dreams of the Green Revolution were crumbling. Economist Prem Shankar Jha wrote in the *Times* of India: "Twenty-five years ago, the country had enough food but no machines. Today it has a technical base and the skills to build most of its own machines but has to import food and fertilizer." [71]

As if to underscore the priorities of India toward industrialization rather than agriculture, he added, "The share of agriculture in successive plans has dwindled from 31.4 percent in the first plan (1951–56) to 20.7 percent in the fourth (1969–74). It has been slashed further in the annual plan for 1974–75." [72]

Because India's population growth—about 15 million per year—requires an annual production increase of several million metric tons of grain just to maintain per capita consumption needs, many food donor countries have inquired about the progress of the family-planning program that India instituted several years ago. In fiscal year 1973–74 India spent about $71 million on family planning; in 1972–73 it spent some $106 million. Ms. Gandhi admitted, "We are aware that our performance has not been very impressive." [73]

Yet, Dr. Karan Singh, the minister of Health and Family Planning, claimed, "The syndrome that India is no longer interested in family planning is totally false. Our foreign critics also say that there is not a political commitment. That is also totally false."

On the other hand, another "expert" told *The New York Times,* "The priority for family planning is just not there. There's high priority for developing industry, for steel mills, for defense. The populaton problem is given lip service, and that's it." [74]

Underlying these conflicting views of the population problem in India, and a fundamental cause of the problem, is the deeply embedded social practice of having many children,[75] partially to ensure the survival of a few sons to care for their parents when they grow old. With sons as their principal social security, it is no wonder that the poorer Indians continue to have many offspring. Clearly, family-planning programs will be most effective when coupled to social security programs to alleviate the economic need for sons. Perhaps with less economic necessity for male children, female children will be more welcome. "A girl is a liability," said Promilla Kalhan, an Indian political journalist and women's activist. "Her family has to feed and clothe her and then provide her with a dowry. But a boy means security in old age. A family will have six daughters, and keep on trying, waiting for a boy." [76]

Lester Brown says of economic and social policies in the developing world: "Making family-planning services universally available, meeting basic social needs, and encouraging new roles for women in society are the most important broad areas for action if birth rates are to be reduced rapidly. In addition, attention must be given to the national structure of economic and social incentives that affect attitudes on family size." [77] Income-tax deductions for an unlimited number of children, which are given in the United States, and child-care allowances, offered in France, are examples of ways in which governments offer no incentive to reduce family size; policies such as those may, in some instances, actually encourage citizens to have even more offspring, Brown points out.

Part of the failure of family planning can be traced to longstanding customs, but those customs also reflect the populations' perceptions of the economic realities of their system. To make a global survival compact that leads to increased food production and eventual economic self-sufficiency and demographic transition, centuries-old social practices will have to be changed; and sufficient progress toward economic security will have to be exhibited so that small-sized families will appear socially practical to the bulk of the people. Will this require a dramatic increase in the general economic standard? If so, then it will undoubtedly take at least a few generations to accomplish; indeed, the momentum of population growth alone would imply a steady-state population well over a billion. On the other hand, there is considerable reason to believe that a small increase in social security, a rapid drive to increase literacy, and some political reforms could get the process begun much faster.

For example, India's island neighbor, Sri Lanka (formerly Ceylon) has made remarkable progress in reducing its birth rate, despite a continuing economic crisis and one of the lowest per capita Gross National Products (GNPs) in the world. According to figures compiled by an Overseas Development Council publication,[78] Sri Lanka has a per capita GNP of only $110 (compared to an identical $110 for India) but a literacy rate of 70 to 80 percent (compared to 28 percent for India), an infant mortality rate of 45 per thousand (compared to 139 per thousand for India) and, most important to our discussion, a birth rate of 28.6 per thousand (compared to 39.9 per thousand for India). Thus, these statistics suggest that literacy might play an important role in determing birth rates and infant mortality rates since India and Sri Lanka have the same low per capita GNP.

The potentially important role of literacy is supported further when Sri Lanka is compared to Brazil, a developing nation striving hard to become a world power; it has already achieved a per capita GNP of five times that of Sri Lanka. But what of Brazil's other statistics? Its infant mortality rate per thousand is 94 (versus 45 for Sri Lanka), its birth rate per thousand is 37.1 (versus 28.6 for Sri Lanka), but the literacy is about 10 to 20 percent less than for Sri Lanka. Clearly, higher per capita GNP *by itself* cannot be proved to lead to lower birth rates, at least not for countries with per capita GNP less than a few thousand dollars.

Is it high literacy, improving health services, culture, or political and land reforms [79] that have given Sri Lanka a lower birth rate than economically similar India or much richer Brazil? Probably, it is some combination of these and other factors, but it is clear that although an increased per capita GNP can ultimately contribute to declining birth rates, many other steps can and must be taken in the interim. The alternatives to these social and political reforms seem to boil down to coercive methods or the horror and political instability of Malthusian collapse.

To recap, the practice of having several surviving sons as social security is not unique to Indians; some have argued that it is the chief common obstacle to economic development and to food production progressing at a substantially faster rate than population growth in most of the developing world. Certainly this, combined with low priority for agricultural development and bureaucratic inefficiency and corrupton, is a chief obstacle.

Bangladesh: The Guernica of South Asia? Bangladesh (formerly East Pakistan) even more than India may be the modern classical example of such obstacles. The great euphoria that followed Bangladesh's bloody fight for independence in 1971 has been replaced by dire worldwide pessimism about the future of the young country. Now referred to as "an international basket case" and the "world's most hopeless nation," [80] Bangladesh has failed to develop a viable system of government, is steeped in corruption, has no agricultural policy nor any plan to deal with its amazingly high birth rate, and has about the lowest per capita income in the world.

Many of the country's myriad problems stem from a relatively large population that cannot be supported by the nation's productivity. Bangladesh comprises some fifty-five thousand square miles—roughly the size of Wisconsin. While 4.4 million people live in Wisconsin, some 75 million are crammed into Bangladesh. Several babies are born every minute. Yet, what can be done to reduce this birth rate? In an article in the *Atlantic Monthly* journalist Claire Sterling noted:

> It is not easy to sell the notion of birth control to people who have no old-age insurance, no money to save against a rainy day, no tractors to work their land, almost no joys in life except sex and their children, and, in some remote rural areas, not even a very clear idea of where babies come from.[81]

In December 1974 Prime Minister Mujibur Rahman, who had led the country (with considerable help from India) to independence three years before, declared a national state of emergency to deal with the lawlessness, corruption, and economic disorder afflicting Bangladesh. Mujib blamed his country's misfortunes on "people who opposed the liberation war . . . collaborators of the Pakistan Army, extremists and enemy agents in the pay of foreign powers." [82] By declaring the state of emergency, Mujib gave himself the latitude to imprison suspects without trial or appeal and to ban any political groups or meetings. It is not surprising that, since Mujib's Awami League party controlled more than 95 percent of the seats in parliament, his declaration was widely supported, but not so widely as to prevent his later overthrow.

A U.N. official commented about Bangladesh: "What may happen in the 1980s in India and the rest of South Asia is happening here right now." [83] At its present rate of growth, the population of Bangladesh will double in two decades. The World Bank reported, "It is impossible to think that this population level will be compatible with anything other than the merest survival on the basis of an international dole." [84]

One senior official in Bangladesh defended his country by stating that the Bengalis never had a chance to develop their own political leaders, bureaucrats, or business people before the war because non-Bengalis were running the country prior to independence. "If there is corruption and lethargy now—if people don't have a vision of society—it's because we never had the experience to build on," [85] he said. However, a different view of the situation, one from outside the government, was expressed by a lawyer and prominent Bengali intellectual, who said of his newly empowered countrymen: "These people have created a society of plunder. They have never made money in their lives; and suddenly they are in charge of factories and giving out licenses and permits. They can write out checks and sell stock. It has gone to their heads." [86]

One Bengali economist, in discussing one of Mujib's cabinet ministers, said, "After the war you could bribe him with two cartons of imported cigarettes. Now it's at least 100,000 takas [$12,000]." [87]

Another factor that is choking Bangladesh's development is its bureaucracy, which is filled with civil servants who are underpaid and often inefficient. Two hundred and four signatures were needed to approve a contract for a foreign consultant, for example. "You talk to them about a project," a foreigner said. "Then they ask you to send a letter. You draft a letter. It takes six months to get a reply." [88]

Although Bangladesh accepts foreign aid, smuggling and black marketeering often make it difficult for the food aid to reach the mouths for whom it was intended. Huge quantities of rice, for example, are smuggled from Bangladesh to India; and India, along with the United States, the United Nations, and the World Bank has been one of the major contributors to its struggling neighbor.

Bangladesh does have fertile agricultural land, but not as much as it needs to feed its burgeoning population—even in good crop years. Furthermore, the region is visited regularly by floods and cyclones. Some agricultural experts say that food production in Bangladesh could be increased significantly with high-yielding crop strains, improved irrigation and fertilization, and mechanization. The greatest obstacle may be the smallness of the individual agricultural holdings; the average farm is a mere 1.7 acres, which is too small for the efficient use of technological production methods.[89]

Perhaps the ultimate irony in Bangladesh's short, troubled history has been that the prime minister's reaction to emergency was to throw out civil liberties rather than the corrupt officials who have been stifling the country's development. And even though Mujib himself has paid with his life, it remains most difficult to be sanguine about the future of Bangladesh.

Development Leads to Social Change

The purpose of the foregoing has been to show by specific example some of the tremendous social, political, and cultural obstacles that stand in the way of improving food production and per capita consumption in the developing countries. I have said that the potential to increase food production in the developing world is the bulwark of optimism of those who see a solution to the world predicament. But any change—even an increase in food production—will cause repercussions in a society.

In the small agricultural village of Sawahrejo in Java, for instance, "progress" is widening the gap between the two kinds of farmers: (1) the poor subsistence farmers who work their own tiny plots with rather primitive traditional methods and, with some luck, also work for one of the large landholders; and (2) the richer farmers who own most of the land and whose profits have increased considerably in recent years through the use of Green Revolution methods—high-yield strains, augmented irrigation, insecticides, and a considerable amount of fertilizer. Historically, there has been a gap between the two kinds of farmers, but the system had stabilized and had worked well for both. The large landholders, assuming paternalistic roles, employed and fed the small landholders and the landless.[90] Now, the expensive Green Revolution methods are substantially increasing the yields (and profits) of the richer farmers, while the vast majority of the population is continuing its use of subsistence farming techniques and rapidly increasing its numbers. With the population increasing so rapidly, the rich can no longer employ and feed all the poor, and the poor are becoming relatively poorer. One prosperous Sawahrejo farmer, who spent his profits on building a house, expressed his insecurities: "In the coming twenty years," he said, "the situation will get worse and worse. If the landless and the poor have nothing to eat, they will become robbers, they will become hoodlums." [91]

The Green Revolution cannot be applied to the developing world without effecting potentially large social changes, particularly if the Green Revolution displaces traditional labor patterns without creating new employment opportunities. Some might use the label "cultural imperialism" for a Global Survival Compromise that ties the delivery of Western technical and capital aid to simultaneously implemented changes in social and political practices, even if these practices stifle hope for self-sufficiency among the poorer countries. Yet, the adoption of Green Revolution techniques and the striving for industrialization are already and everywhere creating profound social and cultural changes and new dependencies in the developing world. These changes are for the most part not imposed from without, but rather reflect the needs and desire for more food common to most peoples. It is to be profoundly hoped that these social changes will be anticipated, not simply experienced in ignorance of their meaning or direction.

Thailand Proves That Agriculture Can Pay

Another example of a country with a rice agriculture is Thailand, Asia's only major rice exporter. Thailand has more land and lower population density than most Asian countries. About 80 percent of the population is agrarian, and land has been plentiful enough so that

Thai farmers have not felt pressed to adopt Green Revolution methods, which require large amounts of petroleum-based chemical fertilizers and pesticides. Thus, when oil prices rose sharply in 1974, the Thais, not dependent on large amounts of petroleum, did not suffer as much as other productive Asian farmers or more industrialized countries. Perhaps Thailand's relative success might serve as an example to other less developed countries with sufficient amounts of cultivable land to concentrate their development more heavily on agriculture. For food is a product that will always be in demand, and, especially in heavily populated areas, there is plenty of labor available to work the soil. It should be pointed out, however, that although Thailand's economy is growing, the country is plagued with problems, such as the bureaucratic inertia and corruption that often exist in countries that are industrializing rapidly. But the example of Thailand shows that maintaining agricultural development pays and suggests that further development could bring on major rewards.[92]

The World Bank, a major contributor to agriculture in developing countries, announced in March, 1975, that it planned to double its investment to $1 billion annually by the year 1979 to help the poorest rural populations. The bank, formally known as the International Bank for Reconstruction and Development, is affiliated with the United Nations but is an independent operation. In a report entitled *Rural Development: Sector Policy Paper*, the bank said that 85 percent of the poor people living in the developing countries had annual per capita incomes equal to fifty dollars or less. About 80 percent of the poor, the report said, live in rural areas. More than 80 percent of the 750 million small farmers own about five acres or less, and many of the land holdings are scattered fragments of land. Graham F. Donaldson, agricultural economist at the bank, noted that the first problem in such cases is to consolidate the land.[93] "You don't realize the enormous difficulties involved until you see a farmer, as I did in Kenya, trying to balance a bag of fertilizer on a bicycle," he said. "That farmer's first need may be a cart." [94] How can Green Revolution methods, which require vast amounts of water and fertilizer for highest productivity, be introduced to most of the world's poor? Obviously, they cannot on a very rapid or large scale, with existing conditions such as small, fragmented land holdings, little government encouragement to small farmers to enter the cash economy, and few social programs to deal with farm laborers displaced by the new methods.

The People's Republic of China: Is This the Only Way for Asia?

In talking about developing countries in Asia, I have left out 20 percent of the world population that lives in a single, developing country: The People's Republic of China. In that vast country, which was infamous for starvation a few decades ago, the people seem healthy and content, according to reports from many returning travelers. The nation is largely agrarian, and hard work from everyone is expected—and rewarded. A family's income is based in part on the success of the crop raised by its production team. Agricultural innovations, often unavailable or resisted by poor farmers throughout the world, are widely used in China, along with traditional, labor-intensive techniques. Intercropping is a technique that some Chinese farmers claim yields 40 percent more food than the usual planting methods. One Chinese farmer described the intercropping methods that he uses: Rows of wheat are planted; when the wheat sprouts, two rows of corn are planted between the wheat rows, and before the corn is high enough to shade the wheat, that grain ripens and is harvested. Two months later the corn is harvested. With advanced energy-intensive technology this method would not work, but with hand labor it seems to be effective.

Another innovation used in China involves the traditional use of compost, manure, and human excrement with inorganic fertilizer, which is produced by some eight hundred "backyard factories" that convert coal and water into nitrogen fertilizer.[95]

Many Americans who think that they are doing their part for ecology by recycling bottles, cans, and newspapers could learn a thing or two from the Chinese farmers, who recycle just about everything, according to Dr. Sterling Wortman, who led a team of Western agriculturalists on a tour of China recently. For example, pigs in China are fed material that is not eaten in many other countries—the leaves and stalks of vegetables and cottonseed hulls. The manure produced by the pigs is spread on the fields.

Wortman described a commune whose specialty is carp and silk.

Dirt dug to make ponds for the carp is used to plant mulberry trees, which house silkworms. The carp, which are scavengers, are fed somewhat like the pigs; their diet includes silkworm droppings, rich in digested mulberry leaf. From time to time, the rich bottom sediments are scraped out of the ponds and spread under the mulberry trees as fertilizer. In addition, the ponds are dug close to the trees so that they can share their moisture.

Despite these visible efforts and the apparent lack of starvation in China, it is difficult to assess accurately the agricultural and economic conditions there because statistics are not readily available. However, the United Nations estimates that the Chinese get only 90 percent of their calorie requirements.[96] Wortman reports that when he and his scientific colleagues asked for exact figures on Chinese crop tonnages or per capita production they were merely told that production was up twofold or threefold. Of course, no one can be sure how representative of general conditions in China the experiences of Western travelers are. Perhaps we know even less about the reality of the situation in China than we do about the causes of climatic change. Nevertheless, Wortman and his group concluded that the long-term outlook was mixed; unless Chinese schools and research centers stabilize further, agricultural progress during the next few decades may be considerably slowed. He feared insufficient attention was being paid to long-term research. A similar feeling was expressed by a team of U.S. meteorologists that visited China in 1974.[97]

Another interesting analysis of China's future prospects was offered by Jon Sigurdson, who had been a cultural attaché at the Swedish Embassy in Peking from 1964–67 and worked on problems of technology transfer. He generally confirmed Wortman's impression that the Chinese had developed an enviable record of self-reliance in food production through ecologically sound resource-recycling techniques achieved by labor-intensive agriculture. However, he, too, expressed concern for their future: "The relatively rapid increase of total population, which is smaller percentage wise than many other developing countries, and the still more rapid increase of working age population pose two major problems with which the Chinese planners have to deal. One is the obvious food production problem and the other is where to employ meaningfully the very significant increase in labor force." [98] To meet these needs to the year 2000, he suggests that grain production will have to increase by another 40 percent, and this, he argues, will require more mechanical and chemical agricultural technologies: "As the amount of organic manures is already approaching the maximum production limit, any expansion of inputs must be achieved by the use of chemical fertilizer." [99] Since this mechanization would reduce the number of agricultural field workers (by about 150 million, he estimates), the Chinese might face the need to employ perhaps as many as 450 million more people in the year 2000 than at present.

Sigurdson forsees that these pressures may force China toward greater industrialization, which also raises the risk of severe environmental damage. However, despite the potentially serious problems ahead for China, he is hopeful that China can overcome these obstacles and still maintain "her concern for fundamental human needs." [100]

Despite some ominous clouds on the horizon, the Chinese seem to have overcome most of the governmental corruption, bureaucratic inefficiency, and ingrained cultural practices that slow development in many other developing (and developed) countries. But they have done all this at a very high price, at least from the perspective of those who cherish writing and speaking their minds publicly and working or living where they choose. But the choice between these individual freedoms and a sufficiency of food brings up a difficult dilemma, assuming that these are mutually exclusive options in the first place. The best hope must be that the developing countries find sufficient inner resources, with adequate external help, to cure some of their problems and also to maintain a freedom of choice consistent with their cultural and political needs. We should not dictate to them how to make the choice, although we can certainly expect them to watch our successes—and failures.

Latin America: Still Well Below Its Potential

Asia is not a unique example of food production problems in the developing world. Latin America is faced with the prospect of permanently losing its status as a net exporter of food, because of its rapidly growing urban populations and strong government biases toward industrialization. The region has one of the highest birth rates in the world; its population almost tripled from 1930 to 1970.

The mass migration to the cities has had repercussions throughout Latin America. In 1950 less than 40 percent of the area's populations lived in urban areas. By 1970 the proportion had risen to 56 percent, and it is predicted that by 1990 two out of three Latin Americans will live in cities.[101]

Rising political demands from urban constituencies have caused governments to keep agricultural prices low, even in inflationary periods. But such political palliatives can create long-term problems. Strict price controls on food in Bolivia, for example, led to contraband trading of thousands of tons of food into neighboring countries, and as a countermeasure Bolivia's president, Hugo Banzer Suarez, almost doubled the price of basic necessities in January, 1973. That action was followed by protest strikes by miners and professional workers and a peasant revolt, all of which seriously shook the regime.[102]

In Argentina, Latin America's most important food exporter, food prices have long been among the lowest in the world. That situation, coupled with high taxes on farmers, has prevented the farmers from realizing the full benefits of rising food prices on the world market. In Argentina, as in Bolivia, there is much contraband trading over the borders—perhaps as much as $500 million per year. Significantly, the low income of the farmers has lessened their interest in increasing investments in technology that might improve production. Some experts feel that Argentina could double or triple its agricultural production, but there have been no notable increases in several decades, even though there were many in other major granaries.

In Chile agricultural production fell 25 percent in 1973, the last year of the Marxist regime of Salvador Allende Gossens. In an article in *The New York Times,* Jonathan Kandell reported: "The deterioration was generally attributed to unrealistically low food prices, inadequate credit, rising production costs, badly administered agrarian reform, lax labor discipline on collective farms and the insecurity of private farmers faced with a wave of illegal land seizures." [103] In 1974, under the new military regime, which has not shrunk from coercive measures, production increased 10 percent, and farming cooperatives are being threatened with a return to private land ownership if production does not improve further. The government has also lifted price controls on most food products, which has worsened inflation and poverty in the cities, but has been a boon to the agrarian areas. One hopes the increased food supply will both reduce prices and improve living conditions, but this remains to be seen.

Prospects for Increasing Food Production: Technologically Optimistic but Politically Bleak

In summary, improving farming throughout the world is clearly possible, but as the examples cited throughout this chapter show, a vast array of complex political, economic, ecological, social, and cultural situations must be dealt with simultaneously in any significant effort to improve the food situation.

Agronomists who spend time in developing countries often comment that the low productivity of farming in those lands, relative to what is possible with existing technology, results not so much from the backwardness of the farmers, but rather from the governments' lack of programs to make the most modern methods and equipment available.

"I have a lot of respect for the small farmer," said Norman Borlaug, one of the key developers of the Green Revolution techniques. "Almost invariably, when you look at what he's doing with his land, you find he's producing the maximum under the situation he has to work with. The thing is that he usually doesn't have much to work with."[104]

Production has increased significantly in the few countries where the governments have set up programs to inform farmers about new methods and allow them to borrow money for fertilizer and equipment, or where big farms have been free to develop and prosper. In most countries, however, farming has been considered a drab occupation, deserving low priority relative to the more glamorous pursuit of industrial development. But recent, worsening food shortages, coupled with the apparent long-term inability or short-term refusal of food-rich countries to bail out the poor ones in times of shortage, may force governments to reassess the importance of agricultural development. Related to that is the growing opinion that the adoption of highly mechanized energy-intensive North American farming methods may not be essential or even desirable in areas where small, labor-intensive farms dominate the land-holding system. Most developing countries are rich in the resource of human labor, and there are scientific techniques that can be employed to make reasonably efficient use of that resource. This has been done in Taiwan and South Korea, for example, where the small farmers' standards of living are rising.

Sterling Wortman cites four prerequisites[105] that he feels must be met before poorer farmers can successfully adopt the new, higher-yielding strains: (1) The new techniques or materials must be demonstrably superior to the traditional ones; those include the new strains, fertilizers, methods for irrigation, and multiple cropping or intercropping. (2) The necessary materials must be available to the farmers where and when they need them and they must be affordable, which often necessitates a credit system. (3) The farmers must be shown how to use the materials and methods, and the instructors must be skilled farmers whom the others will respect. (4) The farmer must be able to sell his products at a price that pays for his expensive investment and gives him a profit. The Chinese, for example, use incentives to stimulate agricultural output.

It is important to note that the world's astounding agricultural progress of the last ten to fifteen years has been largely checked by population growth in developing countries and rising affluence in newly rich nations. So, although much evidence suggests that a *sustained* rise in the standard of living tends to slow the birth rate, the continuing rapid population growth that keeps abreast of production increases in many of the most populous countries only increases the likelihood that these areas will be hit by famine—before a rising standard of living could make enough impact on population growth and, ultimately, self-sufficiency goals. And if massive fertilizer use does prove a substantial threat to the ozone layer, these prospects will grow considerably worse.

Boyce Rensberger, in a series of outstanding articles for *The New York Times* on the food crisis,[106] cites three steps that food experts generally recommend to alleviate food shortages in the short run: (1) Americans and other meat lovers should reduce their meat consumption because the production of meat, especially beef, uses grain very inefficiently (this does not apply, of course, to foraging animals). For example, producing a pound of beef from a grain-fed cow requires from eight to ten pounds of grain, much of which could be fed to humans directly; the poor-quality "feed grains" could be phased out by planting more palatable crops, such as wheat (a point usually disregarded by those who argue in favor of meat diets). (2) Improper storage of grain in many poor countries leads to the loss of as much as half of the food to rodents and insects. Solutions to this problem need not be exotic. If farmers would merely store their grain in old oil drums or in baskets on stilts, instead of leaving it piled up in a corner or in a burlap sack, much more grain could be saved. The U.N.'s Food and Agriculture Organization estimated that "55 million Africans could be fed for a year from the (native) grain finding its way to the 'wrong' consumers—rats, locusts, quelea birds, beetles, moths and weevils, and countless micro-organisms."[107] (3) If rich countries would stop or cut back on fertilizing the grass in their yards, golf courses, and cemeteries, more precious fertilizer could be exported to needy farmers—if, of course, they could pay for it, or special terms could be arranged.

Rensberger concludes,

> Many of the experts think that the resources exist to feed every person well but only if they are shared. There is evidence that if they are shared, the rate of population growth will decline. There is also evidence that when they are not shared, the social and economic effects within and among countries may be profoundly disrupting.
>
> At risk is not only the agony of lingering death for millions, mostly children, but the character of the world left for the survivors.[108]

As for the long-term food outlook, there is ample reason for optimism—unless the fertilizer–ozone crisis or some other ecological or political obstacle proves insurmountable. But we should not delude ourselves into thinking that such progress will be cheap or easy—or guaranteed! It will require massive infusions of capital from the rich nations and often dramatic, unpopular changes in existing social, cultural, and political patterns of the poor and the rich. Nothing short of immediate actions, such as a Global Survival Compromise, seems to have much chance of reducing the probability that severe famine catastrophes, and their unavoidable repercussions on world stability, will occur during the transitional period between now and the twenty-first century.

9

The Genesis Strategy: Suggestions for a Way Out of the World Predicament

We are now faced with the fact that tomorrow's today. Over the bleached bones and jumbled residues of numerous civilizations are written the pathetic words: "Too Late." There is an invisible book of life that faithfully records our vigilance or our neglect . . . This may well be mankind's last chance to choose between chaos and community.

—MARTIN LUTHER KING, JR. (1967)

Ecosystems that are damaged or destroyed in the process of use or development recover at rates in tens to hundreds of years. Therefore it is of prime importance to begin the development of economic policies and resources in the context of local long-term stability rather than immediate goals. This is an enormous and difficult problem because the economic pressures are powerful and demand solution now and the kind of development needed will require cooperation among . . . countries at a level that does not now exist.

Fragile Ecosystems, Report of the Workshop on Tropical Ecology (1974)

One is always reading that the sorts of drastic changes proposed here are infeasible, impractical, and unrealistic, particularly in the economic and political sense. I believe that the alternative of proceeding along our present course

is physically *impractical as well as socially unacceptable.*
Can our political and economic scientists devise ways
to bring the world's institutions into line with
physical reality in time?

—JOHN P. HOLDREN
Energy and Resource Program,
University of California, Berkeley (1975)

Energy–Climate–Development Dilemmas

The main purpose of this chapter is to suggest some mechanisms for helping to reduce the possibilities of worldwide catastrophes. Up to this point, I have primarily called attention to immediate and potential dangers to global security if a number of present world attitudes and trends continue. Climate-related problems have been emphasized for several basic reasons:

(1) Climate change can be both a global and potentially irreversible consequence of human indifference to natural systems.

(2) The climatic system does not conform to the prevailing concept of "national sovereignty"; thus, the possibilities of climatic shifts in one area being connected to related changes elsewhere provide an opportunity to catalyze international cooperation and could even serve as a symbol of global interdependence to encourage greater world unity and movement away from the often selfish and short-sighted goals of nation-states. This interconnectedness of the climatic system could also provide a cause for international conflict.

(3) Climatic processes are not well understood, yet potential climatic changes could be serious; this case exemplifies dilemmas that will arise more and more frequently over critical political issues that contain an important, but uncertain, scientific component. That is, the climate may well be understood "enough" to begin immediate and perhaps extensive actions to prepare for possible but uncertain dangers that present knowledge suggests may be ahead.

(4) Considering their immediate importance, problems of climatic change have been given relatively little detailed attention in most of the writings on the world predicament.[1] For example, Robert

Heilbroner, noted economist/historian from the New School for Social Research, commented in his book *An Inquiry into the Human Prospect* that heating up the earth from energy production is undoubtedly the ultimate limit to industrial growth. Yet he interpreted information available to him on the thermal pollution problem as allowing us "to proceed along our present course for about 150 years before the atmosphere would begin to warm up appreciably—let us say by about three degrees. At this point, however, the enormous multiplicative effects of further exponential growth would suddenly descend upon us." [2]

In that statement, Heilbroner was considering only thermal (heat) pollution on a global scale. But, as pointed out earlier, rapid growth of energy production could cause large local or regional climatic modifications from thermal pollution much sooner. Unfortunately, the present capability of climate theory does not include a definite (quantitative) prediction of the character of possible climatic responses to a given amount and distribution of thermal pollution power density. Nonetheless, I feel that enough is known already to compute rough estimates of climatic sensitivity to energy inputs and to conclude two things: one, that global effects of human activities can certainly be expected to grow to the magnitude of natural climatic variability sometime after the year 2000 and, two, that local, medium-scale, and even regional climatic disruptions can be anticipated even sooner than global ones, although it is still impossible to make precise estimates. Furthermore, the use of fossil fuels to produce energy also leads to increases in the levels of atmospheric carbon dioxide, and all available estimates suggest that this factor is many times more significant for *global* climatic changes than the direct thermal heat release that accompanies all forms of energy production.

Unhappily, since improvements in the standards of living of the world's poorer peoples are related in part to greater energy consumption, and since there is a growing disparity in the per capita energy consumption between peoples in rich and poor nations, a serious dilemma faces world leaders: How can the climatic and other environmental risks of greatly increased energy consumption, which may be both global and irreversible, be weighed against the social and political factions calling for vastly increased energy use as one means to help raise the quality of life for billions living in less developed areas? This dilemma is further complicated by the realization that alterations to present rates of energy use and economic growth would take many years to accomplish smoothly; thus, decisions to alter long-term growth patterns may have to be considered now, despite the uncertainties in predicting the climatic or other environmental consequences of projected energy-use patterns (let alone the social consequences). The problem is clear: What do we do in the face of uncertainty?

Although no solution to this dilemma can be offered without introducing some risks, I believe that in many instances we should consider taking some climatic or other environmental risks in order to narrow the energy gap between rich and poor, even if that means more pollution. Perhaps the wisest strategy is a compromise that calls for long-term limits to total global energy use and population size, but also recognizes that immediate increases in energy use in less developed areas may be necessary (for a limited time, at least); and as the heavily populated countries reduce their population growth rates, the highly energy-consumptive nations must begin to practice energy conservation.

One factor in this compromise is especially critical to its success—timing. If we wait much longer to implement such a plan, the populations of the less developed regions may have grown so large that to increase per capita energy consumption would be climatically dangerous—not to mention the risk of other dangers, such as an ecological catastrophe.[3] The need to raise per capita energy-use levels in the less developed areas as quickly as possible is the crux of the energy aspect of the Global Survival Compromise. The hope is that rapid increases in the level of energy use in the less developed countries would contribute to a decrease in the birth rate (through the demographic transition) so that the final steady-state world population would be much smaller than if only a meager effort to raise standards of living were made over the next decade or two. But, as stressed in the previous chapters, the demographic transition is not a guaranteed consequence of increases in energy-consumption levels. Rather, if it can be accomplished at all, the demographic transition depends on a mix of social, cultural, medical, economic, and political factors. And this mix will vary from one society to another, and also will change in time as each society evolves. Thus, the critical point, and one stated many times in the past, is to work toward leveling off the population growth rate as soon as possible.

The climatic disruption from energy consumption is proportional to the total amount of energy used, which is equal to the population size multiplied times the per capita energy-consumption level. Thus, a greater level of consumption per capita is possible (with no additional environmental risk) if the total population size is smaller. Delay might just mean that there would be too many people on earth to achieve a high per capita energy-consumption level worldwide without the risk of severe climatic changes; in that case the world would face increasingly intense pressures for the redistribution of wealth at a time of fixed levels of energy consumption. The unstable political climate that situation implies is most ominous indeed.

Nuclear Blackmail

What is talked about here is certainly not a new issue. The debate over solutions to the world predicament has been heated since the time of Malthus. However, there is at least one new aspect of the predicament, and that is the specter of nuclear terrorism, as discussed by Heilbroner, for example:

> I do not raise the specter of international blackmail merely to indulge in the dubious sport of shocking the reader. It must be evident that competition for resources may also lead to aggression in the other, "normal" direction—that is, aggression by the rich nations against the poor. Yet two considerations give a new credibility to nuclear terrorism: *Nuclear weaponry for the first time makes such action possible; and "wars of redistribution" may be the only way by which the poor nations can hope to remedy their condition.*[4]

Similar sentiments have been expressed by many others,[5] including Mihajlo Mesarovic of Case Western Reserve University and Edward Pestel of Hanover, Germany. Their recent book, *Mankind at the Turning Point,* is the follow-up study to *Limits to Growth* for the Club of Rome, an interdisciplinary international, nongovernmental organization concerned with the world *problématique*. They conclude that on the basis of a computer model of world dynamics (i.e., the interactions among population, resources, and environment) the means to help the hungry billions find their way out of poverty must be found very soon. "Unless this lesson is learned in time there will be a thousand desperadoes terrorizing those who are now 'rich,' and eventually nuclear blackmail and terror will paralyze further orderly development," they say. "Ten or twenty years from today it will probably be too late, and then even a hundred Kissingers, constantly criss-crossing the globe on peace missions, could not prevent the world from falling into the abyss of a nuclear holocaust." [6]

Mesarovic and Pestel do not say that world problems are unsolvable, only that we are rapidly running out of time in which to solve them. They criticize the model of the world made by the authors of *Limits to Growth*, who argued that only an immediate slowdown of economic growth could prevent a world collapse. Mesarovic and Pestel point out that the earlier work for the Club of Rome was based on a computer model that viewed the world as a single system, while their own model considered that "the world can be viewed only in reference to the prevailing differences in culture, tradition, and economic development, i.e., as a system of interacting regions; a homogeneous view [e.g., as the *Limits to Growth*] of such a system is misleading." [7] The newer model indicates that a collapse on a regional level could occur in the absence of a world collapse (unless, of course, the regional collapse led to nuclear war), and that such regional catastrophes (e.g., widespread famine in Asia) could occur long before the middle of the twenty-first century. Such catastrophes can be avoided, they say, but only through global strategies that promote "organic growth," particularly in developing parts of the world. "The delays in devising such global strategies are not only detrimental or costly, but deadly," they conclude. "It is in this sense that we truly need a strategy for survival." [8]

Economist Kenneth Boulding, who reviewed *Limits to Growth* for *The New Republic* [9] in 1972, also reviewed *Mankind at the Turning Point* for *Science* magazine in 1974. He said about the later book:

> In some ways, for all its modestly optimistic tone, this volume is even more depressing than the first report, perhaps because it is more specific. Its suggestions on what has to be done to repent (population control and large transfers *now*) are extremely plausible. Like Donne's great sonnet on the judgment, it points out that by the time the judgment has arrived it is too late to repent, and we have to repent now for repentance to do any good.

Yet, Boulding complains, "What the report does not say is how we are to be persuaded to repent." [10]

The Genesis Strategy: A Hedge Against Catastrophe

Some mechanisms to help persuade us "to repent" are offered in the remaining pages. Boulding's complaint is addressed by suggesting ways in which the mechanisms may be started. Perhaps, it is as risky for a physical scientist to suggest strategies to reorient world political consciousness as it was for an economic historian to estimate how many years remain before human activities lead to an intolerable heating up of the earth. Nevertheless, the predicament is connected to questions of both climatic change and political consciousness, a connection that will persist even if specialist writers on the subject ignore all aspects outside their specialty. One consensus that clearly emerges from all sides in the debate on the world predicament is that there are no easy answers to be found; even a description of the predicament itself calls for a complex and uncertain mix of insights from the social, scientific, political, ethical, and psychological disciplines.

Despite these difficulties Heilbroner chose to make a connection between potential climatic changes and the human prospect. At the very least, his attempt showed the urgent need for interdisciplinary connections, a need that is often ignored by traditional scholarship.

In *The New Industrial State*, John Kenneth Galbraith, the renowned Harvard economist, recognized this ingrained academic "ostrich" tradition and lashed out at those educators, scientists, intellectuals, and artists, who, he said—

> are handicapped by the belief that their role is professionally passive—that it is to feel and think but not to act. Righteousness, as well as convenience, defends this passivity. Politics is not the business of the intellectual or the artist. Nor of the educator, nor of the scientists. Theirs is the purer domain of the spirit and the mind. This can only be sullied by concern for practical affairs. In the last milli-second before the ultimate nuclear fusion, a scientist will be heard to observe that the issue of nuclear control and military security is really one for politicians and their military and diplomatic advisers. And as the last horizon is lost behind the smoke, gas, neon lights, and detritus of the industrial civilization, men of self-confessed artistic sensitivity will be heard to observe that, unfortunately, none of this is the business of the true artist.[11]

The proposals that follow are offered seriously, largely in the hope that those who "know better" will expand and improve on them in order to correct the difficulties they are designed to address. The common thread running through these ideas and proposals can be traced to the underlying principle of the Genesis Strategy, the idea we have borrowed from the biblical story in the Book of Genesis in which Joseph advises the Egyptian Pharaoh to store food during the years of plenty to ensure Egypt's survival in the lean years to come. At first, the solutions attainable by even the most conscientious application of the Genesis Strategy will only buy time in the race to achieve a harmonious transition to a steady-state world order and the catastrophes that threaten in the interim. All proposed solutions to the world predicament must be designed ultimately to lead toward the rapid establishment of a stable equilibrium world, and the interim steps must be carefully chosen to insure that they do not irreversibly mortgage the future to save the present. However, despite the uncertainties of both the dangers and opportunities that lie ahead, immediate actions are needed, primarily to hedge against the worst possible dangers we can now perceive. For example, although building and maintaining sufficient food reserves to insure against climatic variability has been one of the principal points I have advocated in this book, the margin-of-safety principle, implicit in the Genesis Strategy, extends to other problems as well—particularly when the use of new technologies to sustain the exponential growth trends of economies and populations are at issue.

For example, present policy, which is committing the future economic health of the Western world to growing energy production, has led to widespread pressures for the adoption of nuclear power. Ironically, if atmospheric scientists are successful in convincing the public that fossil fuel consumption is climatically risky (through CO_2 production), then the pressure to develop nuclear power will undoubtedly increase. And although the probabilities of a nuclear accident are thought by many to be extremely low, such an accident, if it did occur, could be extremely dangerous and costly.[12] Thus, the decision on whether or not to "go nuclear" is a very uneasy choice.

In a debate with Dixie Lee Ray, former head of the Atomic Energy Commission, Paul Ehrlich warned that "having a nuclear power plant downwind of your home is basically playing dice with your future in a very tricky game where we really don't know what the odds are." Ray responded that such statements only "prey upon the real fears and worries of the people in difficult times," and asked, "I wonder what Dr. Ehrlich's got against human beings? I thought people were rather nice." [13]

She tried to assuage her audience while pooh-poohing Ehrlich's worry that the greatest public danger from a proliferation of nuclear power plants was the existence of growing amounts of radioactive poisons that could be misplaced, mishandled, or even stolen by malevolent or irresponsible *people*—the true weak link in the safety chain of nuclear technology. But shortly after this debate, Ehrlich's argument received unwitting support from an unexpected source, a supervisor of advanced protective systems for the nuclear industry. Thomas Sellers of Sandia Laboratory in Albuquerque, New Mexico, reportedly told a newspaper reporter who was writing a feature on nuclear safety that as far as prevention of sabotage is concerned, "in the long run we just have to count on the fact that there are more good guys than bad guys." [14] As Ehrlich has pointed out, it only takes a few bad guys to do the trick.

Unfortunately the Ehrlich–Ray debate did not quantify further the probabilities or costs of a nuclear accident, and perhaps made some people even more confused than they had been at the outset. But the "experts" are also confused about this matter,[15] and the difficult question to be faced by society and its decision-makers is whether to risk our future physical and economic well-being on a technology that has not been proven safe, recognizing that once we commit ourselves to economic and physical dependence on that technology, its removal could also be fatal. (The Irish potato famine described earlier is a classic example of a dangerous dependence on an efficient, but undiversified, necessity.)

In the case of nuclear power, the Genesis Strategy suggests that future economic stability be supported on a variety of energy technologies. This would mean that the chances of catastrophic economic collapse would be negligibly small if one of those supporting technologies were to prove too risky to be continued, or were to be removed by some other means (accident, sabotage, earthquake, terrorism, etc.). Present economic consciousness is weighted toward maximizing efficiency by adhering to "economies of scale," such as a few large and efficient power plants rather than a large number of smaller ones, or a few supertankers rather than many smaller ships. The

Genesis Strategy of maintaining diversity can, in some instances, mean lower efficiency than economy of scale considerations would dictate, and thus immediately higher costs. But these costs should be viewed as insurance against the even costlier and sudden demise that would follow the malfunction or sabotage of one of a few highly efficient power production systems, or the destruction of a supertanker. A climatological and ecological example of the Genesis Strategy would be a plan that called for use of a diversity of crop strains rather than a single highest-yielding variety that could be more vulnerable to climatic stress, insect pests, or disease. (This is not meant to imply that diversity necessarily sacrifices efficiency, or that efficiency is an undesirable goal. But in some instances there is conflict between these, and present consciousness is perhaps overbalanced toward efficiency.) These few examples are given mainly to illustrate the kinds of difficult trade-offs between efficiency and diversity that face society in the years ahead. Such dilemmas must be resolved, and often sooner than we might like.[16]

But there are ways to help clarify the options. Several of the institutes proposed in the following text could be formed precisely to gather information, clarify points of knowledge and uncertainty, and then disseminate explanations of the strengths and weaknesses of a variety of technological options for future societies. However, for such information to have a significant impact on the political decision-making processes, a substantial turning of political consciousness away from its slavish dependence on short-term rewards toward recognition of the need for long-term stability seems a prerequisite. Furthermore, since complex decisions that involve scientific issues cannot always be clearly resolved in time to avoid the worst possible effects, many decision-makers who expect clear-cut scientific guidance on the merits of technological disputes must learn to be more accepting of a probabilistic approach.

Therefore, in the interests of identifying some of the *functions* that need to be performed in the future, I will offer a concrete set of proposals aimed at serving those functions. It has been said that when fundamental attitudes need to be adjusted the American response is to propose new institutions. Perhaps this is so, but I offer my proposals mainly in the hope that they can focus debate on the difficulties they expose.

Three categories of concrete proposals are now offered: (1) some new institutions, (2) some new or better agreements, and (3) some new attitudes.

Some New Institutions

World Security Institutes

Institute of Imminent Disasters. Warning signals must be given before the threats to humankind of events ahead can be discerned or their potential impacts reduced by our positive actions. These messages of perceived danger must be sufficiently credible to induce our social and political mechanisms to undertake preventive action, despite uncertainties over when the dangers will occur and how effective the actions to mitigate them will be. To that end I propose the establishment of a multidisciplinary, international Institute of Imminent Disasters, whose responsibility would be to identify, study, and forewarn of potential dangers to world security. These would include natural hazards such as drought or flood, environmental disasters such as the elimination of a natural species or the reduction of ozone from some pollutant, and potential human tragedies such as an Irish potato famine or a Sahelian drought.

To maintain the integrity of the institute, a broad range of qualified specialists from many countries would be needed; and the various disaster scenarios studied by the institute would be chosen in part by the members themselves, with no nation having the right to recall one of its people without the approval of an independent board of advisers. The board would comprise an interdisciplinary, international cross section of individuals of outstanding reputation. Immediately an objection could be raised to the provision for no unilateral recall; opposers might claim it couldn't possibly work since a few national governments would agree to it. Perhaps the solution would be to make the set of institutes a nongovernmental organization. The first hope is to convince enough nations that the value of an independent set of international institutes with a global perspective is in every nation's long-term interests.

This concept of "extranational institutions" is neither new nor radical; a prototype working model is the European Economic Community and its set of commissioners.[17]

A chief function of the Institute of Imminent Disasters would be to assess the probable costs of avoiding any and all perceived disasters impending, and the costs of recovery after their occurrence. The likelihood of possible disasters and their probable times of occurrence would also be examined. By eschewing advocacy roles, institute personnel could serve as impartial experts for testimony before an international spectrum of governmental and nongovernmental bodies and could present valuable supplementary inputs to the normal and often confusing information-transfer systems; in such systems decision-makers usually must balance a parade of seemingly irreconcilable opposing advocates, each appearing confident that he or she is correct and having a plethora of facts to substantiate his or her arguments.

The Institute of Imminent Disasters would rely heavily on the data input of another sister institute, which I will call the Institute of Resource Availability.

Institute of Resource Availability. When the energy crisis of 1973–74 was hot news, many Americans cynically assumed that it was caused not by a resource shortage, but rather by mismanagement of supplies or a deliberate conspiracy of producers to drive up prices.[18] An important aspect of that energy "mini-crisis,"[19] was that information on petroleum supplies, reserves, production, and pipeline amounts was not available from an objective or credible source. That is, most of the petroleum statistics came from the very oil companies that were under suspicion of conspiracy. The need for independent resource data became abundantly clear.

In this connection, the Committee on Mineral Resources and the Environment of the U.S. National Academy of Sciences issued a report in 1975 calling for: (a) establishment of an equitable international agreement for sharing natural resources; (b) creation of national stockpiles of vital materials (i.e., an extension of the Genesis Strategy beyond food); (c) provision of federal support for training of workers in the new and more efficient techniques of production; and (d) establishment of a capability to estimate future resource supplies and energy needs, and to predict the environmental and social consequences of exploiting these resources.

In the report, the committee members admitted that they were unable to agree on the future outlook; that is, when or whether the world would be running out of essential resources. As with most people trying to make a technical analysis of future possibilities, this group fell into two points of view: the "doomsters," who view the future as filled with catastrophic depletions of resources (unless drastic measures are taken to reduce rates of economic growth), and the "cornucopians," who believe that technological advances combined with economic adjustments to changing resource supplies will provide sufficient flexibility in the future to delay the need for regulations that might cause immediate and drastic economic and social disruptions. However, both schools of thought agreed on the four basic needs mentioned above and also were unanimous in the belief that there is an immediate need "(1) to husband resources, (2) to generate information in areas where it is inadequate, and (3) to tackle immediately problems where there is adequate information to form a basis for new action or for augmenting existing efforts."[20] These recommended steps would be compatible with the world survival institutes proposed here.

The urgent need for independent data on world food production could also be satisfied by the Institute of Resource Availability. Despite the active participation of some nations (particularly China and the U.S.S.R.) in the Rome World Food Conference and the Bucharest World Population Conference and their rhetorical advocacy of an improved food situation for the world's poor, many of these same nations are notoriously delinquent in providing data on their own food production or reserve stocks. They tend to regard such data as privileged information. Yet, an accurate inventory of world food production, reserves, and production capabilities is essential both to medium- and long-range planning for world food security, as well as for immediate assessment of famine probabilities. Of course, such planning would be really effective only if surpluses were available on an international basis to those in need.

In addition to food production statistics, accurate demographic accounting is needed. Statistics of current trends of population patterns, devoid of local political massaging, would also be an essential part of the institute's data banks, and could be of enormous help to demographers in understanding population trends.

The Institute of Resource Availability could also collect data on worldwide carbon dioxide release, fertilizer and chlorofluorocarbon use, chlorination of water supplies, energy use and growth rates, nuclear materials inventory, and even on cloud-seeding operations. It could also track the reserves, production rates, and distributions of many natural resources, such as copper, oil, rubber, fish, gold, etc. At the suggestion of Resources Availability personnel, the Institute of Imminent Disasters could then initiate studies of potential disasters and feed back questions requiring the gathering of additional data. Ultimately, the Institute of Imminent Disasters, having warned the world of the inability of existing technologies or institutions to avert potential catastrophe, would need to turn to a third institute to press for alternative technologies.

Institute of Alternative Technologies. The warnings from assessments made by the Institute of Imminent Disasters could be sent to the Institute of Alternative Technologies for further study. By encouraging the efforts of governments and industries and by funding outside research on specific projects, this institute would evaluate and promote the development of supplementary or replacement technologies that might be able to avert or at least mitigate the impact of potential disasters. At the same time, the new technologies conceived or examined there, such as Ernest Schumacher's "intermediate technologies,"[21] for example, could be forwarded to the Institute of Imminent Disasters for advanced technological assessment of their potential impacts on society; those results could then be compared to the probable impacts of a mix of existing technologies. Finally, after the fruits of continuous interplay among the three institutes were gathered, the policy options on a particular issue could be examined by a fourth institute.

Institute of Policy Options. The information gathered from the three other institutes may suggest that policy options be considered for a particular region or nation or for a particular issue, such as wheat production. Since all of the institutes would take a global perspective, the recommendations of the Institute of Policy Options would have to assess the probable costs, benefits, and uncertainties of various options and then present these assessments and alternatives to various world governments and institutions for agreements and actions. Questions of the advisability of centralizing technologies, the location and quantities of food reserves, or the production rate and means of transport of crude oil might be considered along with the urgent matters of planning and implementing technology transfer, emission controls on industrial or agricultural pollutants, and a treaty to regulate climate modification. In essence, the institutes would disseminate information to the public about the critical issues of human survival, giving early warnings for potential disasters, and, perhaps most important, would function as *planning bodies* considering the environmental, political, economic, and social aspects of policy options from a global perspective. The latter function will be the most difficult to manage, because it will often involve political issues that raise conflicts over opposing national interests and outlooks. The extent to which the planning functions could contribute to world security would depend on the ability of the institutes to walk a tightrope between global common sense and clashing national interests.

Thus, the effectiveness of these four institutes in lessening the dangers of world catastrophe would depend on their ability to persuade people to act in their own collective long-term interests—as difficult as these are to define. Therefore, it would be essential that the staffs of the institutes represent a cross section of knowledge and international background as broadly based as possible, and that they be relatively free to act as individuals (much like the secretary general of the United Nations) and not be bound too tightly to representation of the interests of their home nations.

Finally, a brief look at what these institutes might cost is instructive. Annual funding for the operation of the institutes with a staff of, say, a thousand professionals in each would probably run about $60 million per institute (although individual research and development projects to create alternative technologies could be financially open-ended). This indicates a combined total of $240 million or so per year for the four institutes. Is this too expensive a price tag for a new bureaucracy that can by no means be guaranteed to be effective?

Let me give these figures some perspective. Perhaps the best way to characterize the organizations is under the heading of World Security Institutes. As such, they would exist as a nonmilitary analog of the myriad nation-state defense establishments, whose avowed purposes are to preserve myriad national securities. Since the preservation of world security cannot possibly detract from a country's goals of national security, perhaps the most logical way to fund the World Security Institutes would be to ask each nation to contribute a fixed percentage of its national security budget. Since present world expenditures related to armaments are roughly $240 billion dollars per year (a figure that might itself be re-examined by the institutes) the total contribution (or perhaps, world security tax) from each nation would be a mere *one-tenth of one percent* of the world defense budget. It seems highly likely that such a paltry investment in these World Security Institutes holds the promise of far greater security-per-dollar expenditure than investments in weapons of war. The world predicament rightly deserves to be viewed as an urgent security matter. Perhaps then it might receive the attention it demands.

A New American Institution: The Fourth Branch of Government

Realistically, the establishment of a quarter of a billion dollar set of institutes with a global perspective is, at best, a utopian vision, given the present levels of world disunity. Perhaps a less improbable trial run could be attempted within a single nation. I'll begin with my own country, since I am more familiar with its workings and can be more specific on the functions and purposes of my proposed independent institution.

The United States, like the world, needs to assess the probabilities for and timing of future "improbable" disasters. In that vein, it also needs to gather and disseminate credible statistical information on human and natural resources; to come up with alternative technologies; and to be presented with policy options that fairly reflect the costs, benefits, *and* uncertainties surrounding various issues. Of course, the existing branches of government already do many of these things, and the press, private foundations,[22] and universities often try to keep the government agencies "honest." However, the widespread dissemination of independent points of view (and even technical facts) to policy-makers and the public is often less than adequately achieved by present institutions. Most government agencies are charged with the dual responsibility of providing short-term policy advice to the executive branch and, at the same time, are left with the often conflicting task of long-range planning. For example, it was pointed out earlier that some people have challenged the U. S. Department of Agriculture on its selection and interpretation of scientific evidence presented to other government agencies and the public about the vulnerability of U. S. crop yields to weather fluctuations. These critics charge that the USDA cannot provide adequate warnings about the long-term dangers inherent in the present food reserves policy since they are the architects of that policy. This, the critics maintain, is a conflict of interest between the political pressure generated by short-term exigencies and the future imperative of long-range planning. Perhaps a *separation of functions* is needed.

Similarly, other critics have charged that the erstwhile [23] Atomic Energy Commission had a habit of discounting or playing down information on radioactive-materials accidents in nuclear plants, yet another example of a conflict of interests between short- and long-term perspectives. Although counter-opinions on the technical merits of food reserves or nuclear safety issues have been available, along with these agencies' opinions, to interested people willing to dig them out, the former are much more fragmented. More important, the public and its elected representatives who have heard various conflicting positions have generally had little opportunity to weigh the merits of the opposing views, since they are rarely discussed in advocate forums. In addition, very few forums that gather opposing sides together *simultaneously* are held in public view (i.e., prime-time television) or even before the majority of decision-makers.

Regardless of the validity of these charges of conflict of interests, it is clear that the decisions over crucial issues facing the United States in the decades ahead must be based on a set of background facts that considers the uncertainties in knowledge and presents all the sides of an issue. It is often apparent that present government officials and agencies are frequently trapped in an untenable dilemma that forces them to advocate short-term policies *and* provide technical information needed for long-range planning, even if that information weakens the case for the former. Thus, to alleviate that burden, I propose the establishment of a fourth branch of government: Perhaps it should be called the Truth and Consequences Branch. The fourth branch would act as a purveyor of available factual information, even if some of it were conflicting, and would discuss the magnitude and likelihood of potential consequences of policy options offered by governmental officials, agencies, and legislative bodies, as well as by lobbyists. These potential consequences would probably not otherwise be given adequate play by the official advocates of a particular policy. The public and its elected policy-makers should not be afraid of the confusion that might follow an airing of all sides of an argument, especially if large uncertainty characterizes the reality of the state of knowledge.

Of course, it could be argued that the media, the classical Fourth Estate, already serve this watchdog function; but for several reasons the media are insufficient, by themselves, as providers of independent information needed to face the urgent issues of the next few decades. First, the media are fragmented. Second, they haven't, nor are they designed to have, the resources required to obtain sufficient information, digest it, and then evaluate the consequences of all important policy options. Third, they lack the public prestige that an independent branch of government would offer. Fourth, they have a tendency to emphasize news that is concerned with immediate problems. And fifth, the media are profit-making enterprises often subject to the same economic pressures that hound elected officials. However, the crucial survival issues facing the United States and the rest of the world will become most critical many years in the future (when most of today's leaders will be out of power), despite the fact that many decisions that will mold our future well-being need to be made today.

The most important role that the fourth branch would play would be to emphasize the *long-term* implications of our current actions. Political considerations (i.e., re-election) in the American democracy almost always demand that visible decisions be geared toward concerns with time scales shorter than the next electoral period. It has always been this way, and will probably continue to be so, unless we experience a radical change in political consciousness—or decide to limit all elected officials to perhaps one or two terms. The short-term perspective that has evolved into our political system makes this year's employment picture, or next year's inflation outlook, more important political issues than the difficult and often uncertain questions of long-term environmental or even economic catastrophes that might occur sometime later, when different people are in charge of the affairs of government. For example, rather than using our food surpluses now to help poorer developing countries rapidly increase their standards of living (and hopefully become more likely to reduce their population growth), we will continue to sell food to richer countries for immediate economic benefits—unless some credible part of the government convinces the voters to ask their elected officials to recognize that our short-term economic-benefits policies may contain the seeds of long-term misery. The press, private foundations, and universities help a great deal, but are too scattered, too specialized, and often too suspect to convince the public to reorient its political consciousness toward the making of shorter term sacrifices, even for its ultimate good. We desperately need the prestige and resources of a branch of government that is also free of short-term electoral pressures to provide to the public the information required for it to develop a long-range perspective. (Environmental impact statements, now required by Federal law, have been an important, albeit small, first step in this direction.)

The fourth branch would not have any legislative, administrative, or judicial powers; these rightly belong with the original three branches. In fact, to preserve the odd number of branches with decision-making power, the fourth branch would have no power at all, *except* to gather and disseminate information. The primary clout of the fourth branch would be its ability to use persuasion (by the marshalling of evidence, not advocacy), and the best tool of persuasion is the very same one used by government officials, industries, and other political advocates: the media.

Advertising one's products or point of view is an aspect of free expression guaranteed in most cases by the Bill of Rights. But advertising in the media is not merely amusing or annoying, nor is it a mere innocuous side show in a free nation. It is, according to John Kenneth Galbraith, a major prop that supports our industrial state.

> There is an insistent tendency among solemn social scientists to think of any institution which features rhymed and singing commercials, intense and lachrymose voices urging highly improbable enjoyments, caricatures of the human esophagus in normal or impaired operation, and which hints implausibly at opportunities for antiseptic seduction, as inherently trivial. This is a great mistake. The industrial system is profoundly dependent on commercial television and could not exist in its present form without it.[24]

Big industry, he argues, requires big technology, which, in turn, demands expensive and long-term capital investments. Thus, a stable market for the products of big industry—regardless of the ultimate social utility of those products—is essential to the survival of our industrial economy as we have permitted it to evolve. Advertising, particularly on television, provides that instrument for the management of consumer demand; but the purpose of this digression is not to berate advertising, but rather to remind us of its power.

Advertising, in fact, has proved to be so successful in capturing the consciousness of the American public that it is the ideal vehicle to be used by the fourth branch of government to inform the public (with facts, uncertainties, and consequences) about issues of our survival. It might be hoped that an informed electorate would then opt for a new political consciousness, a consciousness that would cause it to reward elected representatives for policies that help to insure future survival, not merely to increase immediate prosperity. We already know that such policies can gain widespread support. Take the American defense budget as a case in point. Americans are willing to accept expenditures of over $100 billion annually to hedge against theoretical threats from potential enemies. Thus, long-term consciousness is already a political reality for issues identified with national security. Perhaps, all we need do is find a way to tie issues of the world predicament, which also deal with theoretical threats from potential sources, to questions of our long-term security—where they rightly belong in the first place.

The budget of the fourth branch would play a similar role in the United States as the proposed World Security Institutes would in the world; but in the context of the United States a fair fraction of the branch's monies would be spent (at least at first) on prime-time television programs that would focus on long-term consequences of our current policies and habits. Perhaps an hour-long weekly program during prime time or a fifteen-minute portion of nightly news programs and selected advertisement spots would be appropriate. (This could be done now even without a fourth branch, of course.) Through repeated exposure of the issues, there is a good chance that the public would question present value systems and adopt a new political consciousness. Such a consciousness could move us away from narrow and immediate economic-interest policies and redirect our efforts and resources toward the creation of a stable, equilibrium world order. Unfortunately, there is a danger that such reorientation could cause a serious jolt to our economic system.

Minimizing the disruption to existing industries, even wasteful or polluting ones, might be essential at first; otherwise the public acceptance of actions aimed toward a longer term view will be limited. On the other hand, the fourth branch would have to help prevent us from prematurely committing our future economic well-being to enterprises that already are or will probably become heavily polluting or unacceptably wasteful of resources. If only ten or twenty years ago we had used more planning in setting up certain goals and industries, we might not have become economically dependent on a wide variety of currently wasteful or heavily polluting activities. And we might not now be faced with the prospect of maintaining those activities in order to prevent an economic catastrophe. For example, repeated warnings that an oil crisis was in the cards have been forthcoming for decades, yet Americans have continued to build and buy large, overpowered, inefficient, and uneconomical cars. Europe and Japan, on the other hand, developed more efficient automobiles. So, in 1974, when the energy crunch surfaced and people finally realized that economy cars had major virtues, many purchased the more economical foreign products. The resulting economic chaos and labor setbacks in the American auto industry from the sudden decrease in sales is well known. But, long-range planning might have averted much of the damage, for the auto industry could have slowly reoriented its product lines without the shock of a precipitous crisis.

Returning to the structure of the proposed fourth branch, it should be pointed out that since its emphasis would be toward our long-term well-being its heads would have to be able to serve for long terms. I recognize that I am on marshy ground when proposing detailed institutional structure for a hypothetical institution, but I will, nonetheless, offer a brief suggestion of how the leadership of the fourth branch might be chosen. A board of governors might be elected for ten- or twenty-year terms. (If that sounds like an excessively long term, one should consider the U.S. Supreme Court, whose members are appointed for life, but can be impeached.) Perhaps, a majority of the board would be chosen in elections held in different regions of the country (the Northeast or the Great Plains, for example). Then, to ensure the participation of the other branches, the remaining governors might be chosen by a vote of Congress from a short list of candidates compiled by the executive branch of the federal government. Then, the board would choose the branch's top administrative officers. Any member of the fourth branch would also be subject to impeachment, but a bill of impeachment would have to be approved by the executive and legislative branches, like any piece of ordinary legislation.

But before going on, it is perhaps instructive to point out that others have proposed similar institutions and given considerable attention to the important details of their structures. For example, Rexford Tugwell of the Center for the Study of Democratic Institutions in Santa Barbara, California, proposed that the United States consider adopting a new constitution,[25] with the first meeting of the constitutional convention (to draft the document) to be held on the U.S. Bicentennial, July 4, 1976. One element of Tugwell's proposal is the creation of a Planning Branch. Dennis Pirages and Paul Ehrlich built on Tugwell's idea in their book, *Ark II,* by suggesting detailed structure and functions for the new institution. They describe one of the most important points, the issue of credibility, raising the question of "how the Planning Branch could be kept independent and prevented from becoming a victim of Washington's four-year syndrome" [26] (what I earlier called the politics of two-, four-, and six-year cycles). *Ark II* suggests that a "number of devices could be used to minimize this danger. First, a minimal budget for the Planning Branch should be constitutionally established and pegged to total government revenues. (Tugwell, for example, suggests 0.5 percent of the federal budget.) This would prevent budgetary attacks from a Congress that might not approve of the critical reports of the Planners." [27]

The fourth branch proposed here, although similar in many important aspects to the Planning Branch, differs perhaps in the emphasis it places on public education. I envision the major function of a fourth branch to be information transfer, not merely to Congress or the executive branch—although they would depend heavily on its expertise—but as an active purveyor of issues relevant to long-term survival. Democracy requires an informed electorate to remain viable, and being informed on the important current issues increasingly requires attention to debates over technical problems, which are often baffling. The question of independence is perhaps the most troublesome part of the proposal. An *independent* fourth branch could provide important help in classifying the truth and consequences surrounding the vital questions of future survival, thereby helping the voters to make informed choices on a variety of issues.

Clearly, serious consideration of a fourth branch of government would first require extensive study and public debate before implementation and, despite the urgency of the issues of survival, this will take time. However, in the interim we could still work to correct the two crucial deficiencies in our present institutions, which I hope the proposal has identified. First, is the need for considerable improvements in the gathering and dissemination of independent factual information on the nature and potential consequences of present policies and habits, and the particular need for technological assessments and long-term economic and demographic growth strategies. And, second, is the need to reorient political consciousness from an overwhelming preoccupation with immediate concerns to a long-range view, in the hope of achieving a catastrophe-free transition to a stable, steady-state world order. Efforts to achieve such an order must be tried, despite the uncertainties inherent in competitive strategies often proposed to accomplish the transition; these problems are so serious that an urgent effort to find some way to implement the *objectives* of the fourth branch into present structures is long overdue.

Some New and Better Agreements

Global Survival Compromise

Throughout the book various aspects of the debate over the world predicament have been addressed, with stress on questions of climatic variability, food production, and the existence of possible climatic barriers to industrial, agricultural, and population growth. Despite the heated debate between optimists and pessimists over the likelihood, magnitude, and timing of potential catastrophes, there is wide consensus that the exponential growth, for both economies and human populations, cannot continue indefinitely. Another general point of agreement is that food production in the developing world must be increased rapidly, although great divisions remain on how to do it, where to do it, and who should pay for it. Both sides of the debate also seem to agree that the highly unequal distribution of human and natural resources on earth suggests that long-term solutions to the world predicament will need to be global in scope, and that the sooner we begin to equalize the resource distribution, the easier it will be to solve the problems before they grow out of hand (if they haven't already). This, I feel, is the crux of the present dilemma: Despite various uncertainties inherent in trying to solve complex global problems, the chances are not at all small that various "improbable" disasters are imminent. Nothing short of a massive and immediate full-scale attack on the world predicament is likely to avert potential disasters; the requisite scale of effort calls for a worldwide plan, a Global Survival Compromise.

Since balancing food production with need is a major component of the world predicament, this case can be used to illustrate how the compromise might work. In chapter 8 it was mentioned that the prospects for increased food production, and for famine, depend on a combination of environmental, economic, social, and political issues. For example, enough arable land remains to feed several times the world's present population (in principle, at least), yet much of this land is marginal for agriculture or lies within fragile ecosystems. In any case, the capital costs of developing such land, to keep food supply abreast of food demand, would be many tens of billions of dollars *per year,* if population continues to grow at present rates. On the other hand, there is much hope for using existing know-how to make considerable improvements in food yields on land that is already tilled in much of Asia, Africa, and Latin America. But such improvement often requires extensive capital investment, as well as some fundamental changes in agricultural practices in developing countries and, in many cases, a reduction in bureaucratic inefficiencies with some resultant cultural adjustments. Furthermore, even if these obstacles to efficient food production could be overcome, growing populations would counteract the gains that would be made and would eventually overtake any foreseeable increase in food production. Fortunately, this Malthusian limit, pointed to by the advocates of lifeboat ethics or *triage*, is not yet a provable global certainty, although it looms closer every year, and is already occurring in a number of seriously food-deficient countries.

Another point often made in connection with the food-population problem is that population growth rates have been reduced without Malthusian catastrophes in some countries where industrial growth raised the standard of living rapidly enough so that people had economic incentives to reduce their family size. This phenomenon, the demographic transition, has taken several generations to approach in the industrialized nations, but may have to occur in as little as one or two generations in today's developing world. And it is not at all clear that a Western industrial model will be effective in bringing about a demographic transition in present developing countries. But if the transition toward a stable, steady-state world is to be smooth and the size of the steady-state population manageable, then the process must begin as soon as possible, even if there is no certainty that a demographic transition will occur in all cases. Some have argued that hope for long-term self-sufficiency for the developing nations will be impossible without massive technical and capital assistance from the developed, industrialized nations. But will the rich sacrifice part of their growing standards of living unless, in return, they perceive some hope that their help will really be effective? For example, they may well expect to see that their aid monies lead to sustained food production increases and perhaps a stable population in the developing world. This is further complicated, since inherent in massive aid is the risk that the desired goal of long-term stability will not result; it is possible that transfers of food and technology now will merely allow the impoverished existence of an even greater number of people, and these people "helped" by massive aid would be even more dependent on the massive aid than the generations before them.

In return for an immediate transfer of wealth, the industrial world may expect to see rapid progress toward such food-population stability, progress that may not be possible without comparably dramatic progress in land tenure policies, social security practices, and lessening bureaucratic inefficiency and corruption in the developing world. No doubt, many will quickly condemn even the thought of such meddling by donor nations in the internal affairs and traditions of recipient nations as "cultural imperialism" at best, and aggression or genocide at worst. For example, have the donors the right to insist in exchange for their aid that other cultures abandon aspects of their religious beliefs, if the donors perceive that these beliefs tend to prevent long-term food–population stability? As a case in point, consider the comment of Karan Singh, Indian delegate to the 1974 Bucharest conference on population as reported in *Time* magazine: "They shout and scream about standing room only! But we believe in reincarnation, my friend, so don't worry. You'll be back." [28] On the other hand, is it the donor's moral obligation to give available food to those who would otherwise perish today, if these recipients will only use the food to maintain or increase their population growth levels, rather than stabilize their numbers and prevent even greater suffering later on? Indeed, if the food aid helps to increase populations, even the rich would soon become unable to support the growing numbers of food-dependent poor.

The urgent need to work out the details of a compromise to the satisfaction of rich and poor is obvious. Fortunately, there are some people groping to set the framework for just such a compromise. The term "Planetary Bargain" has been used often by Harlan Cleveland of the Aspen Institute for Humanistic Studies to describe the process of negotiations prerequisite to the achievement of a Global Survival Compromise. In an article, "The Fairness Revolution," written as an introduction to an Aspen Institute monograph, Cleveland commented on the need for mutual sacrifice:

> If the more affluent peoples are asked to modify their living standards and rearrange their priorities, which for most of them may require important changes in life styles and work ways, their peoples and especially their political leaders will want to know that the changes are worthwhile, that they give promise to meeting the basic needs of the needy—rather than of speculators, brokers, feudal chieftains, and military governors. At the same time in the poor countries, the political courage and administrative drive to be self-reliant (get population growth under control, maximize food production, extend education, assure employment) will also depend on the larger bargain—on assurance that the "advanced" economies don't advance past the limits of environmental prudence, and on large and unpatronizing transfers of resources and technology.[29]

In the summer of 1975, Cleveland directed a workshop at the Aspen Institute on the problems of planetary bargaining. The participants brought with them widely differing views and experiences; some were from developing countries and others from developed nations, some from academic institutions, others from transnational corporations, and there was even a parliament member from a developing country. The participants argued often, and those from developing areas tended to stress the urgency of satisfying "basic human needs," something they felt was essentially impossible with the present international economic order. Some called for nationalization of multinational corporation holdings and others for trade concessions from the rich to make it easier to sell products of the poor. Representatives from corporations responded that such threats would only prove counterproductive, discouraging any transnational corporation from risking an investment in a poor country; thus, they argued, nobody else would aggregate the capital and technical skill needed to set up industries to manufacture marketable products there in the first place. In other words, no incentives, no help.

The response from the other side pointed to recent efforts of Third World intellectuals to develop a new set of economic rules widely known as the New International Economic Order. A preliminary document from the First Conference of the International Development Center held in Algeria, June 24–27, 1975, was discussed.[30] This U.N.-sponsored activity produced ideas that almost surely will dominate much of the deliberations of that world body over the next few years and could well prove a foundation stone in a future planetary bargain. Thus, a brief summary may be valuable here. Ten New Concepts of Development were listed by the participants; and I quote directly:

> 1. Development is a process of profound structural change in the economic, political, social, and cultural fields brought about by the people and for the people, through mass participation and for the benefit of the masses.
> 2. The purpose of development is to develop people, not to accumulate things.
> 3. The development process must be need-oriented, not only growth-oriented. Development involves meeting primary human needs.
> 4. To develop man is to develop all of man, not only to meet basic needs. Particularly important in this connection is the fight to attain cultural identity.
> 5. Development is for all the people, not only a small minority.
> 6. Development is by the people, not by making people into clients.
> 7. The basic strategy of development is self-reliance, not division of labors.
> 8. There are many roads to development, not only one [i.e., improving GNP].
> 9. All countries are in need of development, not only the Third World. . . . The paternalism of yesterday with its idea of donors and recipients has to yield to a global perspective on development.
> 10. Development has to take place in solidarity with future generations [so-called eco-development].

Despite the third world origins of the authors of the Algiers document, it has a decidedly Jeffersonian ring; that is, an emphasis on self-reliance. No one at Aspen strongly disagreed with these concepts; but the Algiers recommendations for concrete steps to help achieve such development touched off considerable disagreement. The Algiers report went well beyond an affirmation of these principles of mass participation and self-reliance. For example, it also called for "militancy and willingness to engage in confrontation," that international stockpiles of food "financed by industrialized and OPEC countries" be set up "to reduce speculation in the food market," and that there should not be any "links between national population policies and international cooperation programs."

Most of the workshop participants agreed that the Algiers report is a rhetorical document, and, despite its often belligerent tones, constituted an opening statement from the Third World to put on the planetary bargaining table. But, as Harlan Cleveland commented, "The growing debate about the content of a 'new international economic order' still mostly consists of everybody talking at once about his own problems, without relating them to the similar problems of others. The facts of interdependence are bypassed in rhetoric about nation-states being 'master in their own houses.' Charges that the affluent consume too much provoke the countercharge that the poor have too many babies." [31]

One important element identified at the workshop was that global bargaining is not only about the foreign policies of countries, but about each others' *domestic* actions as well. Foreign policy is very much constrained by domestic behavior. Harlan Cleveland said that "international affairs are the result of everybody's domestic affairs"; therefore, the overprotective attitude of most countries toward their internal policies will have to be softened if planetary bargaining is to be successful. Nobody at the workshop thought that would be easy to achieve.

The strong consensus that emerged from the Aspen Institute workshop was that the world must quickly move past the rhetorical stage into the negotiating phase. The participants were hopeful that this step is possible and were optimistic, perhaps because the workshop itself was a microcosm of the larger effort needed to hammer out a Global Survival Compromise. After some initial lack of communication between opposing points of view, the face to face contact led at least to an appreciation of the other sides' viewpoints, and thereby raised hope that a similar effort on a larger scale could prove successful.[32]

Yet, the specific terms of a Global Survival Compromise or planetary bargain will be very difficult to negotiate among nations. However, delay in beginning the negotiating process could be more than costly—it could significantly increase the chances for catastrophe to occur before a steady-state world is achieved. For example, in *Mankind at the Turning Point* Mesarovic and Pestel used their world computer model to assess the costs of reducing the per capita income ratio between rich and poor regions of the world to five to one, respectively, by the year 2025. To illustrate the ways that such a transition might occur, they presented several "investment aid" scenarios in which various amounts of money are redistributed within various time frames. They found that not only is it essential to transfer "aid" to achieve the five-to-one income ratio, but it is much cheaper in the long run to concentrate the aid in the beginning. They call this approach the "early action scenario" and say that the total cost of this scenario "is barely one-third of the cost under the continuous aid scenario and little more than one-fifth of the cost under the delayed action scenario. This result is truly stunning: Late action costs nearly five times as much as early action." Perhaps an even more important aspect of the early action scenario is that "the developing regions would become self-sufficient by the year 2000. The globally beneficial political and economic effect of their arrival at the economic takeoff point cannot be overestimated." [33]

When populations remain impoverished and illiterate, they often tend to increase rapidly in numbers, unless that growth is checked by disease or famine. Thus, another risk of delaying the implementation of a Global Survival Compromise aimed at creating long-term food–population stability and improving standards of living is that the growth of human populations can, for example, make per capita energy equality climatically risky. The potential for climatic or other environmental side effects from the use of energy or fertilizer, for example, is proportional to the magnitude of the disruption of natural processes—which is itself proportional to the amount of unnatural energy production or fertilization. Thus, the inescapable conclusion is that the longer we delay implementing the likes of a Global Survival Compromise, the more expensive it becomes to narrow the gulf in living standards between rich and poor; and the more likely it is that technological schemes aimed at improving those standards will have unacceptably large environmental impacts.

John P. Holdren, an energy and resources expert at the University of California at Berkeley (and, incidentally, the only tenured professor at Berkeley outside of a regular academic department), has emphasized the need for a scheme of global cooperation to raise per capita energy consumption rates much faster in poor countries than in rich ones. He warns, however, that this may not be possible without reductions in present population growth rates in *both* rich and poor regions: "The success of such a scheme depends heavily on the success of programs to limit the growth of populations. Only at lowered population growth rates can the high growth in per capita energy use needed and desired in the poor countries be achieved within an economically and environmentally sustainable rate of growth. In the rich countries, the effect of multiplying even small population increments by the very high per capita energy use already prevailing there makes it essential to approach zero population growth as soon as possible." [34] (Fortunately, the United States has already reached below the level of replacement fertility, which means that it will indeed reach zero population growth in a generation or two, assuming the fertility rate does not rise.)

Despite the bitter arguments over how many people the earth could sustain, the obvious point about population growth must be stated and restated: *It is already socially difficult, extremely expensive, and perhaps environmentally dangerous to achieve rapid improvements in per capita living standards for the present 4 billion people on earth; thus, population increases will only dilute the effectiveness of what is already a monumental task.* In debates on population growth, the important question is not how many people could be sustained on earth at some unspecified future time, but rather, can the well-fed afford not to help raise the per capita standards of the poorly fed, when such aid is vital to the early achievement of a demographic transition?

On the other hand, can the poor afford to indulge in political justifications for population increases such as Pope Paul VI's statement that the rich have no right to claim that the solution to poverty lies in "forbidding the poor to be born," or the claim of the Chinese delegate to the World Food Conference that more people are needed to fight "imperialism"? Can we justify such statements simply because some possibility exists that more people than now exist will be able to live comfortably on earth at some unspecified future date—regardless of the suffering that might occur in the interim?

Finally, the food production-and-demand aspect of the world predicament should be viewed as a three-stage problem. Any Global Survival Compromise must be flexible enough to address each in turn. First, the prospect of famine over the next five to ten years (or at least until a "safe" level of food reserves is rebuilt) is likely to be dependent on small fluctuations in food production related to climatic variability, pest infestation, and fertilizer availability. If a food crisis arises before world food security is restored through satisfactory amounts and distributions of food reserves, then the rich may be forced to restrict their diets (especially red meats) temporarily to avert starvation elsewhere.

The second or interim stage of the food-population problem covers roughly the two or three decades after 1980, a period in which an intense effort will have to be made, especially in the developing world, to ensure that food production grows at a faster rate than population. This interim period is crucial to the success of a Global Survival Compromise, for if the world does not begin to achieve a demographic transition and economic self-sufficiency on national bases in the next generation or so, the chances for a smooth transition to a steady-state world in the long term (the third period) will be greatly reduced—along with the likelihood of a peaceful world order.

Although I have no illusions regarding the ease or speed with which present world governments will act in concert to strike a Global Survival Compromise, the sooner we agree to begin the task, the greater will be our chances of success.

Terrorist Abatement Treaty

Hardly a week goes by that we do not hear of some dramatic act of terrorism somewhere in the world. Although some are daring, others vicious and ugly, and many seemingly pointless, suicidal, and doomed to fail, nearly all share two features: (1) Innocent people are often randomly victimized; and (2) virtually no one is immune. Terrorism, in my view, may become one of the main obstacles to a catastrophe-free transition to a steady-state world order, since it directly threatens the supportive use of technology and the desirable relaxation of tensions among nations, both of which are essential to help bring about a Global Survival Compromise.

Terrorism Threatens Centralized Systems. Terrorism endangers the very essence of those benefits we expect to receive from our growing reliance on large-scale modern technologies; that is, the vastly more efficient use of our shrinking supplies of natural resources. Blackmail and terror are not new, of course; but the creation of societies almost wholly dependent on grandiose technologies for their survival is new. These technological props to our economic system—such as electricity, gasoline, refrigeration, fertilizer, reservoirs of drinking water, telephones, transportation systems, and exotic weapons for use as deterrents against foreign aggressors with comparable weapons—provide fertile opportunities to terrorist groups, particularly those with suicidal tendencies, to disrupt seriously the tranquility and productivity of industrial societies, the very conditions needed (in the interim, at least) for the improvement of living standards of both rich and poor alike. Although I have previously argued for more diversity in technologies than practiced at present, the point is that the risk of terrorism further increases the need for less efficient, but more diversified, systems and thus may waste resources needed to fight poverty.

Take a simple example of the terrorist risks inherent in centralization for the sake of efficiency: the telephone system. We have evolved into a society of telephone addicts. Even our computers talk to each other over long-distance telephone lines. The ability of the phone companies to provide reliable and inexpensive service to all private, governmental, and business customers depends on how efficiently this communication system can be operated. Two seemingly sound principles have followed: (1) The number of telephone lines among various regions has been limited to a certain small fraction of all users in each region, a number determined by a statistical analysis of the likelihood that a given fraction of all customers would use their phones simultaneously. (2) Modern electronic equipment has been centralized in switching centers to facilitate the efficient operation of phone service. Under normal conditions both of these practices work very well. However, the extra efficiency introduced by the small number of lines also causes occasional troubles familiar to all of us: One often can't complete a call successfully on those special occasions (Mother's Day, for example) when more people want to use their phones than there are lines available to be used. Merely a temporary and rare inconvenience? Not if there were a mass emergency in which many people needed their phones; then, phone service would be very limited indeed.

An ironic example of the ease in which the statistics of normal phone use can be invalidated by a seemingly innocuous event occurred in the 1974 Christmas season in Denver, Colorado. The public television station, KRMA, announced in early December that Santa Claus would be on the air for an hour or two one evening before Christmas. KRMA set up a few special telephone numbers for Santa's studio appearance in anticipation of the special TV program. However, the producers of the show feared that it would be a big flop, I was told by one of the station's announcers, because today's sophisticated children (used to highly stimulating shows) might ignore something as old-fashioned as a chat with Santa. As it turned out, their anxieties were justified, but for reasons they never expected. The kids of Denver responded enthusiastically to the news that Santa would be in town and available to talk to each of them by phone while he was on TV. (I suppose a Santa on educational TV is far more credible to a child than a Santa in a department store or one on a street corner with a bell and coin box.) In fact, they called KRMA so often and for so long that phone service in much of Denver was completely disrupted for hours before and after the show; and city lines—including police and fire departments, hospitals, businesses, and ordinary house phones—were overloaded for days. As the story goes, a telephone company official, not realizing his emergency call to KRMA had cut into a conversation between Santa and some child, was broadcast all over the city, and his words were not wholly appropriate to the Christmas season. The man from Ma Bell bluntly chided the TV station for screwing up phone service in Denver—as if it were the producers' fault! I understand that he apologized later.[35]

I was at the TV studio several days later and was shown a big white box that should have been called the North Pole, since it contained several phones whose buttons lit up for the incoming calls for Santa. The ringing mechanisms were disconnected, of course, and only the lights worked to indicate a call was incoming. Even though the Santa show had been aired several days earlier, kids were still calling the studio at an incredible rate. Amusing as this seems, it was impossible to answer all of the endless calls and tell the disappointed children that Santa had gone back to the North Pole.

The point of this bizarre example is that a technologically dependent society can operate efficiently only under normal conditions, and that abnormal situations can cause disruptions to the services on which people have become reliant. If long-term disruptions were to occur, we would have to alter our accustomed life styles—which, according to some [36] would be just as well. But some long-term disruptions could be serious to large proportions of the population. For example, if the trucking lines that carry food to our population centers were not operating, the risks of being technologically dependent would become immediately apparent.

Another telephone story is useful here, especially since it brings us back to the trade-off dilemma between centralizing technologies for the sake of efficiency and diversifying them to reduce vulnerability to their failure or to terrorism.

New York City has collected its telephone operations into a number of switching centers, each serving hundreds of thousands of customers in different parts of the city. (It is fortunate, as we shall see, that the phone company has not yet centralized itself in only one building.) On February 27, 1975, there was a serious fire on Second Avenue between 12th and 13th Streets in Manhattan. That would not have been big news in New York, except that the building that burned was the New York Telephone Company's switching center for the Lower East Side, and phone service for some 170,000 telephones was knocked out for weeks. "It's the most severe disaster we've ever had," [37] said the chairman of the board of the American Telephone and Telegraph Company about the loss of service, which affected three hundred blocks. The phone company rushed in twenty-two mobile telephone vans to restore immediately some service to critical customers such as the Beth Israel Hospital, the Department of Health centers, and other welfare and emergency sites. According to *The New*

York Times, Beth Israel Hospital "dispatched one of its telephone operators and a security guard with a two-way radio" to a health center outside of the affected area, so that the hospital could continue making appointments for admissions of patients and could contact families of already admitted patients. "We're just trying to cope,"[38] a hospital spokesperson told the *Times* several days after the fire.

The city fire commissioner who investigated the fire "definitively ruled out the possibility of any sabotage or that any incendiary device was used."[39] However, this example serves to show that centralization of technological services on which our economic livelihood and our well-being depend subjects us heavily to the dangers of large-scale disruption from a single catastrophe, whether its cause is advertent or not. (Imagine the aftermath of that fire if there had been only a *single* switching center for New York City.)

In the same vein, although food transportation and storage technologies have dramatically reduced the vulnerability of people to local droughts, these same technologies have created a dependence on food imports for hundreds of millions. Thus, while the frequency of small disasters has been substantially reduced by the centralization of much of the world's food systems into a few major granaries, even more people have become vulnerable to a less likely, but even more threatening, hazard: a disruption to the food production or distribution systems in a few major granaries because of climatic fluctuation; a pest infestation that attacks a widely used strain; food embargoes; or even terrorist interference with the food systems. In mathematical terminology, it could be said that the historical situation of high-frequency, low-amplitude catastrophes has been changed by technology to a low-frequency, large-amplitude situation—and will remain so, at least as long as reserves remain low.

Returning to questions of possible precautions that could be devised by society against the threats of terrorism, it is important to point out that these threats should be weighed (along with the dangers of accident, miscalculation, or natural hazard) in making present decisions that are likely to determine the size and character of future technologies. Society would be vulnerable, for example, if it were to permit the efficiency-oriented "economy of scale" approach to increase our future dependence on centralized systems. There are many examples of centralization practices or failure to build in diversity that should be re-examined in the light of terrorist hazards *before* committing ourselves to more of them. Here are a few: monoculture agriculture; energy production concentrated in power parks with only a few transmission lines carrying electric power to population centers; abundant stores of deadly plutonium; transporting most of our oil in supertankers; a city dependent on a very few tunnels or bridges; and even urban areas with high population densities that make them more vulnerable to nuclear blackmail than areas with dispersed population.

These kinds of issues, especially the terrorist threats, often prompt people to ask: "Aren't we already subjected to a high vulnerability of terrorism or blackmail, and haven't we made out all right so far? Why do you think the terrorist threat will be so much worse in the future that we should sacrifice our present levels of efficiency and prosperity by decentralizing many of our technologies or economic centers?" The question can be answered in two parts. First, it is not sensible to maintain or even increase our vulnerability to the risk of terrorist threats merely because we have been relatively free of them so far; the longer we wait, the greater will be the chance that terrorists will succeed in the blackmailing and disruption of centralized systems. There is a theorem in mathematical statistics stating that the probability of occurrence for any event that *can* happen becomes unity (i.e., the event happens) as the waiting time for that event's occurrence becomes infinite. In other words, the longer we wait the more likely it is that a particular event will occur.

The obvious steps to insulate us from the potential damage from terrorism are to reduce as close to zero as possible the chances of an incident and to reduce the potential impact of the event—before it does happen. The second step calls for reducing our vulnerability to a terrorist threat by the relatively simple stratagem of diversifying the means of our survival—which, unfortunately, may mean wasted resources. The first step, reducing the probabilities of an incident, is much more difficult to achieve and probably requires a mix of social, economic, technical, and diplomatic strategies. One diplomatic strategy would to be agree to a Terrorist Abatement Treaty.

Terrorism Reduces the Chances for a Successful Global Survival Compromise. Terrorism is a dramatic expression of dissatisfaction. Often, terrorists are driven by both personal and political motives rooted in legitimate injustice, though the victims of terror are rarely the perpetrators of that injustice. Even if we accepted the notion that all terrorists are merely self-professed deities or outright maniacs, and that their avowed identification with political causes is purely a convenience (motives are scarcely ever so simple), we would still have to admit that those few nations or subcultures harboring or supporting terrorists have acceded to the premise that terrorist means may be justified by desired political ends. Thus, a reduction in the political tensions that justify support, or even shelter, for terrorists is bound to reduce the probabilities of terrorism. Furthermore, terrorism, even nuclear blackmail, might be a desperate response to the widening gap in living standards between rich and poor. Heilbroner has even gone so far as to suggest that those in "underdeveloped nations who have 'nothing' to lose might point their nuclear pistols" at the rich, "who have everything to lose." [40] One obvious way to reduce the probabilities of terror attacks would be to reduce that gap.

Unfortunately, if terrorism increases, much of the resources needed to close that gap may be squandered on defense mechanisms to neutralize the terrorism. Thus, it is in the long-term interests of *all* nations, rich and poor, to eschew terrorism as a mechanism for redistribution of wealth or a ploy to focus on a political grievance. The inevitable response of the wealthy nations to terror will probably be proliferation of secret police and political subversion, and increased spending of vital resources on weapons of war—rather than an acceleration of the transfer of wealth. We can readily see, then, that the best policy to assure a reduction in the chances for terrorist acts is to treat all terrorists as international criminals. *There must be no haven anywhere for terrorists regardless of any nation's sympathies for their professed causes.* The world must agree to a Terrorist Abatement Treaty as part (or perhaps even a first step) of any Global Survival Compromise designed to redress the inequities in world living standards. Again, such an agreement is likely to prove essential to the creation of a stable world order. Despite the inability of the United Nations to agree on a policy (let alone a treaty) on terrorists, my hope is that those countries balking at the idea of punishing murderers and kidnapers who share their political beliefs will soon perceive terrorism as harmful to everyone in the long run.

Even threats from small non-establishment terrorist groups can be minimized by collective world action. For example, the signatory nations could agree to accept no aid from countries undergoing blackmail—that is, let the booty rot in harbors if necessary. This would significantly reduce the rewards of terrorist blackmail. However, for underdeveloped nations to agree to such terms, they probably would have to be a part of a Global Survival Compromise with provisions for the orderly transfer of technology and wealth, arrived at through freely negotiated agreements. As the conditions in developing countries then improve, their peoples will have, in Heilbroner's terms, more to "lose" and thus be far less likely to condone terrorist blackmail. Finally, and perhaps most important, it is absolutely crucial to render the ingredients of the most fearsome terrorism, nuclear blackmail, almost impossible to steal or obtain in quantities sufficient to pose a substantial threat. This step suggests another aspect of the Genesis Strategy, a Nuclear Materials Inventory and Control Commission.

Nuclear Materials Inventory and Control Commission

Much is being said these days about the dangers of the element plutonium, a radioactive chemical so deadly that a grapefruit-sized lump of it is thought to have the potential to cause cancer in nearly every human on earth; such a lump could certainly make a nuclear bomb big enough to be an effective blackmail device. Obviously, with the spread of nuclear reactors and even nuclear weapons (which can be built from the by-products of "peaceful" nuclear reactors) across the globe, the prospects for safeguarding this dangerous radioactive element become worse daily. Without further discussion here of the dangers of nuclear blackmail or a nuclear reactor accident (often referred to as a "zero-infinity dilemma"[41] since the probability of such an accident is often assumed to be near zero, while the costs of such an improbable disaster could be "infinitely" large), it is clear that very accurate inventory and control of the dangerous radioactive materials of the nuclear industries need to be implemented on a worldwide basis, regardless of the outcome of the present debate over the safety of nuclear reactors. There can be no exceptions—that is, the global security of plutonium is only as strong as the weakest link in its chain of protective measures. All nations must submit to careful accounting and control of this material, whose misuse could make doomsday a reality for many.

Many have argued that the world would be far better off if plutonium and related products were no longer produced or even had never been discovered! But since plutonium will most likely be around for awhile, the need for strong steps to assure its international accounting and control is both obvious and pressing. The potentially devastating situation demands the strengthening and expansion of current efforts of the Internatonal Atomic Energy Agency to keep track of such dangerous nuclear materials. Every day lost in implementing such a plan increases the chances for a nuclear accident or terrorist incident.

Geophysical–Environment Modification Treaty

Extensive treatment has already been given to the potential dangers of climate modification, both inadvertent and deliberate, and a strong plea for a treaty to control such modification operations has been made. This agreement should also cover all other potential types of geophysical modification (e.g., deliberate tidal wave creation or earthquake activation). Although the danger of these geophysical manipulations may seem remote—from a decade to a century away—and thus a low priority item in terms of the present world consciousness, now is the time to develop a treaty *before* any nation becomes so nearly able to implement such an operation that it might then feel compelled to refuse to sign such a treaty. We still seem to have some time left to steer into constructive channels our growing knowledge of the geophysical sciences, and it would be inexcusable to waste that opportunity merely because the specter of malevolent geophysical modification seems years away.

Some Revised Attitudes and Procedures

In the spirit of the Genesis Strategy and the offering of proposals for new and better institutions and agreements, considerable stress has been placed on the need for taking steps now that promise some chance to diminish the likelihood of future catastrophes. Continuing in that vein, a few more specific areas are brought up where a change in procedures could help to force political, scientific, and intellectual attention on the world predicament. Let us begin with the issue of elections.

Limits to Incumbency

There is much truth to the saying that a political leader seldom sees beyond the next election. And, because such attitudes are widespread, politicians generally do not exhort their constituents to make sacrifices for the sake of future generations—except, of course, for national security issues. Thus, the most obvious procedural change (and one suggested many times in the past) to alleviate the short-term bias forced on politicians by the worry over re-election is to unburden them from that tiresome worry: Eliminate re-elections! No doubt it will be argued that such a change would lead our representatives to vote according to their prejudices—or those of special interests—instead of the wishes of their districts. But they might also vote their consciences more often! However, I would agree that one-term-only officials would put the public in the precarious position of depending too much on the consciences of their elected officials. To reduce the chance for future catastrophe, a balance must be achieved between the understandable political-survival tactic of short-term responsiveness to the electorate and the long-term outlook, which often requires making immediate sacrifices for human survival.

One way to work toward such a balance would be to limit all elected officials to two consecutive terms—as we already do for the U.S. president. (Perhaps the terms of the House members should then be extended to three or four years.) After this procedure had been enforced for a few years, about half of the elected officials would be faced with the prospect of re-election, and the other half, free from re-election pressures, could concentrate on the difficult and perhaps unpopular problems of future well-being. Further revision of a variety of election laws might be needed (e.g., an expanded recall procedure) after implementation of a two-consecutive-term system. In any case, something needs to be done to reduce the preponderant attention our political leaders devote to short-term exigencies. Clearly, without thoughtful attention soon, some of the problems and threats that are being predicted for the future will become horrible realities. And many lie beyond the next election—or even the one after that. If we are to avert these looming disasters, political leaders must begin to face this reality and convince their constituents that to sacrifice some things now will pay off in the long run in the form of a more secure world. Unfortunately, many of our politicians may not feel free to deal with such tricky problems until the next election is no longer an issue; in fact, regardless of the pressures of elections, most people—politicians and their constituents—feel little compulsion to effect major changes whose benefits may only be realized years later.

Science Advice to Government

President Richard Nixon's abolition of the post of presidential adviser (and related staff) in 1972 led to much speculation about his political motives,[42] and considerable pressure to resurrect the post has been exerted since.[43] Undoubtedly, the ready availability of such advice is central to informed planning on the proper mix of technological solutions to future problems, and the office will probably soon be reinstituted. But, regardless of the optimal organizational mechanisms to act as a focus of such advice, I have the nagging concern that the state-of-the-art knowledge (and uncertainties) on scientific questions facing society is not being clearly communicated to policy-makers, planners, and the public—not least because many elements of the present political system would have difficulty accepting the proposition that present policy or planning may have to be devised on uncertain scientific prophecies (such as the likely long-term effects of chlorofluorocarbons on the ozone layer). There is an important political message that often tends to be obscured in hopelessly contradictory "expert" testimony. In the case of continued use of aerosol spray cans, for example, there is a reasonable probability that this technological and economically profitable operation will have a future negative impact on society (skin cancer or ecological damage in this case), and that immediate policy action to regulate its continued production is therefore required. Yet, that immediate policy action can have immediate and obviously unpleasant effects (unemployment in the chemical industry, for example). Even worse, there is no guarantee, given present scientific uncertainties, that such a policy is needed this year or next—or at all. The confusion that results from such complex, uncertain issues often leads policy-makers—and their constituents—to adopt a wait-and-see attitude. That is, "Let's wait for more conclusive scientific information before we take actions that could mean immediate economic hardship for some and a loss of votes and campaign contributions for ourselves." Unfortunately, many scientific problems of this type (climate-related issues are excellent examples) may not be resolved with a sufficient degree of certainty soon enough to prevent the unwanted occurrence or lessen the impact of potential catastrophes—catastrophes that would be inevitable *if* present estimates prove correct or, worse, to be underestimates (*which is just as likely as their being overestimates*). It is essential that scientific advice to the government clearly spells out: (1) what is known; (2) what needs to be known; (3) what might be learned in a reasonable time; (4) the probabilities that present estimates are correct, too low, or too high; (5) the probability that these estimates are reasonable; and, most important, (6) the political and value assumptions of the individual scientists that guide the choice and interpretation of evidence presented in scientific testimony.

The last point is perhaps the key difficulty with present government procedures for making a technical assessment. This process often begins with expert testimony before government forums, usually legislative committees. Individual scientists may agree fairly closely on the uncertainties and sureties in present estimates of the ozone reduction from aerosol cans, for example. Yet one respected scientist might argue that too little is known to regulate the industry, while another who knows quite as much science as the first might plead in the name of sanity to "ban the can." In all likelihood, neither is an "expert" on whether regulations should be imposed or not, and their opinions on the propriety of regulations should be no more or less valuable than those of any other well-informed registered voter. Their expertise lies in estimating possible impacts, and in assessing the probabilities that present scientific estimates are correct. Scientists, as any citizens, are entitled to their views or value judgments on whether such estimates and probabilities justify a ban on spray cans, but as scientists they have a special obligation to declare where their scientific opinion fades into their political philosophy. In the future, society and its representatives will have to make many more political choices based on the scientific judgments of experts, but these judgments must be clearly distinguished from political biases. It is unfortunate that the scholarly credentials of an expert called to testify on an issue can so easily become confused with credibility of the partially political conclusions inherent in the overall testimony. There is need for a procedural modification of the information-gathering system to help legislative commitees (or other assessment bodies) decide precisely where the scientific portion of any expert testimony shades into the political.

To give a well-rounded view of technical issues to the committee members (who are usually not technical experts), I propose that present procedures be altered: Opposing advocate experts should be present before the committee *simultaneously*, and, in addition to responding to the questions that normally come from the committee members, the opposing experts should be quite thoroughly interrogated by special committee scientists capable of drawing from the witnesses distinctions between their scientific and political opinions. To insure objectivity from the committee scientists, perhaps "majority" and "minority" scientists could be recruited from each of the two major political parties for each assessment issue. Although such a system may provoke technical arguments between specialists that bore or confuse some committee members with seemingly contradictory and obscure details, I can see few other reasonable means of defining just where the limits of scientific opinions end and the realm of value judgments begins—other than asking the testifying individual to declare his biases at the outset.

It is tempting here to suggest that more scientists run for public office to assure that at least some legislators sitting on assessment committees will be less naïve about the merits of conflicting scientific testimonies. Yet, I suspect that it would not be long before the scientist legislators grew too far removed from the cutting edge of research to effectively cross-examine conflicting witnesses as efficiently as the specially appointed interlocutors borrowed from the forefront of the issue under debate. (It would also help to increase the size of the staff of Congressional committees by including people with the appropriate expertise.)

Interdisciplinary Integration

Despite a few laudable attempts to create interdisciplinary research institutions, the academic establishment continues, by and large, to reward narrow specialization and attention to detail in promotion and review policies of its members, especially the young ones.

The chief qualification for advancement at nearly all reputable universities or government research laboratories is the ability of the scientist to publish significant work in his or her discipline. This ability is roughly determined by the number of published contributions and their impact in the particular field of specialization. No one would object to this procedure for quality review if it were part of a flexible evaluation that considered many factors in assessing "quality," but more often than not, "high quality" in academic terms means publication of more than a few research papers, each narrowly specialized and mostly the work of the single investigator.

An interesting counterpoint to the publish-or-perish syndrome of traditional scholarship as just described was expressed by Norman Borlaug, head of the agricultural research team in Mexico that developed some miracle wheat strains of the Green Revolution. When asked if his scientists are encouraged to work on publishing their findings, Borlaug scoffed at the idea that publishing should receive a higher priority than transferring new findings into the fields so that farmers could benefit immediately.[44] Perhaps his inversion of the traditional attitude is aptly phrased: "While we publish, they perish."

But the publish-or-perish procedure, crude as it is, is not a bad means of evaluating scholarly work. The problem is it too often discriminates in favor of the narrowly specialized. The problems that tend to suffer neglect because of this traditional institutional evaluation procedure are the multidisciplinary ones, problems that offer high risk to an investigator foolish enough to tackle them. A number of reasons can be given to explain why present promotion policies discourage nontenured researchers from attempting multidisciplinary problems: First, multidisciplinary problems are rarely susceptible to study by any one individual. Second, the originality and creativity that mark quality work on multidisciplinary problems generally come from the clever combinations of existing knowledge from a number of fields, whereas traditional science usually better understands, and thus better rewards, the digging-out of new knowledge from a single field. Third, the tenured members of individual disciplinary departments who must evaluate their "lesser" colleagues are often unfamiliar with much of the content of multidisciplinary problems (except for the part in their own specializations). Fourth, since it is a herculean task for a single individual to be a competent specialist in many fields without weakening somewhat the degree of attention to the basic field, multidisciplinary researchers are often viewed with suspicion by their monodisciplinary colleagues, and are sometimes branded dilettantes who, unable to handle the stiff rigors of advancing knowledge in a single discipline, take refuge in multidisciplinary problems. Fifth, multidisciplinary problems are difficult to solve. The chances of making important progress toward solving a multidisciplinary problem in a few years are much lower than the prospects for contributing significantly to some small aspect of a single field in that time. For all of these reasons it is risky to work on multidisciplinary subjects—at least until one has earned this right after five or ten years' apprenticeship in the mono-discipline. Therefore, the traditional disciplinary structure is, by itself, insufficient for evaluation and promotion of young multidisciplinary workers, and as such discourages them (at a time when they are often highly productive) from tackling the most important multidisciplinary problem of their generation: survival. (A parallel criticism is appropriate for peer review procedures, which often judge multidisciplinary research proposals by the same standards applied to specialized proposals: their potential contribution to the reviewers' fields—far too narrow a perspective in most multidisciplinary cases.)

To correct this imbalance inherent in the traditional system of reviews and promotions in science, I recommend a number of steps: (1) Increase awareness of the intellectually restrictive aspect of mono-disciplinary divisional structures. (2) Encourage the establishment of interdisciplinary administrative units in academic institutions so that evaluation of interdisciplinary research by interdisciplinary researchers is institutionalized and a receptive work environment is provided to those workers who choose to tackle multidisciplinary problems. (3) Encourage multidisciplinary research efforts by expanding research grants for this sort of work, by reconstructing peer review systems to eliminate their disciplinary bias and by assuring the research community that such funding will be sufficiently stable that individual researchers can risk withdrawal from their traditional disciplines for a number of years at least.

One way *not* to encourage multidisciplinary work is the method suggested to the U.S. House of Representatives by Maryland Congressman Robert Bauman. Bauman, using the political tactic of reading obscure-sounding titles from National Science Foundation research-grant lists and strongly implying their irrelevancy, somehow managed to convince the House to pass legislation giving Congress veto power over *individual* research grants. (I was disappointed to see so many opinion-makers support this absurd policy which is little more than a quick and shallow response to deep-seated scientific and managerial problems.)

Perhaps the best statement against this proposal came from freshman Congressman Timothy Wirth of Colorado. On the House floor he argued against the Bauman amendment:

"It should also be noted that the NSF budget, which totals less than one-hundredth of the Defense budget, probably receives more careful scrutiny by the Congress than any comparable sum of money. The Science and Technology Committee and subcommittee proceedings and the floor debate on the less than $1 billion NSF budget totaled about 28 hours. If we subjected the Defense budget to similar scrutiny, we would have to spend 2,800 hours or more than 350 8-hour days at it. And where was Mr. Bauman when the Defense budget came up before this body? Indeed, Mr. Harkin, the gentleman from Iowa, proposed an amendment to the Defense authorizations bill identical to the one we are considering today. Mr. Bauman wisely voted against it as did 345 other Members of this body." [45]

It is not enough for a congressional representative to be "tough" with the research establishment to convince any informed voter that he or she is looking out for the public treasury, when that same representative turns around and with little scrutiny votes for "defense" projects costing tens to hundreds of times more than the "irrelevant" research project he or she has expunged from the taxpayers' burden.

Obviously, Congress has the right and obligation to oversee the granting mission of agencies such as the National Science Foundation; and, frankly, I hope it uses its power to encourage multidisciplinary research and push such agencies to formulate procedures for multidisciplinary peer reviews of multidisciplinary research proposals. Congress, however, has neither the time, nor the objectivity, nor the ability to pass judgment on the importance or quality of individual research grants; that task is tough enough for tens of thousands of experts called upon by the academic peer review system. Reviewing the quality of *individual* research proposals, if done thoughtfully and thoroughly, is a time-consuming task that should continue to be assigned to those most knowledgeable in the particular field (or fields) that the proposal involves. The U.S. congressional representatives do not seem the likely group to carry out such a mammoth and technical task. Congress can and should suggest and fund broad areas of research that it deems important. A representative's harangue against a twenty-thousand-dollar grant to study the mating habits of the African toad may win a few votes from dissatisfied anti-intellectuals in the home district, but contributes nothing to the multifaceted, complex combination of research projects that is needed to solve the world's multiplicity of problems.[46]

The public and Congress must be made aware that although research grants advance knowledge in specific, seemingly irrelevant fields, such studies can be (and often are) the basis of a key link to a major discovery. (The aerosol spray can–ozone problem, for example, was discovered accidentally by researchers who were funded to study obscure-sounding chemical reactions between various unpronouncable atmospheric constituents.) And, even if many grants provide no such dramatic payoff, such grants perform the valuable service of giving research institutions the means to train students and others by working on *real research problems*. And there can be no solutions to the world predicament in the long run if the flow of top quality, fresh talent into research is arbitrarily cut off by the policies of a few shortsighted officials who prey on the worst fears and hostilities of their understandably confused constituents at a time of exceptionally bewildering national and international problems.

In connection with the last point, an article in the *Journal of the American Association of University Women* lamented the state of education in the United States with regard to international problems.

> A survey by the American Association of Colleges for Teacher Education revealed that "only 8 percent of the 225,000 teachers graduated each year in the U.S. have any access to information, analysis, or experiences in the international field." A similar survey by the same group found that "of 900 deans of education polled throughout the U.S., 60 percent said they had absolutely no interest in international education."[47]

This account graphically suggests where Congress might intercede by directing the granting agencies to increase priorities for education and training in the international field, and by then following up with a continuous review of progress in this direction.

Alternatively, Congress might push for "relevant" research by requesting that some research funds be made available for important long-term problems that require short-term answers, such as the absolutely serious fertilizer–ozone problem discussed previously. As an example of one such study, the three-year, 20-million-dollar, multidisciplinary Climatic Impact Assessment Program (CIAP)[48] was surprisingly helpful in reducing the uncertainties surrounding estimates of the potential climatic impact of a fleet of high altitude aircraft. Since fertilizer use is now contemplated to increase by some 500 percent or more over the next twenty-five years, and its use is considered to be essential to the support of the world's growing population, it seems sensible that Congress authorize the establishment of another CIAP-like program to study this question, a problem which is many orders of magnitude more important to society than supersonic airplanes.

Furthermore, Congress might request that a multidisciplinary food–climate study institute be established to perform the kinds of actuarial analyses recommended earlier, since these studies could provide a quantitative basis for setting minimum levels of expensive food reserves.

These examples suggest a few obvious ways in which Congress can appropriately exercise its influence on research funding, ways far more logical than attempting to judge the value of a grant on the basis of its title alone.

Attitudes for the Future

Throughout the book several issues have been emphasized tying climate-related problems to other important issues of the world predicament, such as food and energy production, the relation between technology and natural hazards (e.g., drought), and economic and population growth rates. An attempt has been made to state my personal biases plainly and to spell out where a scientific opinion tends to merge into a political philosophy. That ruling philosophy is best summarized by the Genesis Strategy concept, which, in a nutshell, can be codified into a few fundamental axioms:

(1) Humanity, collectively and individually, is worth preserving; savable lives should be saved.

(2) Humanity can be preserved in the long run; that is, a highly technological steady-state world order is both politically and physically possible if population size and technologies do not create dangerous dependencies or environmental stresses.

(3) The prevention of human catastrophes (e.g., the starvation-related deaths of a hundred million people any year) during the transition period to such a steady-state world order should become the

number one world priority; to permit starvation that is preventable is morally unacceptable.

(4) The world predicament is not likely to be solved without a high level of international cooperation and reduction in the level of belligerency—larger than has ever occurred in history—and these conditions must be attained very soon.

(5) At least some forecast disasters will inevitably occur unless massive preventive measures are taken soon. Such actions can be justified even if there is no certain knowledge of the timing or magnitude of the potential calamities.

(6) Realization of the gravity of the world predicament and recognition that various proposed solutions have no guarantee of success are neither causes for malaise and despair nor reasons to adopt a wait-and-see attitude that discourages attempts to reverse the situation.

This last point is extremely important. Some have lost all hope; they call for the adoption of lifeboat ethics, a policy whose likely outcome would be a nearly immediate and perhaps unnecessary starvation and suffering for hundreds of millions of people—probably followed by political instability and terrorism. In my opinion such a philosophy can be rejected on both practical and moral grounds, given that it may well be preventable with massive, immediate actions, such as those suggested in the Global Survival Compromise. Many others, including so-called pessimists, share the view that timely actions can make a difference. For example, after a rather sobering analysis of potential disasters, John Holdren and Paul Ehrlich concluded an article in the *American Scientist* by saying, "All this is not to suggest that the situation is hopeless. The point is rather that the potential for grave damage is real and that prompt and vigorous action to avert or minimize the damage is necessary." [49]

Similar statements have been made by many of the optimists as well.

On the other hand, there are some who take the opposite view: total laissez faire optimism. This extreme is typically expressed by those whose faith in past dogma or belief that "technology saves" drives them to rest assured that no special social efforts are worthwhile since a timely natural or technological development always turns up—given economic incentive—to save us from catastrophe. They often believe that there is some unnamed and unexplained stabilizing mechanism in the geophysical world that will continue to mitigate all of our impact on the physical and biological environment. Such complacency is comparable to the hard line of the lifeboaters for it results in the same policy—inaction.

Between Lifeboat and Laissez Faire. Obviously the real status of the world predicament and the most effective proposals for its solution almost certainly lie between the lifeboat and laissez faire extremes. But unfortunately the stakes of the global survival game are so high and the likelihood of tragedy sufficiently real that the risk of a wait-and-see attitude needs to be clearly perceived, and may justify such dramatic actions as a Global Survival Compromise or a fourth branch of government. Urgency is essential. The world predicament should become a household word, as Spiro Agnew ultimately did, and not merely remain primarily a matter of intellectual debate to be fought out among a clique of "eggheads" in obscure journals and dingy lecture halls, as the world disintegrates.

What of the future? As of now it appears that the debaters will simply continue to argue the relative merits of lifeboat ethics, transfers of technology, demographic transitions, and climate control treaties. Political leaders will generally opt for business as usual, economic growth, and campaigns for re-election. And the vast bulk of the people, frustrated and seemingly powerless, will wait for more certainty before pressing their leaders for earthshaking changes. After all, something will turn up to pull us out of our mess—it always has, most people rationalize. But has something always turned up in time?

The collapse of civilizations that couldn't accept in time the need for drastic action is, despite popular rationalizations to the contrary, the repeated reality of human history. L. Sprague and Catherine de Camp, authors who have visited and explored the ruins of numerous dead civilizations, summarized common elements in the destructions of once mighty societies in their book *Citadels of Mystery.* A number of their findings still have great relevance today:

> We have learned that . . . if civilizations can arise under varying conditions, they can also be snuffed out by a chance, unforeseen combination of factors, none fatal in itself but all together overwhelming.
>
> We have learned that, when the foe is at the gate, virtue and good will cannot take the place of physical force. We have also learned that unity in the face of danger, although a seemingly simple lesson, is one of the hardest for men to learn—as the Roman Britons, the East African Arabs, and the Mayans all found to their cost.

> We have seen that, when a people in peril can save themselves only at the cost of a quick and drastic change in their habits and beliefs, they usually prefer to perish.[50]

But most people I know would certainly prefer not to perish, although few of them can decide whether the foe is at the gate. Unfortunately, those who do believe that disaster is imminent most likely feel powerless to do anything.

But even recognizing the absolute priority status of the world predicament is a good starting point, and uncertainty over the likelihood of possible catastrophes or the viability of proposed preventive measures need not detract from such recognition. Optimism or pessimism about the human prospect are not the real issues. The world has endured periods of both. Historian Johan Huizinga eloquently tells of epochs of hope and times of despair:

> In the fifteenth century, as in the epoch of romanticism, it was, so to say, bad form to praise the world and life openly. It was fashionable to see only its suffering and misery, to discover everywhere signs of decadence and of the near end—in short, to condemn the times or to despise them.[51]

That this attitude created a stifling atmosphere of pessimism is well established, Huizinga relates:

> If in all that regards the things of this world there is no hope of improvement and of progress, however slow, those who love the world too much to give up its delights, and who nevertheless cannot help aspiring to a better order of things see nothing before them but a gulf. We will have to wait till the eighteenth century—for even the Renaissance does not truly bring the idea of progress—before men resolutely enter the path of social optimism, only then the perfectibility of man and society is raised to the rank of a central dogma, and the next century will only lose the naïvete of this belief but not the courage and optimism which it inspired.[52]

Admittedly, the resolve of people to act does depend on hope, but the problems facing the world today are not *solved* by clarifying the human prospect in terms of hope versus despair. Problems are solved best by actions. To be sure, history will record whether the optimists or pessimists were better analysts of the future, but it will judge our era by its actions. I hope that we, unlike the Romans or the Mayans, take corrective steps in time. To do otherwise would negate the characteristics that exempt humans from the cruelly efficient world of natural selection, a world where individual survival is always subordinate to the survival of the species. Humans *can* have it both ways. They have the ability to use their hands and minds to shape their environment and to preserve both individuals and the species—if only they would act in concert to prevent their handiwork from damaging both.

Part IV

Appendices, Illustrations, Bibliography, Notes, and Index

A

Climatic Effects of Deforestation of the Amazon[1]

Change in either the albedo or moistness of the land surface could affect the local surface temperature. If the albedo of the surface were to be increased, leaving the moistness of the surface unchanged, then less heat would be absorbed at the surface and the temperature would be decreased locally (assuming there were no local change in cloudiness resulting from the diminished surface heating). On the other hand, if the surface albedo were to remain unaltered while at the same time the moistness of the surface were to vary, then there would be no change in the amount of solar energy absorbed by the surface; but there would be some change in the proportions of absorbed energy used to evaporate water to absorbed energy directly added to the local environment as sensible heat. If the surface moistness is increased while the albedo remains unchanged, then more liquid water is evaporated at the surface and less sensible heat energy is available to warm the local region, and the climate is cooled near the surface. At the same time, the evaporated water vapor is carried upward into the atmosphere and is transported downstream until it eventually condenses back into droplets in the clouds, releasing the latent heat energy needed originally to evaporate the liquid water at the surface and convert it to water vapor. Thus, there is a net energy loss at the original evaporation site (the surface) and an equal net energy gain at the condensation site (the atmosphere). Of course, the overall global energy input remains unaltered since the albedo was unchanged.

If very large artificial bodies of water were to be constructed, they would likely warm the global climate and moderate the local climate, since it is probable that an artificial body of water will both decrease the surface albedo and increase the surface moistness. Deforestation, as explained below, is likely to do just the opposite (see Figure 24).

Professor Reginald Newell of the Massachusetts Institute of Technology was concerned with this question:

"What will happen to the large-scale atmospheric general circulation if the tropical forests are removed over Brazil and Indonesia and perhaps over Central Africa? I think that the answer . . . is that we do not know, but we can speculate on the kind of influence it will have." Newell then speculates on the possible effects of deforestation: "Reduction of the water cycling between the forests and the air, which provides the latent heat high in the column, would clearly alter the pattern of the latent heat forcing function." [2]

Although Newell has argued that tropical changes could affect the global energy budget, he has also recognized that predictions of any such possible effects must be approached with extreme caution. Newell states: "I should emphasize here that the linkages between the various parts of the general circulation are not well understood, even diagnostically." [3]

An important question arises: How can we learn more quickly about the potential climatic consequences of large-scale deforestation? One possible way is to simulate a deforested Amazon in a large computer model of the climate. Although experiments with these "general circulation models" are worth trying for this case, a strong caution is needed in light of recent work by my colleagues, Robert Chervin and Warren Washington, and myself. We have demonstrated that unambiguous interpretation of the results of general circulation models for such experiments can be a difficult problem.[4]

Finally, a point on the danger of the potential irreversibility of climatic changes due to deforestation should be made. Although no one can be certain that a perturbed climate would return to its normal state, even if a deforested Amazon were allowed to regenerate, the main risk of permanent damage comes not so much from the possibility that changes in the climatic system are irreversible, but rather from the prospect that changes in the ecological systems of the forest are irreversible. Daniel Janzen, a distinguished ecologist from the University of Wisconsin, said at a recent conference that tropical forests, when destroyed, are much less likely to regenerate than mid-latitude forests. Thus, any climatic shift induced by tropical deforestation could be irreversible for the simple reason that the deforestation process in itself may well be irreversible.

All of this suggests adherence to the Genesis Strategy principle of conducting large-scale technological projects like deforestation of the Amazon basin with caution since they have the capacity to create global and irreversible changes.

B

Natural Power Densities in the Earth-Atmosphere System

Before discussing power densities in the climatic system, an unrigorous and brief set of definitions is given. For a fuller explanation of energy concepts and terminology, the reader is referred to the first chapter of the very readable volume on energy by John Holdren.[5]

Energy is often defined as the capacity to do work. It can take on many forms: heat, light, chemical, elastic, pressure, velocity or radiation, among others. Power is the *rate* of doing work, that is, a measure of how much energy is expended in a given time. In the metric system this is most often measured in watts; and in the English system, in horsepower (1 horsepower= 746 watts).

When energy flows continuously and is in the form of radiation (from the sun, for example) it is often convenient to define it in terms of a power density. Power density in the context of the climate is often used as a measure of the amount of energy released at or impinging on a unit of surface area per unit of time. At the earth's orbit the natural power density of solar radiation (i.e., the amount of radiant energy that passes each second through an area of one square meter perpendicular to the rays of the sun) is about 1,350 watts per square meter (W/m^2). This fundamental natural power density is a useful gauge against which potential thermal pollution power densities can be measured.

The earth's equilibrium temperature is determined by a balance between the unreflected portion of the sun's radiant energy (that is, the part absorbed by the earth-atmosphere system) and the outgoing infrared radiation emitted by the system to space. Figure 25, redrawn from Rotty and Mitchell,[6] shows the complex disposition of natural energy flows as a percentage of the incoming solar radiation Q (1,350 watts per square meter), the "solar constant" or solar irradiance. About 28 percent of Q is reflected back into space, and the remainder drives the atmospheric and oceanic motions that determine the climate. About 47 percent of Q is absorbed at the earth's surface (Q_s). However, the important natural power densities for future comparison to artificial power releases are the *global average* densities (denoted by overbar—e.g., $\bar{Q}$—since the solar energy intercepted by the earth's disk is spread rapidly (by a rotating earth) over the entire earth's surface (which has four times the area of the earth's disk). Thus, the important natural power densities to remember for comparisons to unnatural power generation densities are the global average solar irradiance, $\bar{Q} \approx 340$ W/m^2, and the global average solar energy absorbed by the earth's surface, $\bar{Q}_s \approx 160$ W/m^2.

Temperature–Energy-Sensitivity Analysis [7]

An order-of-magnitude estimate can be made of the *change* in surface temperature, ΔT_s, caused by an incremental change in surface power density, $\Delta \bar{Q}_s$. The calculations of a one-dimensional climate model show that a 1 percent increase in global average solar power density (i.e., $\Delta \bar{Q}_s = 0.01 \bar{Q}_s \approx$ 1.6 W/m^2) would give an 0.8°K increase in global average surface temperature (see the Schneider and Dennett article for details). This figure, however, could be an under- or overestimate by severalfold, as explained in chapter 6, particularly in polar regions or the midlatitudes in winter, where it is likely to be an underestimate.

C

Illustrations

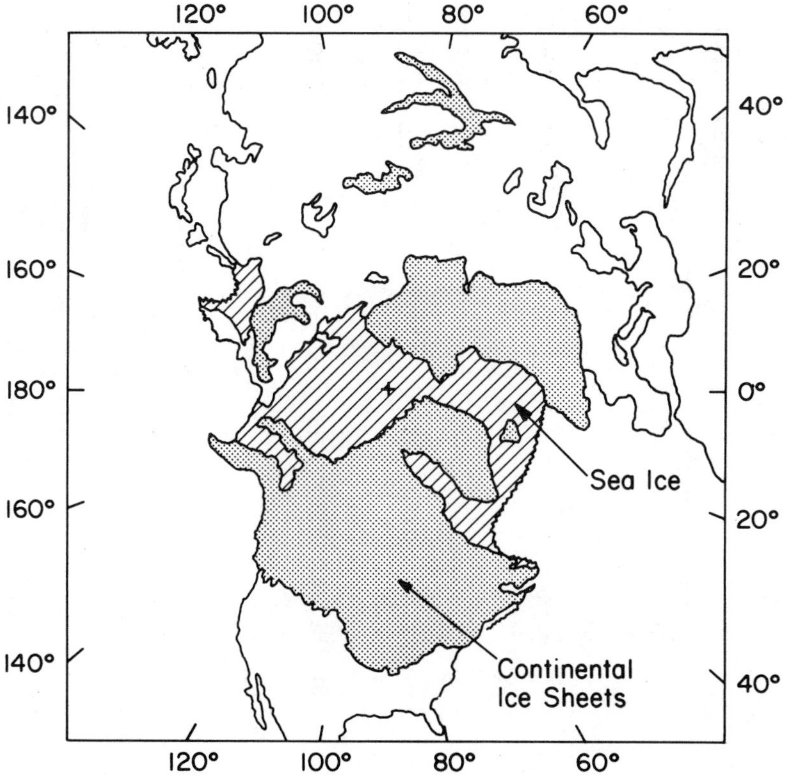

Figure 1. Maximum extent of Northern Hemisphere ice cover during the present glacial age. The southernmost extent of the ice sheets represents the furthest the sheet has moved over the past million years. If this pattern were to recur today, ice would cover most of the territory that is now the northern United States, Canada, and northern Europe (after U.S. National Academy of Science).

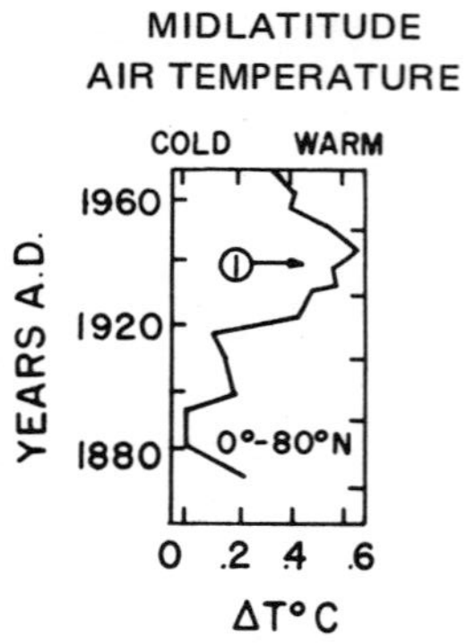

(a) THE LAST 100 YEARS

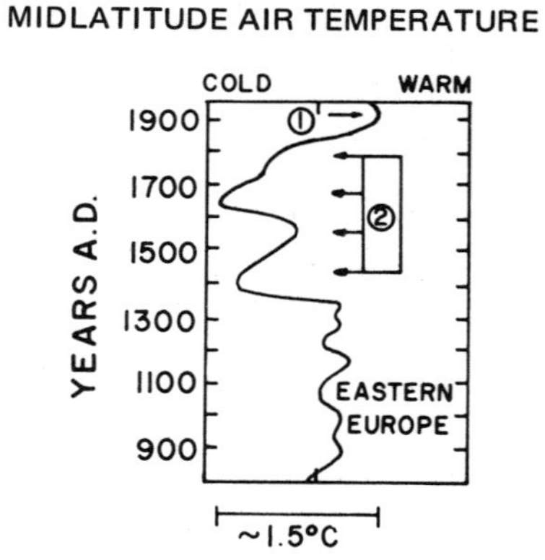

(b) THE LAST 1,000 YEARS

LEGEND

1. Thermal Maximum of 1940s
2. Little Ice Age
3. Cold Interval
4. Present Interglacial (Holocene)
5. Last Previous Interglacial (Eemian)

SOURCE: U.S. National Academy of Sciences

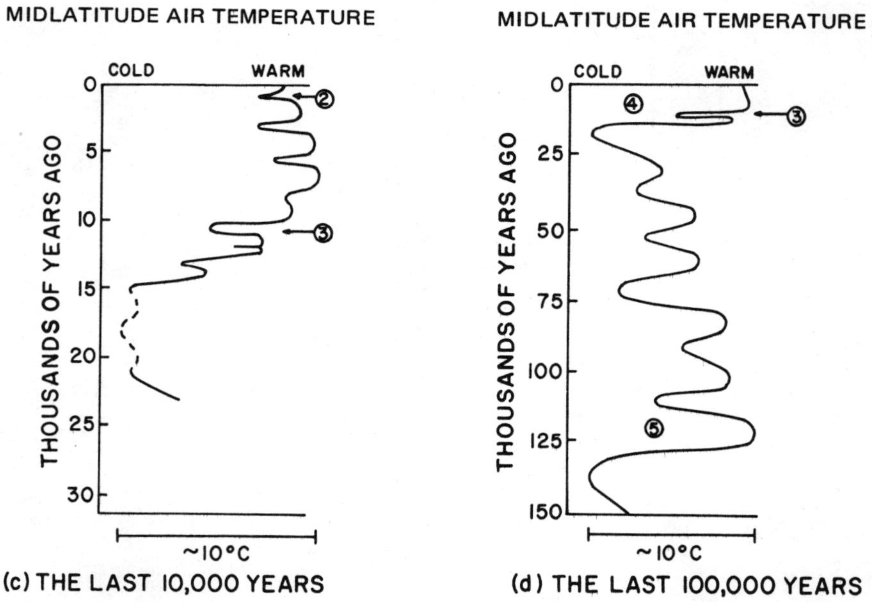

Figure 2. Generalized trends in global climate represented by approximate surface temperature patterns that prevailed over a variety of time scales for the past 150,000 years.

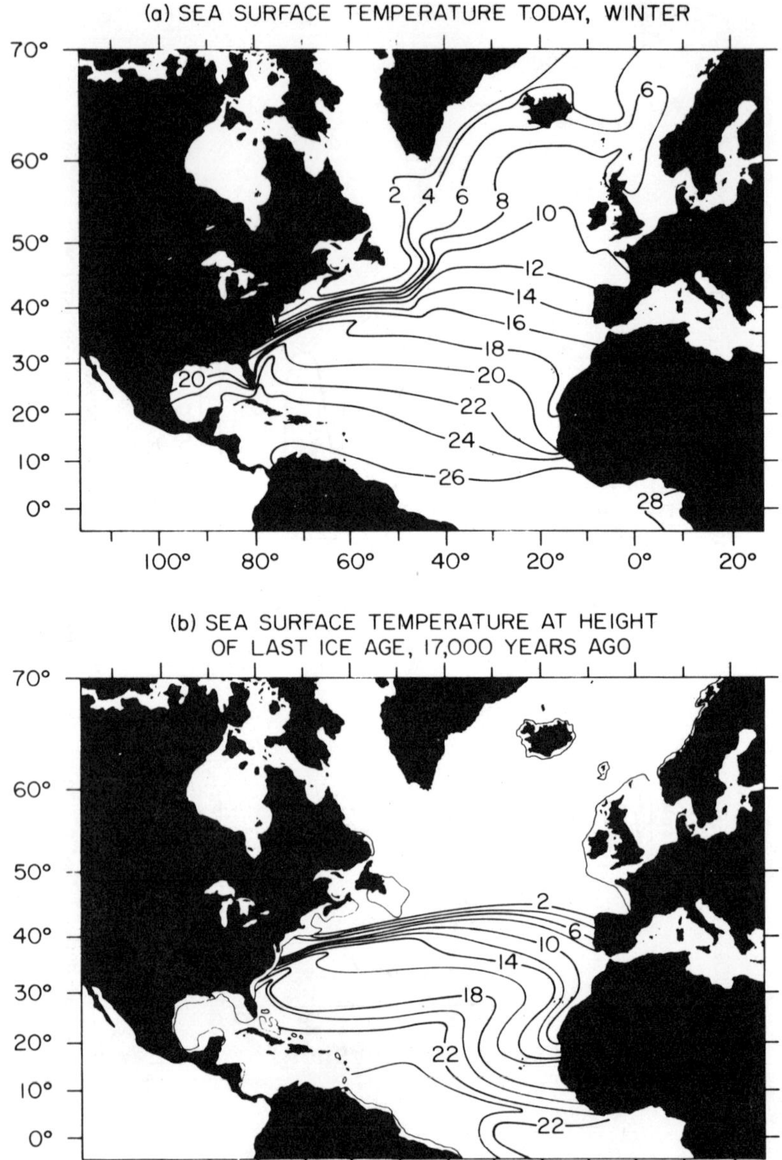

Figure 3. Maps of the average surface temperature (°C) in the North Atlantic Ocean (a) today and (b) as it was 17,000 years ago at the height of the last Ice Age. Notice that the warm waters carried by the Gulf Stream ocean current that today keep the England, Ireland, and Northern Europe region relatively mild were displaced to the south of Spain in the Ice Age.

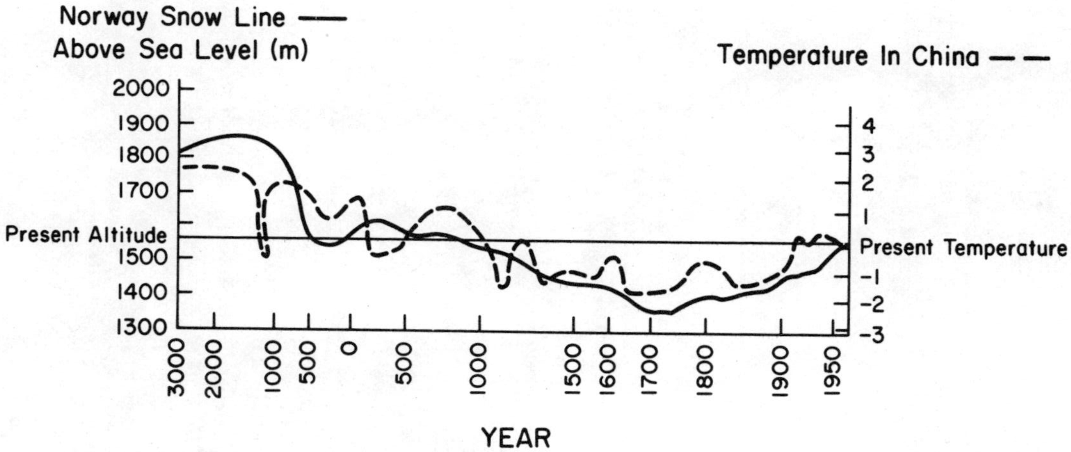

Figure 4. Changes in the Norway snow line (solid line) suggesting that general climatic patterns in Europe exhibit some similar features to the temperatures in China (dotted line) over the past 5,000 years, as reconstructed by Chu, Ko-Chen.

Figure 5. Engraving of the Argentière Glacier (French Alps) made about 1850–1860 and showing its front still close to the plain and the village. (From Emmanuel Le Roy Ladurie, *Times of Feast, Times of Famine*, published by Doubleday and Company, Inc., 1971. Translated by Barbara Bray.)

Figure 6. Photograph of the town of Argentière in the French Alps, taken in the mid-1960s. The view is essentially the same as that in the engraving shown in Figure 5, which was made about 100 years earlier when the mean hemispheric temperature was less than 0.5°C cooler. The terminus of the glacier can now barely be discerned in the upper part of the picture. (From Ladurie.)

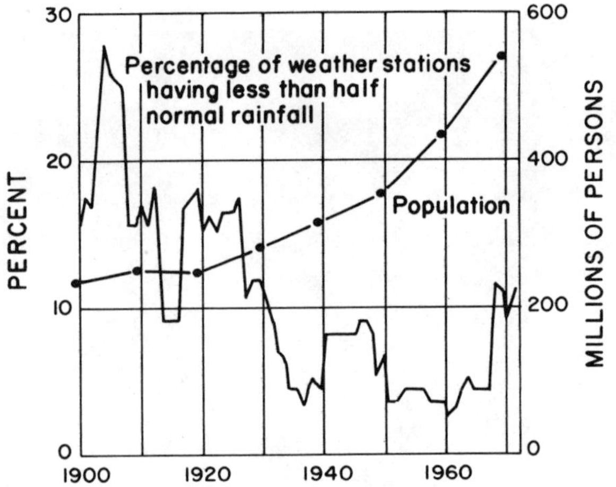

Figure 7. The wiggly line gives the percentage of weather stations in northwest India having less than half-normal rainfall, and the line with dots gives the approximate population of India over the same period. The graphs show that from the mid-1930s to the mid-1960s, a period of rapid population growth, the likelihood of droughts was as much as 4 times less at these weather stations than in the decade after 1900 (after Bryson).

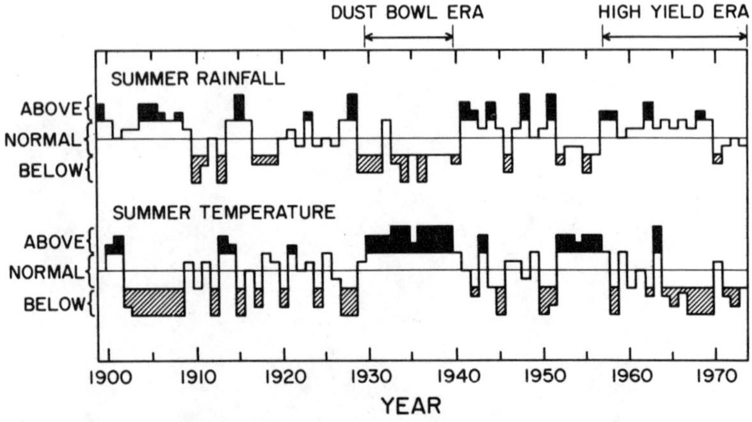

Figure 8. Seventy-five year record of summer average temperature and rainfall in the five major wheat-producing states of the United States, compiled by Donald Gilman of the National Weather Service and showing the 10-year drought period (i.e., high temperatures and low rainfall) of the dust bowl era and the 15-year recent high-yield era (i.e., above-average rainfall and below-average temperature).

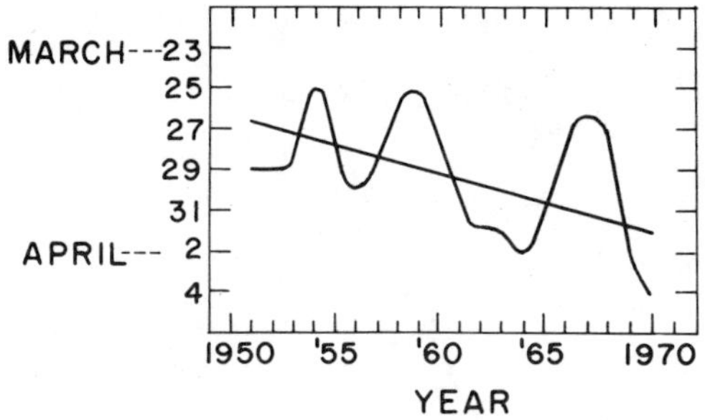

Figure 9. The opening dates of the blooming of the cherry blossoms in Tokyo, from Asakura. Ostensibly, the cherry blossoms bloom early if the average weather conditions are favorable, but the graph suggests a trend of unfavorable weather since 1950.

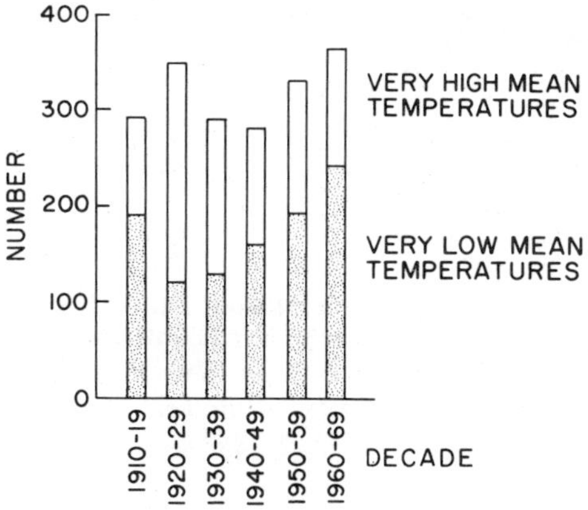

Figure 10. A record of very high and very low temperature extremes (i.e., extreme means higher or lower than 90 percent of the recorded temperatures) from 65 long-term stations around the world as compiled by Tadashi Asakura of the Japan Meteorological Agency. The graph suggests that although the total number of extremes has not changed much since 1910, the number of extreme low temperatures at these stations has roughly doubled since the 1920s.

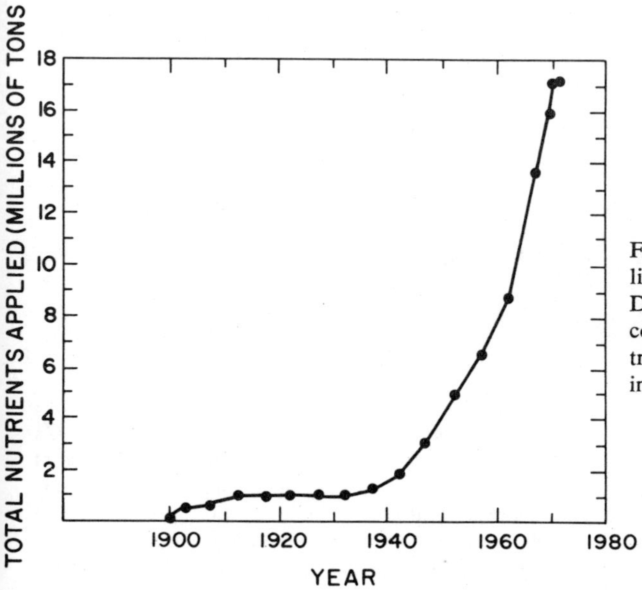

Figure 11. Trend of fertilizer use in the U.S., from Decker. It is interesting to compare this trend to the trend in crop yields shown in Figure 12.

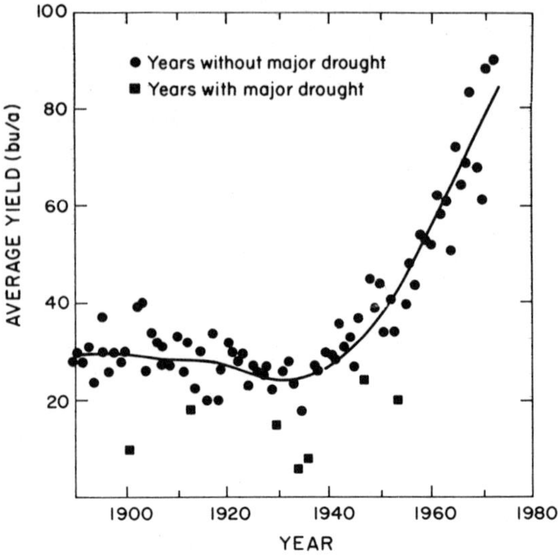

Figure 12. Solid line is the trend in corn yield per acre in Missouri, after Decker. The circles represent individual yearly yields for years without major drought and the squares are for years with drought. Note that since 1956 not only have yields increased significantly, but variability in yields from one year to the next has been reduced.

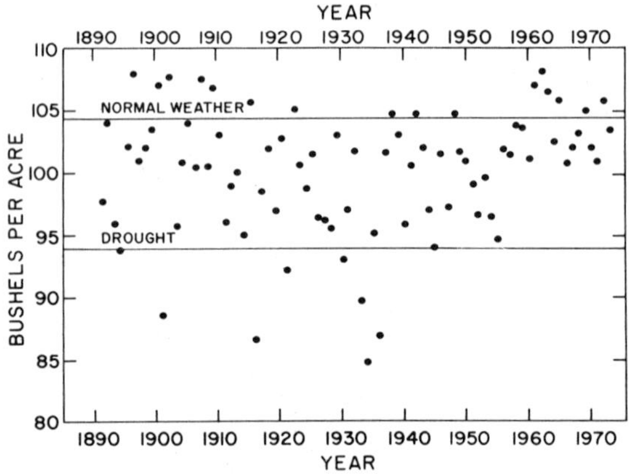

Figure 13. Simulation with Louis Thompson's crop model of corn yields in the five major corn-producing states under the assumptions that technology is fixed at 1973 levels and the only influence on corn yield variability is the weather patterns that prevailed from 1890 to 1973. This technique isolates the impact of weather on year-to-year crop variability, and demonstrates that the weather in the corn belt from 1956 to 1973 was abnormally good; this, not improved technology, largely explains the low *yield variability* seen in Figure 12 for this 17-year period (after McQuigg, LeDuc, Lockard, and McKay).

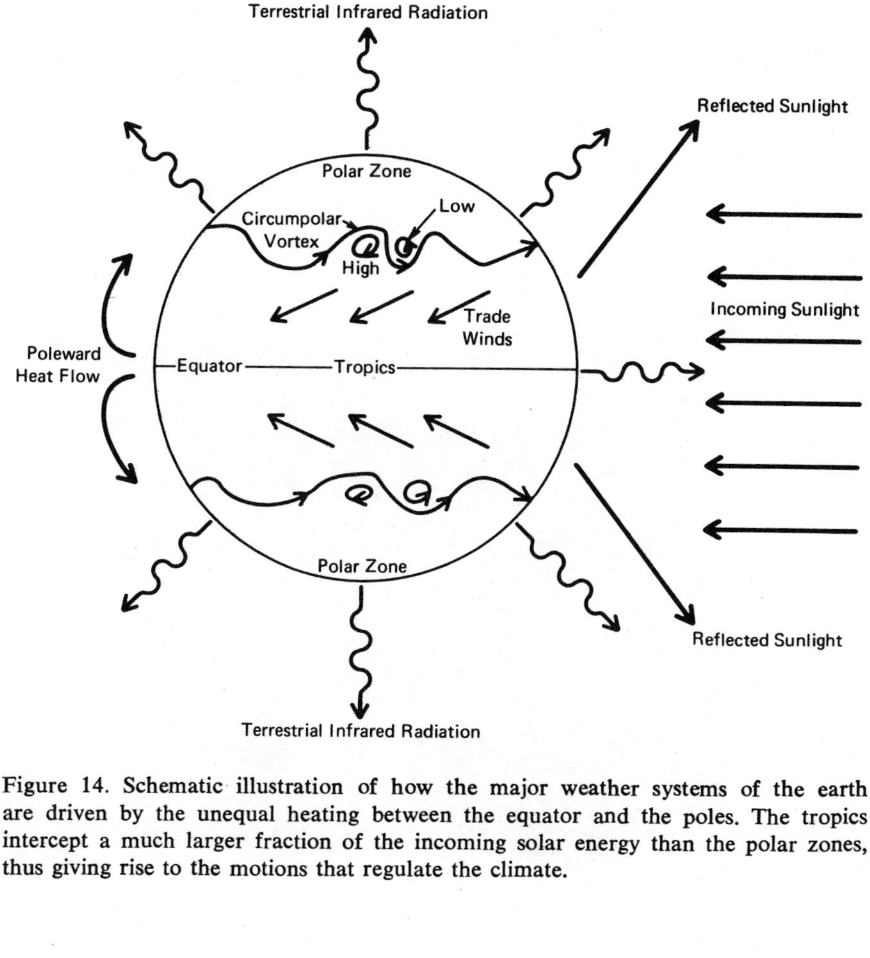

Figure 14. Schematic illustration of how the major weather systems of the earth are driven by the unequal heating between the equator and the poles. The tropics intercept a much larger fraction of the incoming solar energy than the polar zones, thus giving rise to the motions that regulate the climate.

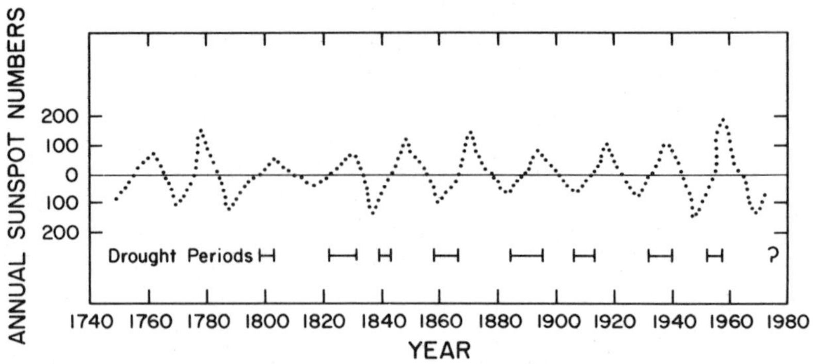

Figure 15. A plot of the double sunspot cycle versus drought in Nebraska. The graph suggests that droughts in the U.S. Great Plains tend to occur in a 22-year cycle centered near the minima of that cycle. If this relationship holds, then the next such drought is "due" in the last half of the 1970s (after Thompson).

Figure 16. A remarkable photograph of the Gaza Strip region taken by the NASA ERTS satellite showing a visible *political border*. The 1948 armistice line between the Israeli Negev (darker region) and the pre-1967-war Egyptian Sinai (lighter region) is easily seen. The Sinai side is lighter because Bedouin Arabs' animals have grazed off much of the vegetation cover up to the fence at the border.

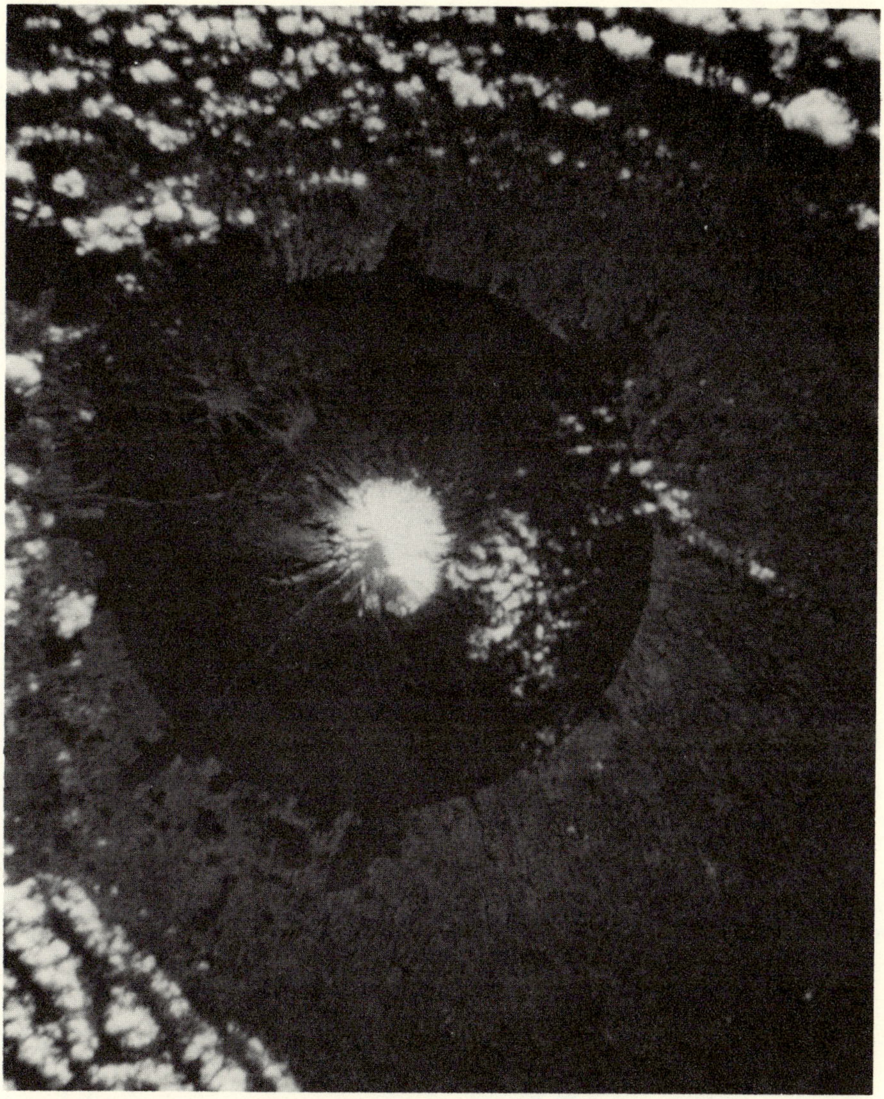

Figure 17. Another example of the effects of overgrazing on the land surface is visible from space in this NASA Skylab photo of Mt. Egmont, an extinct volcano in New Zealand. The nearly perfect circular shape marks the line where a fence keeps grazing sheep out of the heavily vegetated slopes (darker color) of the mountain.

Figure 18. A NASA ERTS photograph of a part of the drought-stricken Sahel region taken at the height of the drought in 1973. Despite the severity of this recent dry spell, sufficient vegetation was able to grow inside the pentagonal region to make this zone clearly discernible to the satellite. The darker vegetated region inside the pentagon was protected from hungry foraging animals by a fence, which prevented the interior from being overgrazed.

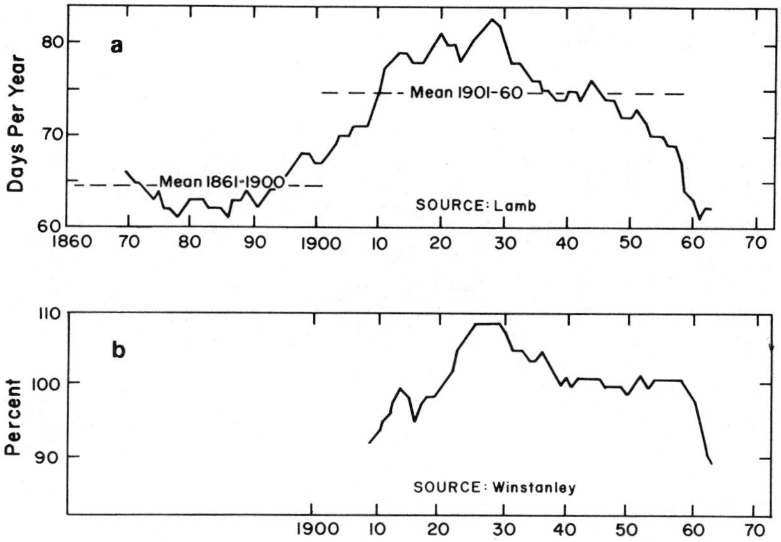

Figure 19. In (a) the frequency of westerly-type weather over Great Britain can be seen to coincide closely with the rainfall in the Sahel shown in (b). This figure was prepared by meteorologist Derek Winstanley to suggest that the Sahelian drought may have been related to other climatic patterns thousands of miles from Africa.

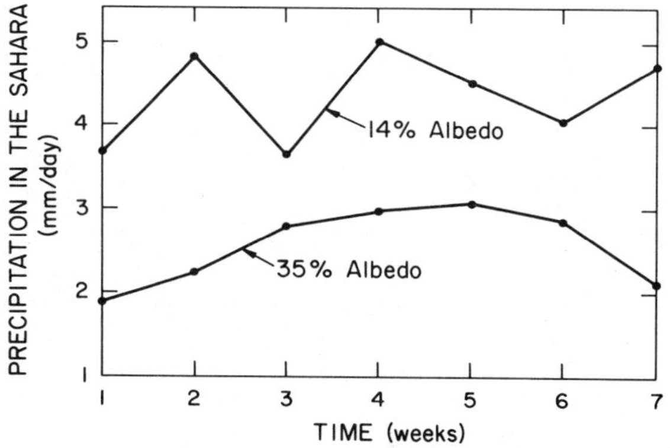

Figure 20. Rainfall in the Sahara predicted by a computer model of the atmosphere for two possible values of land-surface albedo (reflectivity). Meteorological theorist Jule Charney has used this calculation to demonstrate the possibility that overgrazing of the Sahel (which would increase the albedo) may have contributed to the Sahelian drought.

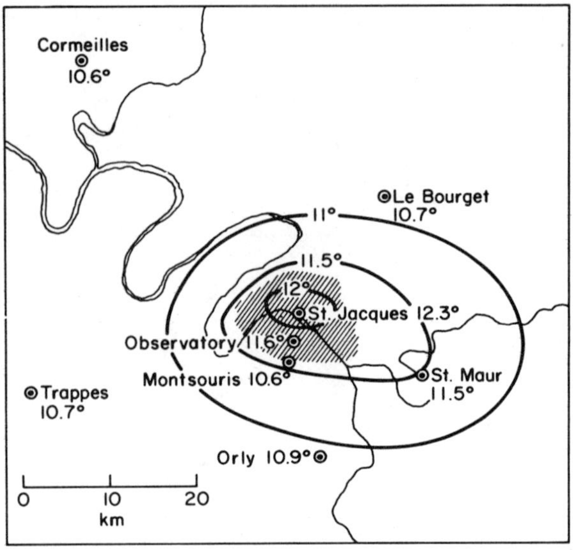

Figure 21. The influence of an urban area in raising the temperature of the city center is clearly seen in this figure, which shows this urban heat island effect for Paris (after Dettwiller in Landsberg).

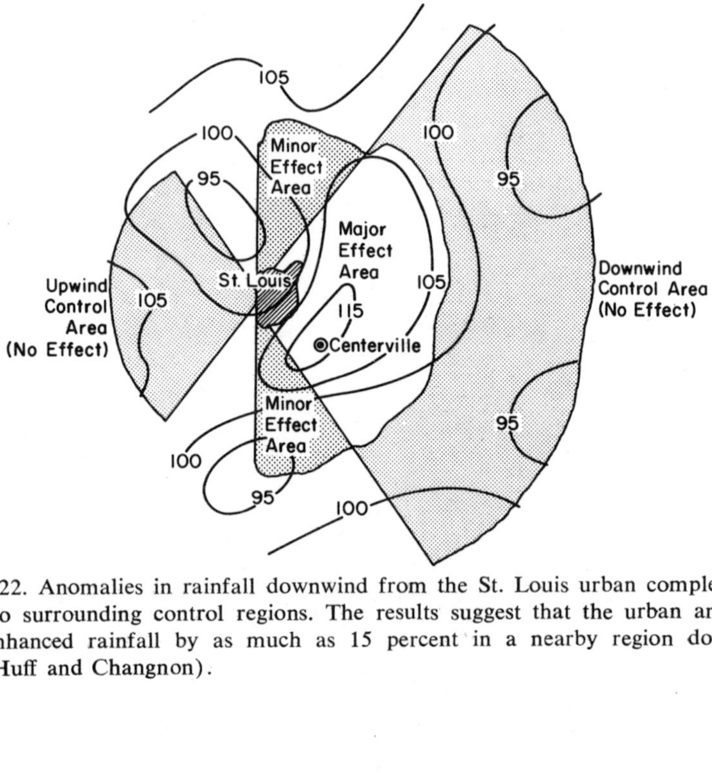

Figure 22. Anomalies in rainfall downwind from the St. Louis urban complex compared to surrounding control regions. The results suggest that the urban area may have enhanced rainfall by as much as 15 percent in a nearby region downwind (after Huff and Changnon).

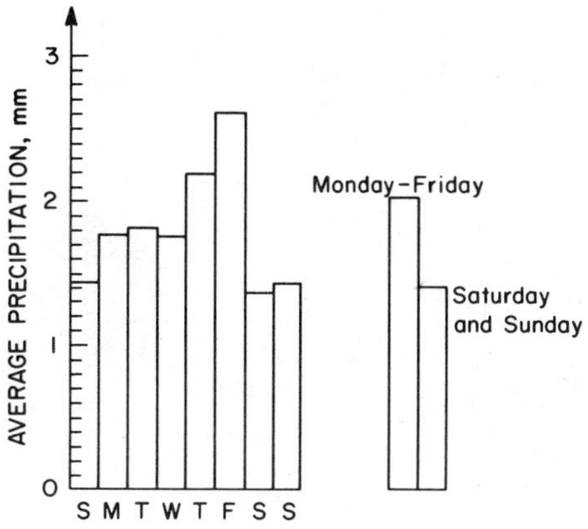

Figure 23. Average precipitation in Paris broken down into day-of-the-week categories. These data support the hypothesis that urban activities increase precipitation, since precipitation is significantly greater on the weekdays, when urban activities are most intense (after Dettwiller in Landsberg).

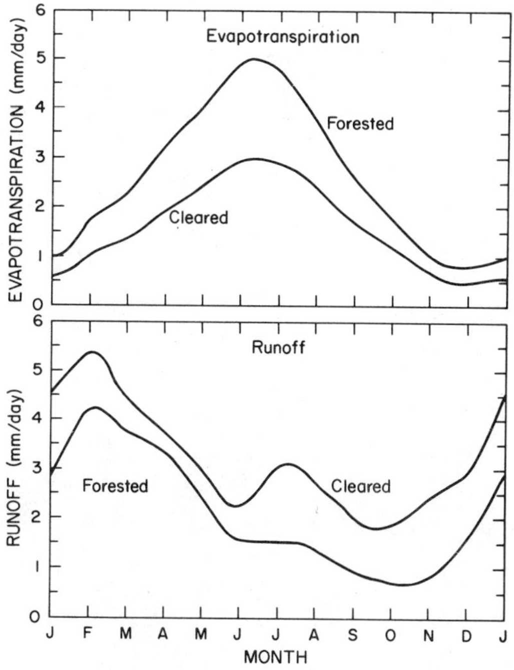

Figure 24. (Upper) evaporation from the land and transpiration from plants (together called evapotranspiration) of water vapor and (lower) runoff of water from forested and cleared land in North Carolina. Clearing land increases the runoff and decreases the evapotranspiration in comparison to forested land in all seasons. Thus, deforestation on a large scale can change the water-bearing characteristics of the land and perhaps even the climate (after Sellers).

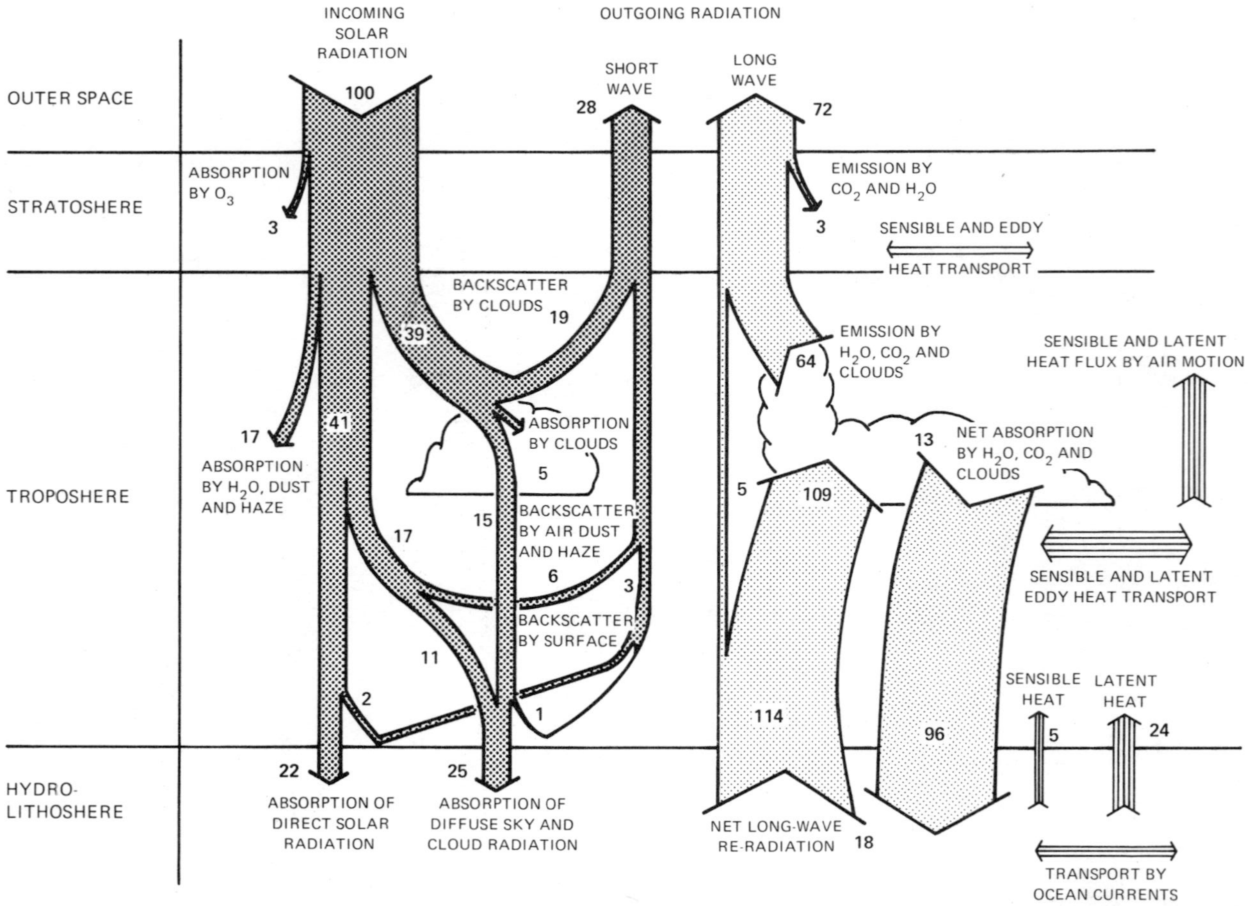

Figure 25. Schematic diagram of the complex disposition of absorbed solar energy in the earth-atmosphere system. The potential climatic consequences of any human activities will be proportional to the relative disruption these might cause to the natural energy flows shown on the diagram (after Rotty and Mitchell).

Figure 26. An Air Weather Service satellite photograph of the eastern half of the United States taken at night with a sensitive camera. Note that the urban regions literally glow in the dark as a result of the high energy-consumption densities in these population centers.

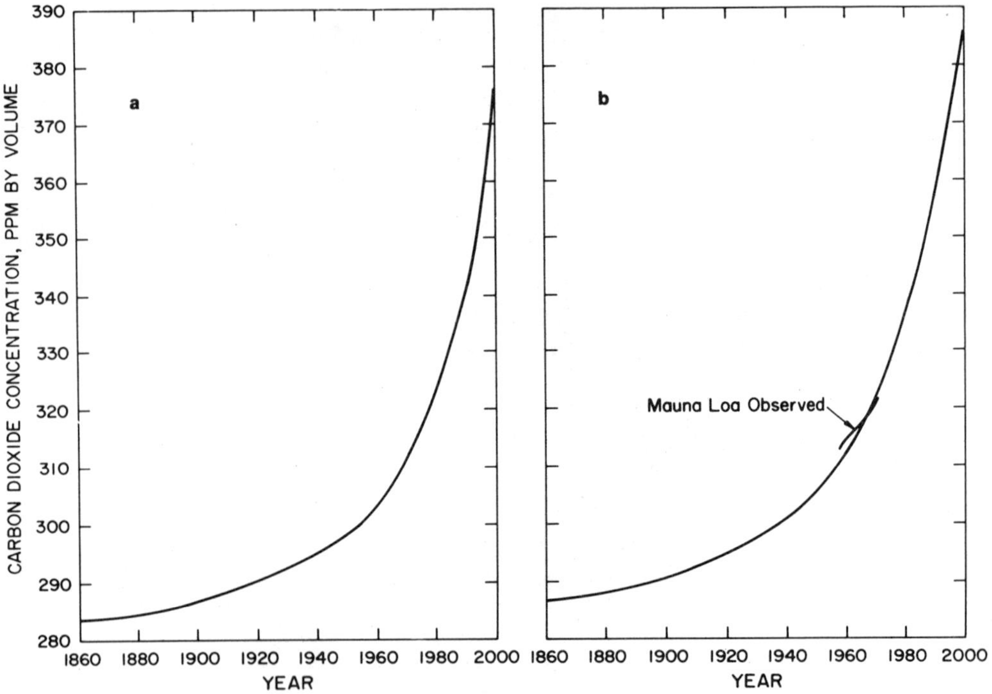

Figure 27. Projections of atmospheric carbon dioxide (CO_2) concentration from fossil fuels calculated by a model of the carbon dioxide cycle. Projection (a) shows an early estimate of 375 parts per million (ppm) CO_2 concentration by the year 2000, whereas an updated model (b) predicts a CO_2 concentration of near 390 ppm, proving that not all "doomsday" projections are overestimates. This prediction implies a global warming of roughly 1°C, which is as large as any natural climatic changes in the past several centuries. Projection (a) is from Machta (1971) and Projection (b) from Machta and Telegadas (1974).

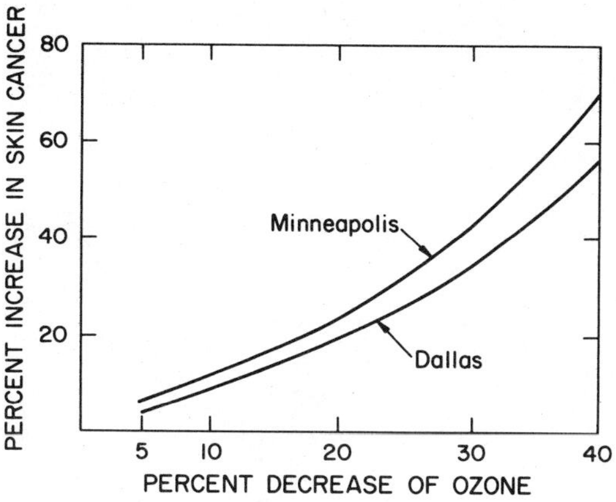

Figure 28. An estimate of the percent increase in skin cancer in light-skinned humans as a result of a decrease in ozone at two locations (after Epstein, Urbach, Lee, Forbes, and Scotto in "The Report of Findings," Climatic Impact Assessment Program).

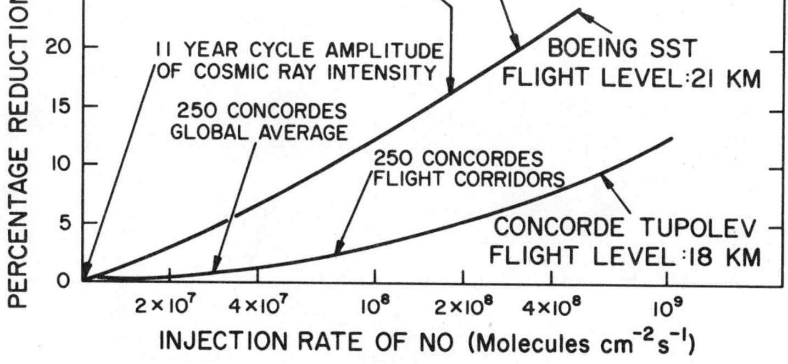

Figure 29. A computer model calculation of the percentage reduction in ozone from an injection of nitrogen oxides (NO) into the stratosphere. The figure shows the relative effect of many natural and human-induced phenomena on the ozone layer (after Crutzen).

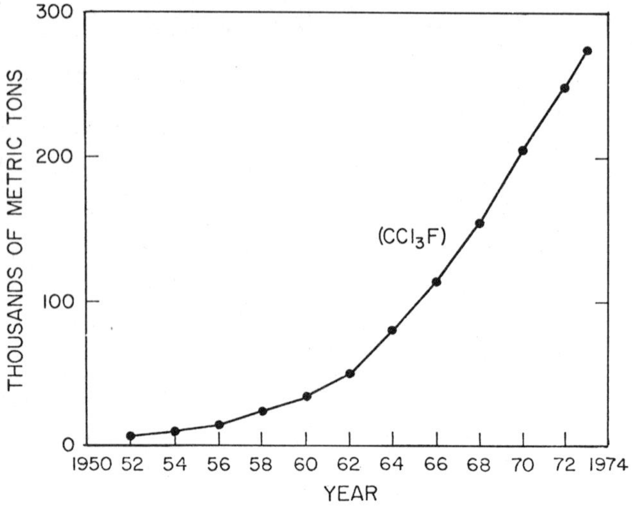

Figure 30. Annual production of one chlorofluorocarbon (CFC) shows that up to 1974 this has been a growth industry. Since chlorofluorocarbons may be able to destroy the ozone shield that protects life on earth from the sun's ultraviolet rays, regulation to inhibit this growth rate is under debate (after Crutzen).

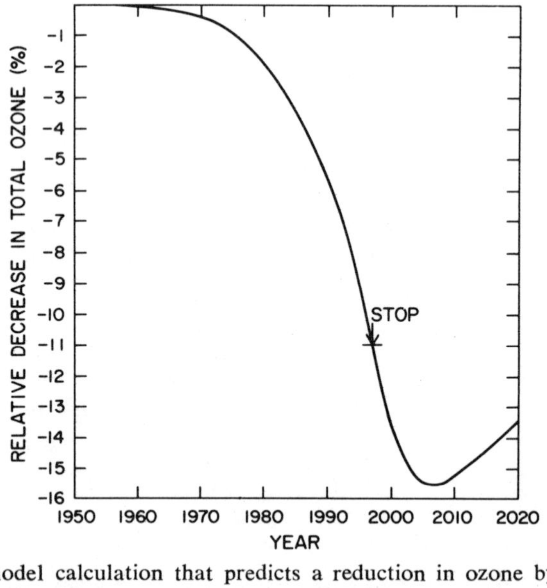

Figure 31. A model calculation that predicts a reduction in ozone by 11 percent in 1998 if CFC use continues to grow at a rate of 10 percent per year. Notice that ozone keeps on decreasing for about 10 years *after* the CFC use is cut off completely (at the "stop" point on the diagram). (After Crutzen and Isaksen.)

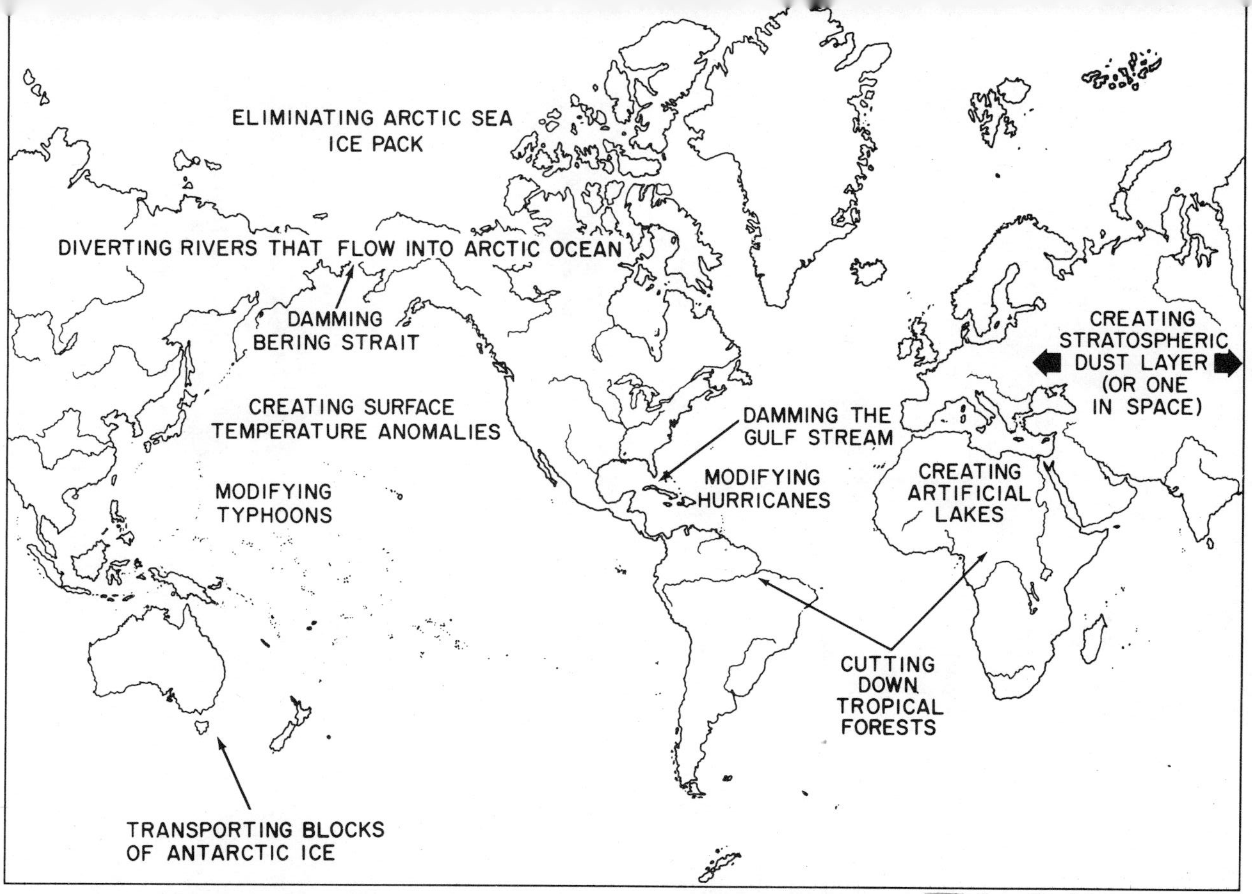

Figure 32. Schematic map from Kellogg and Schneider showing a number of deliberate climate modification projects that have been or are likely to be proposed in the near future.

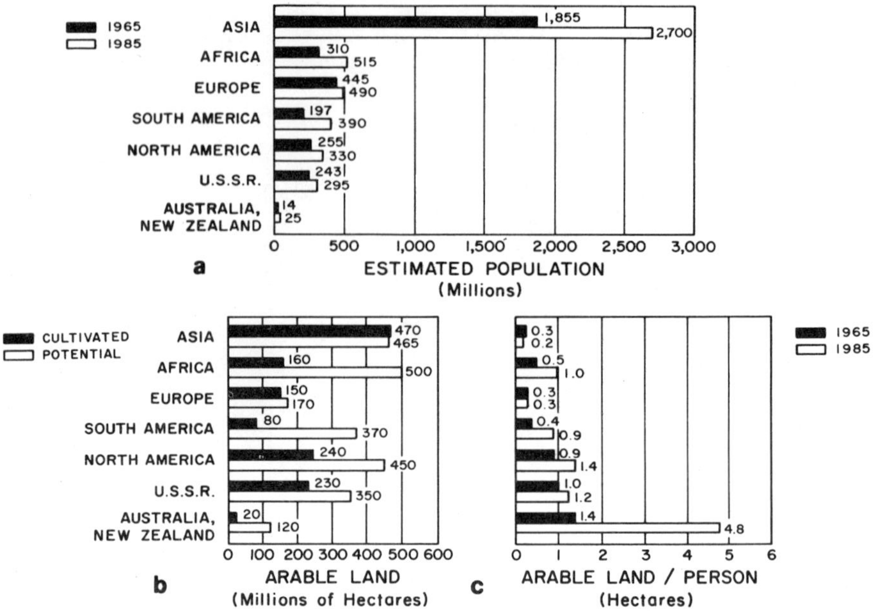

Figure 33. A comparison of estimated world population and arable land, from Revelle. The figure suggests that food self-sufficiency for Asia is unlikely in the near future, but that Africa and South America have large quantities of arable land. The problems of exploiting those lands are explained in chapters 6 and 8.

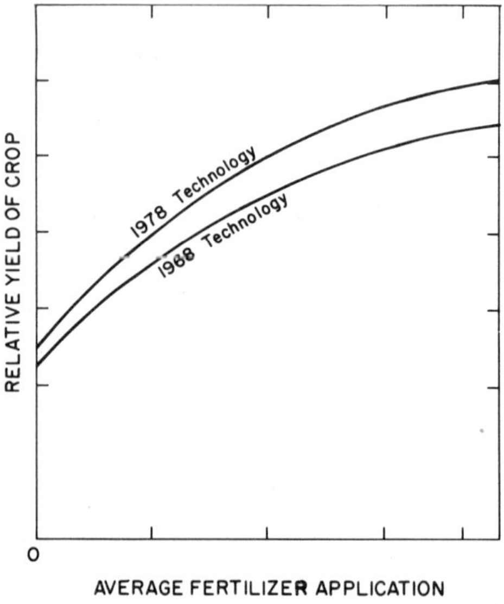

Figure 34. Conceptual relationship between U.S. crop yield and U.S. average fertilizer application, from Decker. The figure shows a law of diminishing returns for large applications of fertilizer.

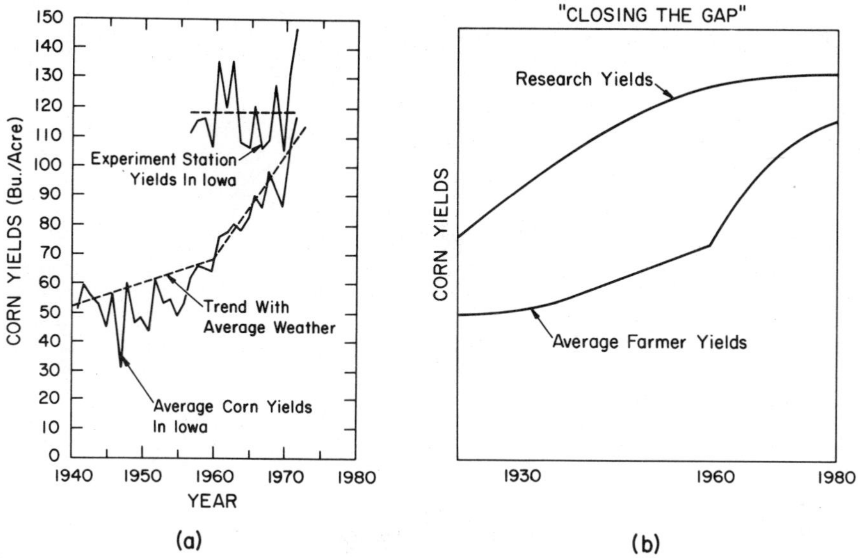

Figure 35. The gap between corn yields at an Iowan experimental station and those of the average Iowan farmer, from Thompson. Clearly, in the absence of a technological breakthrough, the average farmer will soon catch up to the productivity of the researchers.

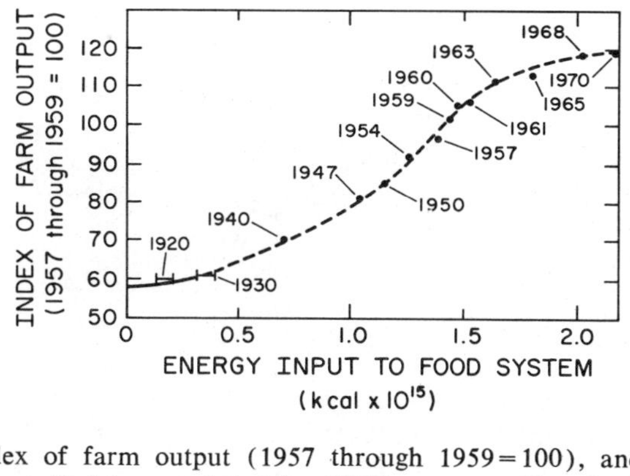

Figure 36. Index of farm output (1957 through 1959 = 100), and energy input to food system, from Steinhart and Steinhart. This curve typifies the law of diminishing returns and suggests that increasing farm output in the United States is becoming an inefficient process.

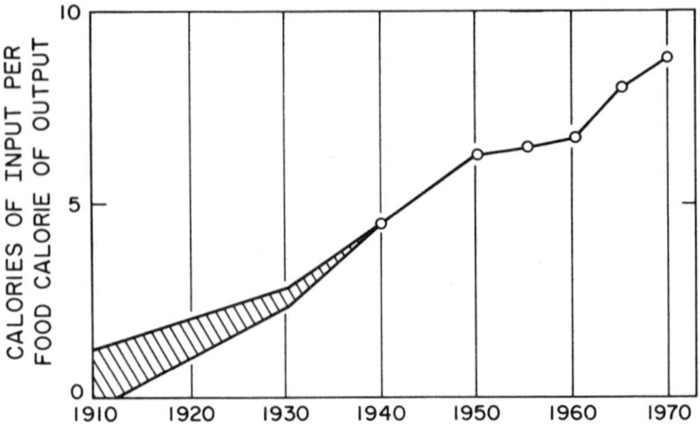

Figure 37. Calories of input per food calorie of output, from Steinhart and Steinhart. Figure points out that the U.S. agricultural system has become highly energy-intensive over time.

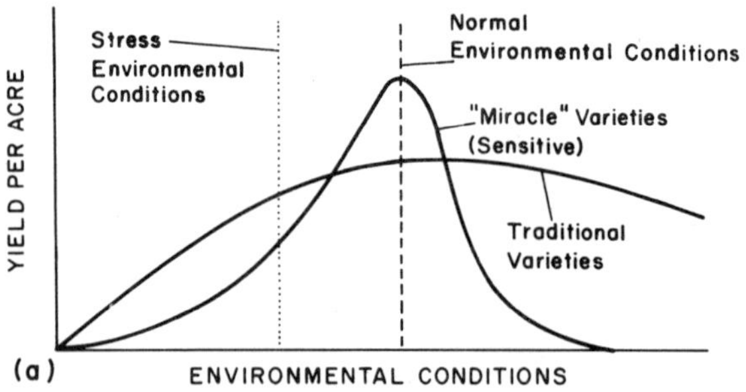

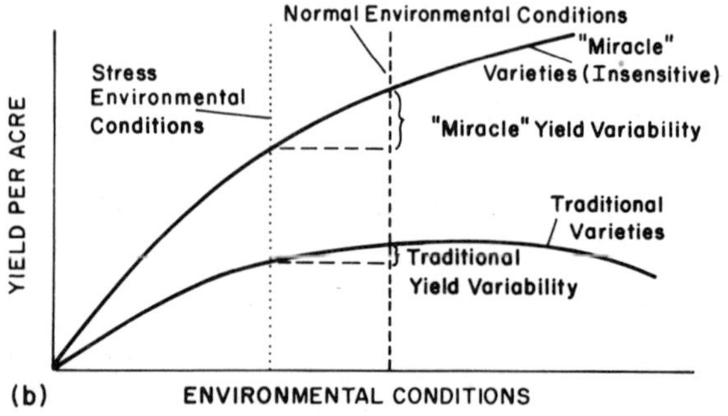

Figure 38. *Variability* in crop yield as a function of environmental factors (e.g., soil moisture, fertilizer input or sunshine) is seen to be greater for miracle crops than conventional varieties, as illustrated by this schematic diagram. This increased sensitivity to changing environmental conditions remains even if the insensitive varieties shown in (b) are used. Figure suggests that large food reserves are essential to hedge against the potentially high yield variability of miracle crops.

D

Suggested Readings

A short list of further references is given for those readers who may wish to dig deeper into a number of aspects of *The Genesis Strategy* subject matter.

Very few books are listed, and preference is given to readable and recent books since these provide a useful entry point into the subject, and contain up-to-date citations of the latest literature as well as lists of more classical works.

Climatic Change

Books

Glantz, M. H., *The Politics of Natural Disaster: Case of the Sahelian Drought,* Praeger, New York, 1976, is an interesting volume that documents a number of aspects of the Sahelian disaster. As a multidisciplinary work combining physical and social sciences, it is relatively unique, but, more significantly, it presents a background to the Sahelian catastrophe, which should be studied as an example of potential future disasters.

Gribbin, J., *The Climatic Threat,* Walker, New York, 1976, presents an interesting challenge to our understanding of the climate and how it can change, written from the viewpoint of a scientist and with the skill of a science writer.

Hare, F. K., *The Restless Atmosphere,* Harper Torchbook, New York, 1961, is a very readable introduction to meteorology and climatology.

Hess, W. N., ed., *Weather and Climate Modification,* Wiley, New York, 1974, is a good and comprehensive reference intended primarily for scientists.

Ladurie, E. L. R., *Times of Feast, Times of Famine,* translated by Barbara Bray, Doubleday, Garden City, New York, 1971, is one of the most interesting treatments of climatic history.

Lamb, H. H., *Climate: Present, Past and Future,* Methuen, London, 1972. No one can study climatic change properly without reading a book or two by Hubert Lamb.

Panel on Climatic Variation, *Understanding Climatic Change,* National Academy of Sciences, Washington, D.C., 1975, is an up-to-date review of progress in climate research.

Rasool, S. I., ed., *Chemistry of the Lower Atmosphere,* Plenum, New York, 1973, although written primarily for technically trained people, requires little meteorological background and provides a useful introduction to problems of atmospheric chemistry and pollution.

Report of the Study of Man's Impact on Climate (SMIC), *Inadvertent Climate Modification,* MIT Press, Cambridge, Mass., 1971, is still one of the best and most understandable treatments of the subject.

Articles

Alexander, T., "Ominous Changes in the World's Weather," *Fortune,* **LXXXIX,** 90, February, 1974, is one of the first and best popular articles warning of potential climate-related crises.

Climatic Change (D. Reidel, Dordrecht, Holland) is a journal that publishes interdisciplinary papers on any subject related to climatic change.

Food

Books

Brown, L. R., with Eckholm, E. P., *By Bread Alone,* Praeger, New York, 1974, is probably an ideal starting point for reading about world food problems, and includes an excellent list of selected readings.

Paddock, W. and E., *We Don't Know How,* Iowa State University Press, Ames, 1973, is an iconoclastic follow-up to the now famous *Famine 1975!,* which correctly predicted the food crises of the 1970s.

Articles

Science's May 9, 1975, issue is devoted entirely to food and contains an excellent set of articles on the subject. In particular, see Louis M. Thompson's landmark article.

Energy

Books

Ehrlich, P. R., Ehrlich, A. H., and Holdren, J. P., *World Science,* Freeman, San Francisco, 1976, contains a large section on energy that brings up-to-date the major developments in energy since Holdren's 1971 book, and contains myriad references to the most recent works.

Holdren, J. P., and Herrera, P., *Energy: A Crisis in Power,* Sierra Club, San Francisco, 1971, 252 pp. Part I, "Energy: Resources and Consumption," by John Holdren is probably the most readable overview of energy concepts, prospects, and problems available to a lay audience.

Lovins, A. B., *World Energy Strategies,* Ballinger, Cambridge, Mass., 1975, 131 pp., is an intelligently written discussion of a number of aspects of future energy strategies.

Articles

Ambio, A Journal of the Human Environment, published by the Royal Swedish Academy of Sciences. See Volume II, No. 6, 1973, which is devoted to energy issues.

The *Bulletin of the Atomic Scientists* has devoted considerable attention to energy issues, particularly nuclear safety questions, throughout 1975.

Schipper, L., *Energy Conservation: Its Nature, Hidden Benefits and Hidden Barriers,* Report No. UCID-3725, Energy and Resources Group, University of California, Berkeley, 1975, is a lively and informative discussion of the strengths and weaknesses of the energy conservation alternative. It is well worth a stamp to write to Lee Schipper for a copy (and for anything else he has written since).

Science's April 19, 1974, issue is devoted to energy problems. The article on energy and food by John and Carol Steinhart is especially important.

Ecology

Books

Commoner, B., *Science and Survival,* Viking, New York, 1967, is a book that can help to show the thinking that got the environmental movement started. Commoner, whose name has almost been synonymous with the movement, has written a number of books in addition to this important early work.

Eckholm, E. P., *Losing Ground,* Norton, New York, 1976, presents a fresh look at questions of environmental stress and the human prospect. The discussions of deforestation and a firewood crisis are especially important.

Ehrlich, P. R., and Ehrlich, A. H., *Population Resources Environment,* Freeman, San Francisco, 1972, is the 2nd edition of their pioneering text on environmental issues, and is one of the most readable books on the subject. It is a free-swinging, yet authoritative, account of the environmental crisis, and really is a handbook on ecology.

Ehrlich, P. R., Ehrlich, A. H., and Holdren, J. P., *World Science,* Freeman, San Francisco, 1976, will probably define a new field. This book expands greatly on *Population Resources Environment* by including details on the energy, climate, and political dimensions of the environmental crisis, thus combining into one volume most of the important issues of the world predicament. It contains extensive documentation and discussions of various opposing points of view. No doubt, it is the most up-to-date and comprehensive work available. A long overdue addition to the literature on the *problématique*.

Farnworth, G. A., and Golley, F. B., eds., *Fragile Ecosystems,* Springer-Verlag, New York, 1974, is a remarkably important book since it is the consensus of a distinguished group of ecologists, including many Latin Americans who, despite their proximity to poverty and deprivation, recommend extreme caution in development plans for the tropics. It should be read by anyone who looks to the tropics as the world's future breadbasket.

Report of the Study of Critical Environmental Problems (SCEP), *Man's Impact on the Global Environment,* MIT Press, Cambridge, Mass., 1970, provides a good overview of environmental problems.

The World Predicament

Books

Ehrlich, P. R., and Ehrlich, A. H., *The End of Affluence,* Ballantine, New York, 1974, in which they predict what the title suggests, is a nonspecialist, gloves-off follow-up to the now classic *Population Bomb.*

Heilbroner, R. L., *An Inquiry into the Human Prospect,* Norton, New York, 1974, is a short and disturbing book, in which a very knowledgeable futurist warns of serious problems ahead.

Mamdani, M., *Myth of Population Control,* Monthly Review Press, New York, 1972, is recommended for its interesting insights into the difficulties of population control. It clearly shows the complexity of population control issues by recounting experiences of a U.S. team of researchers in a poor village in India.

Articles

For a new approach to solutions to the world predicament, write to the Aspen Institute for a copy of Harlan Cleveland's workshop report: *The Planetary Bargain,* A Policy Paper/Aspen Institute for Humanistic Studies/Program in International Affairs, Aspen, Colorado, 1975.

An important article, still relevant 20 years after it appeared, is Margaret Mead's "Christian Faith and Technical Assistance," *Christianity and Crisis,* **XIV,** 179–182 (January 10, 1955). She reminds us that feeding the entire world had never been considered a realistic possibility until after World War II.

E

Notes and References

Quotations: Notes and Sources

Book title page

"Some questions can be decided, even if not answered." This quote from Justice Louis Brandeis is from Merle Miller, *Plain Speaking,* a Berkeley Medallion Book, Berkeley, Calif., 1973, p. 407.

Preface

John Kenneth Galbraith, *The New Industrial State,* Signet Books, New York, 1968, p. 392.

Harvey Brooks, "Scientific Response to Public Concerns About the Atmosphere," in *Modifying the Weather,* edited by W. R. Derrick Sewell, University of Victoria, Victoria, British Columbia, 1973, p. 86.

Part I

Robert L. Heilbroner, *An Inquiry Into the Human Prospect,* Norton, New York, 1974, p. 13.

The Valery Giscard d'Estaing quote is from: Anthony Lewis, *The New York Times,* Section IV, November 10, 1974, p. 5.

Chapter 1

John Maddox, *The Doomsday Syndrome,* McGraw-Hill, New York, 1973, p. 146.

The International Federation of Institutes for Advanced Study, The Nobel House, Sturegatan 14, Box 5344, S-102 46 Stockholm, Sweden.

Chapter 2

The Secretary Butz quote is from: *Denver Post* (AP), September 22, 1974, p. 7.

The Reid Bryson quote is from: Tom Alexander, "Ominous Changes in the World's Weather," *Fortune,* **LXXXIX,** February, 1974, p. 92. The original quote referred to "a billion people starving," but Reid Bryson told me he prefers "a half billion."

Part II

Report of the Panel on Climatic Variation, *Understanding Climatic Change,* National Academy of Sciences, Washington, D.C., 1975, pp. 1–2.

Chapter 3

Report of the Panel on Climatic Variation, *Understanding Climatic Change,* National Academy of Sciences, Washington, D.C., 1975, p. 127.

F. K. Hare quote is from: Tom Alexander, "Ominous Changes in the World's Weather," *Fortune,* **LXXXIX,** February, 1974, p. 152.

Louis M. Thompson, "Weather Variability, Climatic Change, and Grain Production," *Science,* **188,** 1975, p. 535.

Chapter 4

Lester R. Brown, with Erik P. Eckholm, *By Bread Alone,* Praeger, New York, 1974, p. 61.

Louis M. Thompson, *Iowa Agriculture, World Food Needs and Educational Response,* Center for Agricultural and Economic Development, Report No. AAC156, Iowa State University, Ames, Iowa, 1965, p. 2.

James McQuigg is quoted from: Tom Alexander, "Ominous Changes in the World's Weather," *Fortune,* **LXXXIX,** February, 1974, p. 146.

Chapter 5

Report of the Study of Man's Impact on Climate, *Inadvertent Climate Modification,* M.I.T. Press, Cambridge, Mass., 1971, p. 14.

Chapter 6

William W. Kellogg, "Mankind as a Factor in Climatic Change," in *The Energy Question: An International Failure of Policy. Volume 1.*

The World, edited by Edward W. Erickson and Leonard Waverman, University of Toronto Press, Toronto, 1974, p. 253.

Resolution No. 75–5, January 17, 1975, North American Interstate Weather Modification Council, Vern W. Butler, Chairman. The resolution was proposed and adopted at the annual meeting, Denver, Colorado, January 17, 1975.

Allen L. Hammond and Thomas H. Maugh II, *Science,* **186,** 1974, pp. 335 and 337.

Chapter 7

The Michael Reinbold quote is from: *Rocky Mountain News,* November 20, 1974, p. 17.

Howard Taubenfeld, Professor of Law at the Southern Methodist University, made this remark to the Weather Modification Council in January, 1975, at their meeting in Denver, Colorado.

Robert G. Fleagle, "Preface," in Robert G. Fleagle, James A. Crutchfield, Ralph W. Johnson, and Mohamed F. Abdo, *Weather Modification in the Public Interest,* University of Washington Press, Seattle, 1974, p. viii.

Part III

Walter Orr Roberts, *A View of Century 21,* Claremont University Center, Claremont, Calif., 1969, p. 27.

Chapter 8

Pope Paul VI is quoted from: *Boulder Daily Camera* (AP), November 10, 1974, p. 14.

Secretary of Agriculture Butz is quoted from: *Chicago Daily News,* November 29, 1974, p. 2.

Chapter 9

Martin Luther King, *Where Do We Go From Here: Chaos or Community?,* 1967.

Report of the Workshop on Tropical Ecology, *Fragile Ecosystems,* edited by E. A. Farnworth and F. B. Golley, Springer-Verlag, New York, 1974, p. 226.

John P. Holdren, Energy and Prosperity, *Bulletin of the Atomic Scientists,* January, 1975, p. 28.

Notes and References

Preface

1. Goodfriend, A., *The Only War We Seek,* Farrar, Straus, and Young, Inc., for Americans for Democratic Action, Washington, D.C., 1951, p. 1.
2. Schneider, S. H., "Specter at the Feast," *The National Observer,* July 6, 1974, p. 18.

Chapter 1

1. *Bartlett's Quotations* (14th ed.) says that in actual fact this famous quote was said by Charles Dudley Warner in an editorial in the *Hartford Courant,* August 24, 1897, p. 733.

2. No one knows exactly how many people suffer from malnutrition. U.S. Secretary of State Kissinger mentioned that "one billion people suffer from malnutrition" in his April 15, 1974, address to the United Nations General Assembly. In *The End of Affluence* (Ballantine Books, New York, 1974), p. 259, Paul and Anne Ehrlich cite a number of references on this issue. Also, an article by Harold Schmeck in *The New York Times,* October 6, 1974, is a good update.
3. A point he has made at numerous talks and public appearances.
4. Report of the Study of Man's Impact on Climate (SMIC), *Inadvertent Climate Modification,* MIT Press, Cambridge, Mass., 1971, Fig. 3.8.
5. Reported by R. A. Bryson, *Institute of Environmental Studies Report 27,* University of Wisconsin, November, 1975.
6. The issue of increased variability is discussed by J. M. Mitchell, Jr., in *GARP Publication Series 16,* World Meteorological Organization, Geneva, 1975. However, the issue of a "cooling trend" is still controversial, and it may well turn out that as recent data are analyzed, no continuation of the cooling trend will be apparent, as discussed in chapter 3.
7. Gilman presented this result to the 140th meeting of the American Association for the Advancement of Science in San Francisco, February 27, 1974.
8. John Steinbeck's *Grapes of Wrath* is one example of a fictional account of hardship related to the Dust Bowl.
9. Brown, L. R., with Eckholm, E., *By Bread Alone,* Praeger, New York, 1974, p. 60, was the original source, but these figures have recently been revised by Brown, using new USDA data. The figures indicate that as of early 1976 the grain reserves situation was still about the same as the precarious mid-1974 figures and, if anything, a bit worse. However, the revised USDA figures have been challenged as showing too much reserves.
10. Another danger of low food reserves is wildly fluctuating prices, which are particularly difficult for food-needy, poor countries with little foreign exchange cushion.
11. The USDA confidently predicted a record corn crop in the U.S. in 1974 up through mid-July, despite the well-publicized difficulties farmers experienced in planting in the spring. As severe drought set in during early July, the USDA continued to predict record corn crops. For example, according to an Associated Press story in the August 13, 1974, *International Herald Tribune,* the USDA projected a "minimum" crop of 5,949,950,000 bushels on July 25, but admitted by August 1 that the crop would be 984 million bushels less than their previous week's estimate. Similarly optimistic projections for soybeans were also drastically reduced in August, and again in September, as described in *The New York Times (AP),* November 9, 1974, p. 41.
12. A term coined by Ehrlich and Ehrlich, *End of Affluence.*
13. A number of references to this question can be obtained in S. H. Schneider, *Journal of the Atmospheric Sciences,* "On the Carbon Dioxide–Climate Confusion," **32,** 2060–2066, 1975.

14. Report of the Climatic Impact Committee, *Environmental Impact of Stratospheric Flight,* National Academy of Sciences, Washington, D.C., 1975.
15. This has been suggested by Harvard photochemist Michael McElroy.
16. See the discussion in chapters 6 and 7 for details and references.
17. Some estimates even range as high as 1,000 percent, as discussed later.
18. See in this connection the book *Small is Beautiful* by E. F. Schumacher, Harper and Row, New York, 1973.
19. Holdren, J. P., *The New York Times,* Op. Ed. column, July 23, 1975; see also Hirst, E., and Moyers, J. C., *Science,* **179,** 1299–1304, 1974.
20. Light, not heat, is what is seen from space at night. However, regions where many lights burn are also usually regions of high energy consumption.
21. A. M. Weinberg and R. P. Hammond in a paper presented before the Fourth International Conference on the Peaceful Uses of Atomic Energy, Geneva, September 7, 1971, suggest that in 100 years a population of 15 billion "is quite likely" and 20 billion is "not impossible." W. Häfele, *American Scientist,* **62,** 438–447, 1974, uses similar numbers for a steady-state world where per capita energy consumption could be as high as 20 kilowatts.
22. Ehrlich, P. R., Ehrlich, A. H., and Holdren, J. P., in *Human Ecology,* W. H. Freeman, San Francisco, 1973, p. 214, present a clear discussion of the arithmetic of growth.
23. Private communication.
24. See the extensive discussion in chapters 6 and 7.
25. Roberts, W. O., *National Geographic,* April, 1972, p. 545.

Chapter 2

1. Magazine pages are often used as wall decorations throughout the world, and as such can serve repeatedly and colorfully as reminders of industrially produced affluence.
2. Rensberger, B., *The New York Times,* July 26, 1974, p. 35.
3. Thomas R. Malthus's *An Essay on the Principle of Population,* excerpted in *Introduction to Contemporary Civilization in the West,* Columbia University Press, New York, 1961, 3rd ed., p. 296.
4. Meadows, D. H., Meadows, D. L., Randers, J., and Behrens, W. W., III, *The Limits to Growth,* Potomac Associates, Washington, D.C., 1972.
5. Maddox, J., *The Doomsday Syndrome,* McGraw-Hill, New York, 1972, p. 83.
6. Ehrlich, P. R., *The Population Bomb,* Ballantine Books, New York, 1971.
7. *Ibid.,* p. xi.
8. Maddox, J., *Doomsday Syndrome,* p. 3.
9. Ehrlich, P. R., and Ehrlich, A. H., *The End of Affluence,* Ballantine Books, New York, 1974, p. 11.
10. A list of their books can be found in *The End of Affluence.*

11. See the arguments in chapter 8 for an explanation of this crucial detail.
12. Paddock, W., and Paddock, P., *Famine 1975!*, Little, Brown, and Co., Boston, 1967, p. 230.
13. Paddock, W. and Paddock, E., *We Don't Know How,* Iowa State University Press, Ames, 1973, p. 250.
14. *Ibid.,* p. 248.
15. Estimates range considerably as pointed out by Paul and Anne Ehrlich in *Population Resources Environment,* W. H. Freeman and Co., San Francisco, 1972, 2nd ed., p. 114.
16. Farnworth, E. G., and Golley, F. B., eds., *Fragile Ecosystems,* Springer-Verlag, New York, 1974.
17. Brown, L. R., with Eckholm, E. P., *By Bread Alone,* Praeger, New York, 1974, p. 51.
18. *The New York Times,* June 21, 1974.
19. Gofman, J. W., and Tamplin, A. R., *Poisoned Power: The Case Against Nuclear Power Plants,* Rodale Press, Emmaus, Pa., 1971. Several thousand scientists and nearly a dozen Nobel prizewinners signed an antinuclear statement prepared by John T. Esdall, James D. Watson, Henry W. Kendall, Harold C. Urey, and George B. Kistiakowsky, which was circulated under the name "Union of Concerned Scientists" and published in the *Bulletin of the Atomic Scientists* in late 1975. See also the collection of articles contained in the background report: *The Plutonium Economy: A Statement of Concern,* National Council of Churches of Christ in the U.S.A., New York, September, 1975.
20. Hans Bethe of Cornell University prepared a well-publicized defense of nuclear power that also was published in the *Bulletin of the Atomic Scientists,* **31,** 4–5, 1975, and also was signed by 32 scientists, including 11 Nobel prizewinners. If the fact that distinguished nuclear physicists cannot agree on the safety of nuclear power plants leaves you a bit uneasy, then the message of this issue has been clearly perceived!
21. Ehrlich and Ehrlich, *End of Affluence,* pp. 73–74.
22. *Science,* **186,** 906–907, 1974.
23. See the reference list in John P. Holdren, "Hazards of the Nuclear Fuel Cycle," *Bulletin of the Atomic Scientists,* October, 1974, p. 23, and the entire September, 1975, issue of the *Bulletin of the Atomic Scientists.*
24. Schneider, S. H., and Dennett, R. D., "Climatic Barriers to Long-Term Energy Growth," *Ambio,* **4,** 65–75, 1975.
25. Malthus, T. R., *Principle of Population,* p. 296.
26. Ehrlich and Ehrlich, *End of Affluence,* p. 92–93.
27. A term coined in "The Specter at the Feast" by S. H. Schneider, *The National Observer,* July 6, 1974, p. 18.
28. Bendersky, D., Park, W. R., Shannon, L. J., and Franklin, W. E., "Resource Recovery from Municipal Wastes" in *Pollution Engineering and Scientific Solutions,* Barekette, E. S., ed., Plenum Publishing Co., New York, 1973, pp. 357–379. For a discussion of questions relating to recycling

see Glenn Seaborg, "Science, Technology and Development: A New World Outlook," *Science,* **184,** 13–19, 1973.
29. A number of individuals and groups (e.g., the Agribusiness Accountability Project) have made this charge, especially after the 1972 Russian wheat deal. See, for example, Trager, J., *The Great Grain Robbery,* Ballantine Books, New York, 1975.
30. Gelb, L. H., *International Herald Tribune,* July 29, 1974, p. 6 (condensed from a *New York Times* story).
31. This is exemplified by the Jeffersonian–Federalist feuds.
32. Tugwell, R., *Model for a New Constitution,* J. E. Freel and Associates, Palo Alto, Calif., 1970.
33. Pirages, D. C., and Ehrlich, P. R., *Ark II: Social Response to Environmental Imperatives,* W. H. Freeman and Co.,, San Francisco, 1974, p. 174.
34. That is, it would *seek* to inform people rather than just respond to requests.
35. Hardin, G., "Living on a Lifeboat," *Bioscience,* **24,** 561–568, 1974.
36. Boulding, K. E., "The Utterly Dismal Theorem" in *Population, Evolution and Birth Control,* Hardin, G., ed., W. H. Freeman, San Francisco, 1969, p. 81.
37. *Bartlett's Quotations* (14th ed.), p. 732.
38. *Ibid.*
39. Hardin, G., "Living on a Lifeboat," p. 563.
40. *Ibid.,* p. 562.
41. Rensberger, B., *The New York Times,* October 25, 1974.
42. Lewis, A., *The New York Times,* October 31, 1974, p. 4.
43. *Time,* November 11, 1974, p. 75.
44. Paddock, W., and Paddock, P., *Famine 1975!,* pp. 206–7.
45. The Global Survival Compromise proposed here is related to the "world survival bargain" urged by Columbia University Law Professor Richard N. Gardner, as described in an article by Clyde H. Farnsworth in *The New York Times,* November 11, 1974, p. 1.
46. Revelle, R., "Food and Population," *Scientific American,* **231,** September, 1974, p. 168.
47. Garrett Hardin, in "The Tragedy of the Commons," *Science,* **162,** 1243–1248, 1968, describes a metaphorical situation where overstressed commons could collapse.
48. An expression often used in this context by Lester Brown.
49. Cleveland, H., "The U.S. vs. the U.N.?" *The New York Times Magazine,* May 5, 1975.
50. That the poor nations should not follow the development patterns of the rich is one of the key principles of the debate over the New International Economic Order that took place at the U.N. in September, 1975. This is discussed at length in chapter 9.
51. Paddock, W., and Paddock, P., *Famine 1975!,* p. 229.
52. Dostoevsky, F., *The Brothers Karamazov,* Modern Library, New York, 1950, pp. 288 and 283.

53. *Ibid.*, p. 291.

Chapter 3

1. Walcott, R. I., *EOS: Transactions of the American Geophysical Union,* **56,** 62–72, 1975.
2. Dansgaard, W., Johnsen, S. J., Clausen, H. B., and Langway, C. C., *The Late Cenozoic Glacial Ages,* symposium edited by K. K. Turekian, Yale University Press, New Haven, 1971.
3. The reconstructions of Ice Age temperature from the "CLIMAP" project are described in detail in the Report of the Panel on Climate Variations, *Understanding Climatic Change,* National Academy of Sciences, Washington, D.C., 1975, mostly under the names John Imbrie, Andrew McIntyre, Robert Matthews, and James Hays.
4. Calder, N., *The Weather Machine,* Viking Press, New York, 1975; Kennett, J. P., and Huddlestun, P., "Abrupt Climatic Change at 90,000 Yr BP," *Quaternary Research,* **2,** 384–395, 1972.
5. "The Weather Machine," a two-hour science special produced by the British Broadcasting Corporation, was broadcast in the United States on February 24, 1975.
6. The Emiliani hypothesis is explained by Walter Sullivan in *The New York Times,* September 24, 1975, and is based in part on J. P. Kennett and N. J. Shackleton in *Science,* **188**, 147–150, 1975. The original Antarctic surge hypothesis is put forth by A. T. Wilson in *Nature,* **201**, 147–149, 1964, and more recent evidence comes from T. Hughes, *Journal of Geophysical Research,* **78,** 7884–7910, 1973, and H. Flohn, *Quaternary Research,* **4,** 385–404, 1974.
7. Bryson, R. A., *World Climate and World Food Systems III: The Lessons of Climatic History,* Institute of Environmental Studies, University of Wisconsin, Report 27, November, 1974, p. 3; also, Bryson, R. A., Irving, W. N., and Larsen, J. A., *Science,* **147**, 46–48, 1965.
8. In addition to the works of Lamb and Ladurie, which are later referred to extensively, other books document climatic history. For example, C. E. P. Brooks's *Climate Through the Ages,* Benn, London, 1949; also, Robert Clairborne's *Climate, Man and History,* Norton, New York, 1970.
9. Bryson, R. A., *Lessons of Climatic History,* p. 5.
10. Lamb, H. H., *The Changing Climate,* Methuen and Co. Ltd., London, 1966, p. 7.
11. Ladurie, E. L., *Times of Feast, Times of Famine,* translated by Barbara Bray, Doubleday and Company, Inc., Garden City, New York, 1971, p. 265.
12. Lamb, *Changing Climate,* p. 8.
13. Ladurie, *Times of Feast and Famine,* p. 47.
14. Lamb, *Changing Climate,* p. 10, and J. P. Williams Freeman in Lamb, *Antiquity,* p. 208, Gloucester, 1928.

15. *Ibid.*, p. 11.
16. *Ibid.*
17. Shen Wen-hsiung, Changes in China's Climate, *China Pictorial,* **4,** 30–33, 1974; reprinted in *Bulletin of the American Meteorological Society,* **55,** 1974, pp. 1347–1350.
18. Ladurie, E. L., *Times of Feast and Famine,* p. 67.
19. Hoskins, W. G., "Harvest Fluctuations and English Economic History (XVIth/XVIIIth Centuries)," *Agricultural History Review,* 1964, pp. 28–46; and 1968, pp. 15–31, quoted in Ladurie; see ref. 13.
20. Ladurie, E. L., *Times of Feast and Famine,* p. 285.
21. *Ibid.*, p. 90.
22. Trevelyan, G. M., *English Social History,* Longmans, London, 1942, p. 432, quoted in H. H. Lamb's *The Changing Climate,* p. 5.
23. *Ibid.*
24. Lamb, *Changing Climate,* p. 5.
25. Ladurie, E. L., *Times of Feast and Famine,* p. 92.
26. G. Utterström (1955), quoted in H. H. Lamb's *The Changing Climate,* p. 12.
27. Shelley, M., *Introduction to Frankenstein, or the Modern Prometheus,* quoted in E. L. Ladurie, *Times of Feast and Famine,* p. 65.
28. *Ibid.*, p. 65.
29. Text published by J. M. Desbordes, about 1965, as quoted by E. L. Ladurie, *Times of Feast and Famine,* p. 66.
30. Bryson, R. A., *World Climate and World Food Systems III: The Lessons of Climatic History,* Institute for Environmental Studies Report 27, November, 1974, p. 9.
31. Ladurie, E. L., *Times of Feast and Famine,* plates nos. XXI and XXII.
32. Work by Paul Damon (private communication) of the University of Arizona, Tucson, suggests that, if anything, the Southern Hemisphere has been *warming* up in the past decade. Furthermore, very recent evidence was presented to a 1975 Climate Symposium in Norwich, England, by British climatologist Robert Ratcliffe to the effect that warming has been replacing cooling in the Arctic.
33. The Atlantic cooling is reported by M. Rodewald in *Beilage zur Berliner Wetterkarte* (119/73 SO 29/73), Berlin, Inst. f. Met., Fr. Univ., October 23, 1973, as shown by H. H. Lamb in *The Current Trend of World Climate—a Report on the Early 1970s and a Perspective,* Climatic Research Unit, School of Environmental Sciences, University of East Anglia, Norwich, England, CRURP3, p. 5.
34. Stella Melugin Coakley, Department of Biology, Denver University, private communication.
35. The epoch analysis was performed by Bryson's colleague, John Kutzbach, as reported in Bryson, *Lessons of Climatic History,* p. 14.
36. Walter Orr Roberts interviewed in *U.S. News and World Report,* March 18, 1974, pp. 41–44.

37. Wade, N., "The Sahelian Drought: No Victory for Western Aid," *Science,* **185,** 1974, p. 234.
38. *Ibid.,* p. 235.
39. Glantz, M., "The Sahelian Drought: No Victory for Anyone," *Africa Today,* **22**, no. 2, April–June, 1975, pp. 57–61.
40. Sterling, C., *Atlantic Monthly,* May, 1974, p. 98.
41. Glantz, M. H., "Nine Fallacies of a Natural Disaster: Case of the Sahel" in *The Politics of a Natural Disaster: Case of the Sahel,* Glantz, M. H., ed., Praeger, New York, 1976. Glantz, now resident in Boulder, Colorado, is responsible for a commissioned study, "The Moral, Legal, Political, and Economic Implications of a Credible and Reliable Climate Forecast," sponsored by the International Federation of Institutes for Advanced Study (IFIAS) in Stockholm, Sweden.
42. Brown, L., with Eckholm, E., *By Bread Alone,* Praeger Publishers, New York, 1974, pp. 48–49.
43. Brooks, R. R., "People Versus Food," *Saturday Review,* September 5, 1970, p. 10, as quoted in Brown, with Eckholm, *By Bread Alone,* p. 48.
44. Tadashi Asakura, *I., Jokisho to Kankyo-osen,* Kyoritsu Shuppan, Tokyo, 1972. Quotes taken from the English translation, *Unusual Weather and Environmental Pollution,* U.S. Joint Publications Research Service, 1974, p. 14.
45. *Ibid.,* p. 30.
46. *Ibid.,* p. 28.
47. Reid A. Bryson, private communication.
48. Lydia Dotto, science writer for the *Toronto Globe and Mail* wrote an excellent series of articles on climate-related issues in later 1974. These articles have been put together into a pamphlet, *Climatic Change,* Atmospheric Environment Service, Downsview, Ontario, p. 3.
49. Interdepartmental Committee for Atmospheric Sciences, *Report of the Ad Hoc Panel on the Present Interglacial,* Federal Council for Science and Technology, National Science Foundation, August, 1974, report no. ICAS 18B-FY75, p. 19.
50. *Ibid.*
51. Brown with Eckholm, *By Bread Alone,* pp. 152–153.
52. *Ibid.,* p. 156.
53. Oceanographer Dana Thompson, of JAYCOR Inc., Alexandria, Virginia, pointed this out to me.
54. Brown with Eckholm, *By Bread Alone,* p. 157.
55. Kukla, G. J., and Kukla, H. J., "Increased Surface Albedo in the Northern Hemisphere," *Science,* **183,** 709–714, 1974.
56. See ref. 11, chap. 1.
57. Report to the National Oceanic and Atmospheric Administration, *The Influence of Weather and Climate on United States Grain Yields: Bumper Crops or Drought,* U.S. Dept. of Commerce, Washington, D.C., December, 1973; see the discussion in chapter 4 for more details.

58. "Report on 'The Influence of Weather and Climate on United States Grain Yields,' " enclosed in a communication to U.S. Secretary of Commerce Fredrick B. Dent from U.S. Secretary of Agriculture Earl Butz, February 1, 1974, as reported in L. M. Thompson, *Science,* **188**, p. 535, 1975.
59. See the discussion of this in chapter 4.
60. For example, see LaMarche, Valmore C., Jr., in "'Paleoclimatic Inferences from Long Tree-Ring Records," *Science,* **183**, 1043–1048, 1974; also, see Jeffrey S. Dean, *Chronological Analysis of Tsegi Phase Sites in Northeastern Arizona,* papers of the Laboratory of Tree-Ring Research Number 3, University of Arizona Press, Tucson, 1969.
61. Contained in the Japan Meteorological Agency "Report of a Study of Recent Unusual Weather and Climatic Trends in the World and the Outlook for the Future," as summarized in *GARP Publication Series* 16, World Meteorological Organization, Geneva, 1975.
62. There was a severe drought in the Cornbelt in 1974, and a moderate one in Iowa in 1975. However, two years do not constitute a strong drought cycle, and the weather in the late 1970s will determine if there is to be another strong showing of the 22-year cycle.

Chapter 4

1. Brown, L. R., with Eckholm, E. P., *By Bread Alone,* Praeger Publishers, New York, 1974, p. 61; and Brown, L. R., "The Global Politics of Food: Role and Responsibility of North America," Marfleet-Falconer Lecture, University of Toronto, September 24, 1975, p. 5.
2. *Ibid.,* p. 61.
3. In "The Global Politics of Food" (p. 3), Lester Brown revised his and Eckholm's estimates given in *By Bread Alone,* p. 60, using new data from the USDA. The new estimates are given here.
4. See ref. 2 of chap. 1.
5. Brown with Eckholm, *By Bread Alone,* p. 59.
6. Robbins, W., *The New York Times,* September 5, 1974, p. 1.
7. Since there is no real "surplus" of food in a world with a billion or more malnourished people, it is preposterous to call the economic excess grain in the rich countries a world "surplus."
8. For example, Brown with Eckholm, *By Bread Alone,* pp. 35–44.
9. Workshop Conference, "World Food Supply in a Changing Climate," Sterling Forest, New York, December 2–5, 1974. Details and conference proceedings can be obtained from Dr. George Kukla, Lamont-Doherty Geological Observatory, Columbia University, Palisades, New York; James McQuigg, paper prepared for the conference.
10. Brody, J. B., *The New York Times,* October 11, 1974, p. 41.
11. Wayne Decker, Professor of Atmospheric Science at the University of Missouri, Columbia, in a paper, "The Climatic Impact of Variability in World Food Production," prepared for the 1973 Annual Meeting of the

American Association for the Advancement of Science, San Francisco, reprinted in the *American Biology Teacher,* **36**, 534–540, December, 1974.
12. *Ibid.,* p. 6.
13. *Ibid.,* p. 8.
14. Wallace, H. A., "Mathematical Inquiry into the Effects of Weather on Corn Yields in Eight Corn Belt States," *Monthly Weather Review,* **48,** 439–446, 1920.
15. Decker, "Variability in World Food Production," p. 14.
16. Committee Report to the National Oceanic and Atmospheric Administration, *The Influence of Weather and Climate on United States Grain Yields: Bumper Crops or Droughts,* U.S. Dept. of Commerce, Washington, D.C., 1973.
17. *Ibid.,* pp. 1–2.
18. *Ibid.,* p. 23.
19. *Ibid.,* p. 4.
20. *Ibid.,* p. 23.
21. "Report on 'The Influence of Weather and Climate on United States Grain Yields,'" enclosed in a communication dated February 1, 1974, from U.S. Secretary of Agriculture Earl L. Butz to U.S. Secretary of Commerce Fredrick B. Dent.
22. *Ibid.*
23. Letter from Louis M. Thompson to Earl L. Butz, February 13, 1974.
24. Letter from Bruce M. Graham to Louis M. Thompson, March 1, 1974.
25. Letter fom Louis M. Thompson to Bruce M. Graham, March 14, 1974.
26. *Ibid.*
27. According to *The New York Times* (AP), November 9, 1974, p. 41, original USDA projections for 1974 corn yields were 6.7 billion bushels, but surveys as of November 1, 1974, showed a harvest of only 4.62 billion bushels, more than a 30 percent reduction!
28. *The World Food Situation and Prospects to 1985,* Foreign Agricultural Economic Report No. 98, U.S. Dept. of Agriculture, Washington, D.C., 1974, p. 73.
29. Thompson, L. M., "Weather Variability, Climatic Change, and Grain Production," *Science,* **188**, p. 535, 1975. This article is a landmark as a prototype of how food–climate problems can be studied quantitatively. Furthermore, this issue of *Science* (May 9, 1975) is a gold mine of information on current world food problems.
30. *Ibid.,* p. 536.
31. Gordon McKay, paper prepared for Sterling Forest Conference (see ref. 9).
32. *Ibid.*
33. *Ibid.*
34. *Ibid.*
35. Winstanley, D., Emmett, B., and Winstanley, G., *Climatic Changes and the World Food Supply,* preprint, p. 25 (to be published as Report No. 5, Environment Canada, Planning and Finance Service).

36. See, for example, James Trager's *The Great Grain Robbery*, Ballantine Books, New York, 1975.
37. Ehrlich, P. R., and Ehrlich, A. H., *Population Resources Environment*, 2nd ed., Freeman, San Francisco, 1972, p. 114.
38. Decker, "Variability in World Food Production," pp. 9–10.
39. Paddock, W., and Paddock, P., *Famine 1975!*, Little, Brown, and Co., Boston, 1967.
40. Anthony S. Rojko, U.S.D.A. analyst, in "Food Production and Demand Prospects to 1985," working materials presented to the Sterling Forest Conference (see ref. 9) as background on the USDA study (ref. 28).
41. Their economic assumptions that food supply and demand would be balanced were strongly challenged by RAND Corporation economist Cheryl Cook.
42. L. M. Thompson, "Weather Variability," p. 538.
43. Mark Twain's *Life on the Mississippi*, 1883.
44. L. M. Thompson, "Weather Variability," pp. 538–539.

Chapter 5

1. Thorarinsson, S. and Vonnegut, B., *Bulletin of the American Meteorological Society*, **45**, p. 440, 1964.
2. A more detailed description of the "greenhouse effect" can be obtained in S. H. Schneider and W. W. Kellogg, "The Chemical Basis for Climate Change," in *Chemistry of the Lower Atmosphere*, edited by S. I. Rasool, Plenum Press, New York, 1973, pp. 203–249.
3. A. T. Wilson, in "Origin of Ice Ages: an Ice Shelf Theory for Pleistocene Glaciation," *Nature*, **201**, January 11, 1964, first proposed an ice age theory based on the rapid slippage of a part of the huge Antarctic ice sheet into the oceans. See also the article by W. W. Kellogg, "Climatic Feedback Mechanisms Involving the Polar Regions," *Proceedings of Climate of the Arctic*, University of Alaska Press, Fairbanks, 1975, for more details and references to the role of the poles in climate theory.
4. Taken from the discussion in S. H. Schneider, "A New World Climate Norm?: Implications for Future World Needs," *Bulletin of the American Academy of Arts and Sciences*, **XXVIII**, 20–35, December, 1975.
5. In fact, one of the chief problems with most solar-climatic change theories is that they are based on statistical correlations rather than physical theories. Obviously, more emphasis on the latter is essential if these correlations are to be taken fully seriously by the scientific community.
6. Schneider, S. H., and Mass, C., "Volcanic Dust, Sunspots and Temperature Trends," *Science*, **190**, 741–746, 1975.
7. The classical work on this theory is M. Milankovitch's *Kanon der Erdbestrahlung und Seine Anwendung auf das Eiszeit problem* ("Cannon of Insolation and the Ice Age Problem"), Royal Serbian Academy (Program for Scientific Translations, Jerusalem, 1941). More recently, A. L. Berger

has improved Milankovitch's calculations, as presented in the proceedings of World Meteorological Organization Symposium on Climate Fluctuations in Norwich, England, August, 1975.

8. Reid Bryson, Hubert Lamb, and J. Murray Mitchell, Jr., are among the most prominent of these.
9. Schneider, S. H., and Mass, C., "Volcanic Dust,"; Pollack, J. B., Toon, O. W., Sagan, C., VanCamp, W., Summers, A., and Baldwin, B., "Role of Volcanic Explosions in Climatic Change," *Journal of Geophysical Research,* 1976, presents detailed calculations on the subject.
10. Lorenz, E. N., "Climatic Determinism," *Meteorological Monographs,* **8,** p. 28, 1968.
11. Syukuro Manabe and R. T. Wetherald in *Journal of the Atmospheric Sciences,* **24**, 241–259, 1967; other results are discussed by S. H. Schneider in *Journal of the Atmospheric Sciences,* **32,** 2060–2066, 1975.
12. Hugh W. Ellsaesser, meteorologist at the Lawrence Livermore Laboratory, often bristles at what he considers the one-sidedness of those who point to all climatic modification as being detrimental when, clearly, some changes could be beneficial. See his article, "Where Are We Now in Air Pollution," UCRL-76394 preprint, Lawrence Livermore Laboratory, University of California, Livermore, Calif., February, 1975, p. 8.
13. Project ASTRA at the University of Washington, Seattle, had obtained such results in the early 1970s; see, for example, Paul W. Hodge in *Nature,* **229,** 549, 1971.
14. W. W. Kellogg, J. A. Coakley, Jr., and G. W. Grams, in "Effects of Anthropogenic Aerosols on the Global Climate," in the *Proceedings of the World Meteorological Organization Symposium on Climatic Fluctuations,* Norwich, England (World Meteorological Organization, Geneva, August, 1975), discuss the extent of aerosols from industry. Wayne M. Wendland and Reid A. Bryson, in "Atmospheric Dustiness, Man and Climatic Change," *Biological Conservation,* **2**, 125–128, 1970, point out the role of dust from overgrazing but P. B. Russell and G. W. Grams, *Journal of Applied Meteorology,* **14,** 1037–1043, 1975, point out that even these kinds of dust could warm the climate.
15. See the discussion in chapter 6 for details and references.
16. Otterman, J., "Barring High Albedo Soils by Overgrazing: A Hypothesized Desertification Mechanism," *Science,* **186**, 531–533, 1974.
17. This photo is described by Nicholas Wade in "Sahelian Drought: No Victory for Western Aid," *Science,* **185,** 234–237, 1974. It was sent to me through the courtesy of W. Bandeen and W. Nordberg, Goddard Space Flight Center, NASA, Greenbelt, Maryland.
18. Bryson, R. A., *World Climate and Food Systems III: The Lessons of Climatic History,* IES Report 27, University of Wisconsin, Madison, 1974.
19. Winstanley, D., Emmett, B., and Winstanley, G., *Climatic Changes and the World Food Supply,* Environment Canada, Planning and Finance Report No. 5, 1976, preprint pp. 4–14.

20. Wendland and Bryson, "Atmospheric Dustiness."
21. Charney, J., Stone, P. H., and Quirk, W. J., "Drought in the Sahara: A Biogeophysical Feedback Mechanism," *Science,* **187**, 434–435, 1975.
22. *Ibid.,* Fig. 1.
23. A point Schnell has emphasized at numerous colloquia. Formal publications are in process.
24. In this connection, philosophers Morris R. Cohen and Ernest Nagel, in *An Introduction to Logic and Scientific Method,* Harcourt, Brace, New York, 1934, pp. 391–392, said: "It is idle to collect facts unless there is a problem upon which they are supposed to bear." The expense of collecting data on the atmosphere suggests adherence to this idea. The most troublesome problem in this connection occurs since we are not yet sure exactly what data are needed to understand climatic change.
25. Schneider, S. H., and Dickinson, R. E., "Climate Modeling," *Reviews of Geophysics and Space Physics,* **12**, 447, 1974.
26. Ellsaesser, "Air Pollution."
27. This is a point made often by Carl Sagan, an astronomer at Cornell University, and is discussed in his popular book *The Cosmic Connection,* Dell Publishing Co., New York, 1975.

Chapter 6

1. Holdren, J. P., "Energy and Prosperity," *Bulletin of the Atomic Scientists,* **30**, 26–28, 1975.
2. Landsberg, H., "Inadvertent Atmospheric Modification through Urbanization," in *Weather and Climate Modification,* edited by Wilmot N. Hess, Wiley, New York, 1974, pp. 726–763.
3. Report of the Study of Man's Impact on Climate (SMIC), *Inadvertent Climate Modification,* MIT Press, Cambridge, Mass., 1971, pp. 58–59.
4. *Ibid.*
5. A recent survey article on the toxicity of particles in the urban atmosphere is by David F. S. Natusch and John R. Wallace: "Urban Aerosol Toxicity: The Influence of Particle Size," *Science,* **186**, 695–699, 1974.
6. SMIC, *Inadvertent Climate Modification,* chapter 8.
7. *Man's Impact on Climate,* edited by W. Matthews, W. W. Kellogg, and G. D. Robinson, MIT Press, Cambridge, Mass., 1971, chapters 24 and 26.
8. Schneider, S. H., and Kellogg, W. W., "The Chemical Basis for Climate Change," chapter 5, in *Chemistry of the Lower Atmosphere,* edited by S. I. Rasool, Plenum Press, New York, 1973, p. 218.
9. Guccione, E., *The New York Times,* August 28, 1971, p. 25; Ellsaesser, H. W., "The Upward Trend in Airborne Particles That Isn't," in *The Changing Global Environment,* edited by S. F. Singer, D. Reidel, Dordrecht, 1975, pp. 235–269.
10. Letter to the editor of *The New York Times* in response to E. Guccione (ref. 9): S. H. Schneider, September 16, 1971, p. 42.

11. Monitoring and Air Quality Trends Report EPA 450/1-73-004, Environmental Protection Agency, Research Triangle Park, North Carolina, 1972.
12. Schneider and Kellogg, "The Chemical Basis."
13. Kellogg, W. W., Coakley, J. A., Jr., and Grams, G. W., "Effect of Anthropogenic Aerosols on the Global Climate," Fig. 1, p. 329, in *Proceedings of the WMO/IAMAP Symposium on Long-Term Climatic Fluctuations,* Norwich, England, August 18–23, 1975, WMO-No. 421 (Secretariat of the World Meteorological Organization, Geneva, 1975).
14. See the figure from cloud physicist Sean Twomey on p. 229 of *SMIC, Inadvertent Climate Modification;* P. V. Hobbs, L. F. Radke, and S. W. Shumway in *Journal of the Atmospheric Sciences,* **27**, 81–89, 1970; and P. V. Hobbs, H. Harrison, and E. Robinson in *Science,* **183,** 909, 1974.
15. Changnon, S. A., Jr., "The La Porte Weather Anomaly—Fact or Fiction," *Bulletin of the American Meteorological Society,* **49**, 4–11, 1968.
16. Huff, F. A., and Changnon, S. A., Jr., "Climatological Assessment of Urban Effects on Precipitation," *Conference on Air Pollution Meteorology,* Raleigh, N. C., April 5–9, 1971 (American Meteorological Society, Boston, 1971), pp. 98–103; also published as Huff, F. A., and Changnon, S. A., Jr., "Climatological Assessment of Urban Effects on Precipitation at St. Louis," *Journal of Applied Meteorology,* **11**, 823–842, 1972.
17. Landsberg, "*Inadvertent Atmospheric Modification.*"
18. *Ibid.*
19. *Ibid.,* pp. 759–760.
20. See the discussion of E. N. Lorenz's concepts of climatic determinism in *Understanding Climatic Change,* National Academy of Sciences, Washington, D.C., 1975, pp. 29–33.
21. See Ehrlich, P. R., and Ehrlich, A. H., *The End of Affluence,* Ballantine Books, New York, 1974, pp. 128–136, for additional references.
22. Sellers, W. D., *Physical Climatology,* University of Chicago Press, Chicago, 1965, p. 179.
23. Revelle, R., "Food and Population," *Scientific American,* **231**, 168, September, 1974.
24. *Ibid.*
25. The importance of heating anomalies at the earth's surface in the tropics on the climate both in the tropics and elsewhere has been shown quantitatively in the computer experiments of Robert M. Chervin and Paul Julian at The National Center for Atmospheric Research, Boulder, Colorado; and by Peter R. Rowntree in *Quarterly Journal of the Royal Meteorological Society,* **98**, 290–321, 1972.
26. *Fragile Ecosystems,* edited by E. A. Farnworth and F. B. Golley, Springer–Verlag, New York, 1974.
27. Ehrlich, P. R., and Ehrlich, A. H., *End of Affluence,* p. 133. The problem of poor soils in tropical forests motivated the Ehrlichs' remark, as discussed in chapter 8.

28. Schneider, S. H., and Dennett, R. D., "Climatic Barriers to Long-Term Energy Growth," *Ambio,* **4**, 65–74, 1975.
29. SMIC, *Inadvertent Climate Modification,* chapter 6; Schneider and Kellogg, "The Chemical Basis."
30. Weinberg, A. M., and Hammond, R. D., "Global Effects of Increased Use of Energy," in *Proceedings of Fourth International Conference on Peaceful Uses of Atomic Energy,* Geneva, September 7, 1971; Häfele, W., "A Systems Approach to Energy," *American Scientist,* **62**, 438–447, July–August, 1974.
31. Weinberg and Hammond, "Increased Use of Energy."
32. Häfele, "Systems Approach."
33. Both Alvin Weinberg and Wolf Häfele have been commendably active in attempting to evaluate the climatic consequences of energy production as part of their analyses of future energy prospects, even though their optimistic projections are not at all universally agreed to be feasible or desirable.
34. A sobering analysis of the staggering problem of water resources needed to meet optimistic projections for future food and energy consumption for a growing world population is presented by Malin Falkenmark and Gunnar Lindh in "How Can We Cope with the Water Resources Situation by the Year 2015?," *Ambio,* **3**, 114–122, 1974.
35. Häfele, "Systems Approach," p. 445.
36. *Ibid.*
37. Zener, C., "Solar Sea Power," *Physics Today,* 48–53, January, 1973. Perhaps one might minimize disruption to the heat flow in the Gulf Stream if heat dumped by power parks into the current were later removed downstream by solar sea power generators. Of course, very careful studies would be an *absolute prerequisite* to any such operational scheme.
38. This idea has been discussed and analyzed for many years by meteorologist Jerome Namias, now at the Scripps Institution of Oceanography.
39. Kutzbach, J. E., unpublished manuscript.
40. See the review article by Joseph Smagorinsky, "Global Atmospheric Modeling and the Numerical Simulation of Climate," in *Weather and Climate Modification,* edited by Wilmot N. Hess, Wiley, New York, 1974, pp. 632–686.
41. Warren M. Washington, *Journal of Applied Meteorology,* **11,** 768, 1972, performed a prototype numerical experiment on thermal pollution. R. M. Chervin and S. H. Schneider have prepared a series of papers on the statistical design of GCM experiments, and these papers appear in the *Journal of the Atmospheric Sciences* in 1976.
42. Manabe's model has been used to estimate the response of the surface temperature to energy increases outside the earth's atmosphere. Although this can be made proportional to the effects of heat released at the earth's surface for mean global conditions, it will, as pointed out by Schneider

and Dennett in "Climatic Barriers," seriously underestimate the effects of thermal pollution in regions where near-surface atmospheric stability conditions persist.

43. Thorarinsson, S., and Vonnegut, B., *Bulletin of the American Meteorological Society,* **45**, 440, 1964.
44. See "Atmospheric Implications of Energy Alternatives," National Center for Atmospheric Research, Boulder, Colorado, February, 1975, pp. 11–13.
45. The need for tax help and research funds for "exotic" energy sources is an applicable principle for most alternative energy technologies. However, such help is certainly no guarantee for success, but probably is essential if any success is to be achieved.
46. Odum, H. T., "Energy, Ecology, and Economics," *Ambio,* **2**, 220–227, 1973.
47. Schneider and Dennett, "Climatic Barriers"; footnote 29 contains a private communication from Harvey Brooks.
48. Manne, A. S., and Marchetti, C., *Hydrogen: Mechanism and Strategies of Market Penetration,* International Institute for Applied Systems Analysis Research Report No. RR-74-4, Schloss Laxenburg, Austria, March, 1974.
49. Daniels, F., *Direct Use of the Sun's Energy,* Yale University Press, New Haven, Conn., 1964.
50. Reed, J. W., "Wind Power Climatology," *Weatherwise,* **27**, 237–242, 1974; also Clark, W., *Energy for Survival,* Anchor Books, Garden City, New York, 1974.
51. See notes 32 and 33 in Schneider and Dennett, "Climatic Barriers."
52. Kellogg, W. W., and Schneider, S. H., "Climate Stabilization: For Better or for Worse?," *Science,* **186**, 1163–1172, 1974.
53. Machta, L., and Telegadas, K., "Inadvertent Large-Scale Weather Modification," in *Weather and Climate Modification,* edited by Wilmot N. Hess, Wiley, New York, 1974, p. 697.
54. Bacastow, R., and Keeling, C. D., in *Carbon and the Biosphere,* edited by G. M. Woodwell and E. J. Pecan, USAEC CONF-720510, National Technical Information Service, Springfield, Va., 1973, p. 86.
55. Hoffert, M. I., "Global Distribution of Atmospheric Carbon Dioxide in the Fossil-Fuel Era: A Projection," *Atmospheric Environment,* **8**, 1225–1249, 1974. Of course, all these models depend on one very *nonscientific* assumption: How much fossil fuel will humans burn in the future?
56. Schneider, S. H., "On the Carbon Dioxide Climate Confusion," *Journal of the Atmospheric Sciences,* **32**, 2060–2066, 1975.
57. SMIC, *Inadvertent Climate Modification,* p. 189.
58. Kellogg, Coakley, and Grams, "Anthropogenic Aerosols."
59. Schneider and Kellogg, "The Chemical Basis."
60. See ref. 14.
61. Crutzen, P. J., "Estimates of Possible Variations in Total Ozone Due to Natural Causes and Human Activities," *Ambio,* **3**, 201–210, 1974; Stolar-

ski, R. S., and Cicerone, R. J., *Canadian Journal of Chemistry,* **52**, 1610, 1974; Wofsy, S., and McElroy, M. B., *Canadian Journal of Chemistry,* **52**, 1569, 1974.

62. Crutzen, P. J., "Variations in Total Ozone," p. 201.
63. Ramanathan, V., Callis, L. B., and Boughner, R. E., in the *Journal of the Atmospheric Sciences,* 1976.
64. Ramanathan, V., "Greenhouse Effect due to Chlorofluorocarbons: Climatic Implications," *Science,* **190,** 50–52, 1975.
65. Report of Federal Task Force on Inadvertent Modification of the Stratosphere (IMOS), *Fluorocarbons and the Environment,* Council on Environmental Quality/Federal Council for Science and Technology GPO Stock No. 038–000–00226–1, Washington, D.C., 1975, pp. 9–10.
66. The possibility that some agent other than UV can account for this latitudinal dependence cannot yet be ruled out. For example, atmospheric chemist Edward Martell, in a private communication, has suggested the possibility that radioactive particles on plants could account in part for this trend. Obviously, more research is urgently needed.
67. Bronowski, J., *The Ascent of Man,* Little, Brown, and Company, Boston, 1973, p. 50.
68. Halstead Harrison published one of the first articles on potential SST exhaust dangers, *Science,* **170**, 734–736, 1970.
69. Johnston, H. S., "Reduction of Stratospheric Ozone by Nitrogen Oxide Catalysts from SST Exhaust," *Science,* **173**, 517–522, 1971.
70. Crutzen, P. J., "The Influence of Nitrogen Oxides on the Atmospheric Ozone Content, *"Quarterly Journal of the Royal Meteorological Society,* **96**, 320–325, 1970.
71. Crutzen, P. J., "Variations in Total Ozone," p. 207.
72. Grobecker, A. J., Coroniti, S. C., and Cannon, R. H., Jr., *Report of Findings: The Effects of Stratospheric Pollution by Aircraft,* Department of Transportation/Climatic Impact Assessment Program, Washington, D.C., December 1974.
73. Donahue, T. M., "The SST and Ozone Depletion," *Science,* **187**, p. 1142, 1975.
74. Grobecker, A. J., *Science,* **187,** p. 1145, 1975.
75. Quote appeared in Michael Drosnin's "Not with a Bang but with a Psssst!," *New Times,* March 7, 1975, p. 28.
76. Crutzen, P. J., "Variations in Total Ozone," p. 708.
77. "Human Impact on Ozone Studied at NCAR," *NCAR Staff Notes,* **10**, 1, January 24, 1975. A later calculation (see ref. 65, p. 27) of Crutzen's suggests that this curve may overestimate the effect somewhat, but it is not far off and illustrates the major point: Maximum ozone decrease will occur about ten years *after* CFC production is cut off.
78. Hammond, A. L., and Maugh, T. H., II, in *Science,* **186**, 335–338, 1974.
79. Climatic Impact Committee, *Environmental Impact of Stratospheric Flight,* National Academy of Sciences, Washington, D.C., p. 49.

80. Chemical Specialties Manufacturers Association, *Chemunicator,* No. C004-74, November 25, 1974.
81. Wilkniss, P. E., Swinnerton, J. W., Lamontagne, R. A., and Bressan, D. J., *Science,* **187,** 832–834, 1975.
82. Ref. 80, p. 3.
83. *Ibid.,* p. 4.
84. F. S. Rowland has suggested Freon 22 as a possible substitute, but it is quite doubtful that he shares Sobelev's "wait-and-see" attitude. Perhaps Ralph Cicerone's statement *(Science News,* **107,** 324, 1975) summarizes best the feeling of many prominent "scientists in the field": "Complete scientific proof to everyone's satisfaction will take years, so we are faced with a benefit-risk analysis. I have come to the reluctant conclusion that the risks are greater than the benefits, and the evidence is already strong."
85. Drosnin, "Not with a Bang," p. 29, quoted McElroy.
86. Crutzen, P. J., *Pure and Applied Geophysics,* **106–108,** 1385–1399, 1973.
87. Crutzen, P. J., "Variations in Total Ozone," p. 206.
88. Hahn, J., *Tellus,* **26,** 160, 1974.
89. Hardy, R. W. F., and Havelka, V. D., "Nitrogen Fixation Research: A Key to World Food," *Science,* **188,** 633–643, 1975.
90. Thompson, L. M., and Troeh, F. R., *Soils and Soil Fertility,* 3rd ed., McGraw-Hill, New York, 1973, chapter 10, excellent background material on the nitrogen cycle in the soil.
91. Hammond and Maugh, in *Science,* p. 338.
92. McElroy, M. B., *Chemical Processes in the Solar System: A Kinetic Perspective* (preprint to be submitted for publication), August, 1975, p. 87; McElroy discusses a class of potential chlorine sources in this comprehensive work on geochemistry.
93. *Ibid.,* p. 88.
94. Ehrlich, P. R., and Ehrlich, A. H., *Population Resources Environment,* 2nd ed., Freeman, San Francisco, 1972, pp. 243–245, a chilling account of various effects of thermonuclear warfare.
95. Committee to Study the Long-Term Worldwide Effects of Multiple Nuclear-Weapons Detonations, *Long Term Worldwide Effects of Multiple Nuclear Detonations,* National Academy of Sciences, Washington, D. C., 1975.
96. Kellogg, W. W., and Schneider, S. H., "Climate Stabilization."
97. Fletcher, J. O., *The Heat Budget of the Arctic Ocean and Its Relation to Climate,* Report R-444-PR, RAND Corporation, Santa Monica, Calif., 1965.
98. M. Warshaw and R. R. Rapp, in *Journal of Applied Meteorology,* **12,** 43, 1973.
99. Budyko, M. I., "The Future Climate," *EOS, Transactions of the American Geophysical Union,* **53,** 868, 1972.
100. Maykut, G. A., and Untersteiner, N., *Journal of Geophysical Research,* **76**, 1550, 1971.

101. This and literally dozens of other schemes to alter the climate are discussed seriously in N. Rusin and L. Flit, *Man Versus Climate,* translated from the Russian by Dorian Rottenberg, Peace Publishers, Moscow, no date given. This amazing pamphlet (written sometime after 1940) states "that if we want to improve our planet and make it more suitable for life, we must alter its climate." Hopefully, Soviet leaders today will be more prudent, although current rumbles about diverting northward-flowing rivers southward are not encouraging.
102. Aagaard, K., and Coachman, L. K., "Toward an Ice-Free Arctic Ocean," *EOS, Transactions of the American Geophysical Union,* **56**, 484–486, 1975, a strong case for concern over diverting northward-flowing rivers away from the Arctic basin.
103. Rosenthal, S. L., "Computer Simulation of Hurricane Development and Structure," in *Weather and Climate Modification,* edited by W. N. Hess, Wiley, New York, 1974, pp. 522–551.
104. Gentry, R. C., "Hurricane Modification," in *Weather and Climate Modification,* edited by W. N. Hess, Wiley, New York, 1974, pp. 497–521.
105. *Ibid.,* p. 519.
106. *Ibid.*
107. Sundquist, H., *Tellus,* **XXIV,** 6–12, 1972.
108. Roberts, W. O., "We're Doing Something about the Weather!," *National Geographic,* April, 1972, p. 545.
109. Budyko, M. I., *Climatic Changes,* Hydrometeorological Publishers, Leningrad, 1974, p. 269 (in Russian). An English summary in Budyko's book points out that the long-range warming effects of human activities "can result in drastic changes of global climate which are of great practical importance. In this connection the development of methods of climate modification for the purpose of climate control would be essential. One such method can be based on maintaining certain aerosol concentrations in the stratosphere."
110. For example, see the article by Janet H. Weinberg, "Decision at Asilomar," *Science News,* **107**, 194–196, March 22, 1975.
111. The most recent perhaps is Paul R. Ehrlich's "The Benefits of Saying Yes," *Bulletin of the Atomic Scientists,* 49–51, September, 1975.

Chapter 7

1. Cadle, R. D., "The Chemistry of Smog," in *Man's Impact on Climate,* by W. H. Matthews, W. W. Kellogg, and G. D. Robinson, MIT Press, Cambridge, Mass., 1971, pp. 338–359, an overview.
2. Ottar, B., "The Long-Range Transport of Air Pollutants," in *Proceedings of the Third International Clean Air Congress,* Verlag des Vereins Deutscher Ingenieure, Düsseldorf, 1973.
3. Likens, G., and Bormann, F. H., "Acid Rain: A Serious Regional Environ-

mental Problem," *Science,* **184,** 1176–1179, 1974, the case for concerns over acid rain.

4. Hardy, R. W. F., and Havelka, V. D., *Science,* **188,** 633–643, 1975.
5. McElroy, M., *Chemical Processes in the Solar System,* August, 1975, preprint, submitted for publication.
6. *Ibid.,* pp. 36–45. See also the text by L. M. Thompson and F. R. Troeh, *Soils and Soil Fertility,* 3rd ed., McGraw-Hill, New York, 1973, for an introduction to soil chemistry.
7. Delwiche, C. C., *Scientific American,* **223,** September, 1970, p. 136.
8. J. O. Frohliger and R. Kane, in "Precipitation: Its Acid Nature," *Science,* **189,** 455–457, 1975, present the case against concern over acid rain.
9. Chróściel, S., and Nowicki, M., *New Concepts in Air Pollution Research, Interdisciplinary Contributions by an International Group of 20 Scientists,* edited by Jan-Olaf Willums, Birkhauser Verlag, Basel und Stuttgart, 1974. This interesting book deals with a variety of air pollution problems.
10. See the discussion in chapter 9 on the new international economic order.
11. Gilbert White made his remark on the Public Broadcasting Corporation/American Association for the Advancement of Science program "Nova," aired in 1974.
12. The south coastal area of southern California imports about 59 percent of its water, of which about 66 percent is from the Colorado River, according to *The California Water Plan Outlook in 1974* (Calif. Dept. of Water Resources, 1974), p. 165.
13. Weisbecker, L. W., *Snowpack, Cloud-Seeding, and the Colorado River: A Technology Assessment of Weather Modification,* University of Oklahoma Press, Norman, 1974.
14. Ehrlich, P. R., Breedlove, D. E., Brussard, P. F., and Sharp, M. A., "Weather and the 'Regulation' of Subalpine Populations," *Ecology,* **53,** 1972.
15. Cooper, C., and Jolly, W., "Ecological Effects of Silver Iodide and Other Weather Modification Agents: A Review," *Water Resources Research,* **6,** 88–98, 1970. See also Charles Cooper's "Ecological Opportunities and Problems of Weather and Climate Modification," in W. R. D. Sewell, ed., *Modifying the Weather: A Social Assessment,* Western Geographical Series, v. 9, University of Victoria, Victoria, B.C., 1973, 99–134.
16. Lynne Mesirow, in an article written for *Colorado Business Magazine,* published in early 1976.
17. Committee on Water, *Water and Choice in the Colorado Basin,* National Academy of Sciences, Washington, D.C., 1968, pp. 18–19.
18. Stockton, C. W., *Long-term Streamflow Records Reconstructed from Tree Rings,* The University of Arizona Press, Tucson, 1975; also Stockton, C. W., and Jacoby, G. C., Jr., *Long-Term Surface-Water Supply and Streamflow Trends in the Upper Colorado River Basin based on Tree-Ring Analysis: Lake Powell Research Project Bull.,* Institute of Geophysics and Planetary Physics, U.C.L.A., Los Angeles, Calif., 1976.

19. Committee on Water, *The Colorado Basin,* p. 34.
20. See the article by Merry Maisel in *Atmospheric Implications of Energy Alternatives,* Proceedings of the October, 1974, UCAR Forum, National Center for Atmospheric Research, Boulder, Colorado, February, 1975, pp. 11–13.
21. Burns, J., special to the *Christian Science Monitor* in the *Toronto Globe and Mail,* September 9, 1974.
22. Hidalgo, H., *Some Considerations on the Feasibility of Advertent Climate Modification,* Institute for Defense Analyses, Paper P-856, October, 1972, p. 5.
23. Ehrlich, P. R., and Ehrlich, A. H., *The End of Affluence,* Ballantine Books, New York, 1974, p. 60.
24. Remark made at the Annual Meeting of the North American Interstate Weather Modification Council, Denver, Colorado, January 16, 1975.
25. Glantz, M. H., and Parton, W., "Weather and Climate Modification and the Future of the Sahel," in *The Politics of A Natural Disaster: Case of the Sahelian Drought,* edited by M. H. Glantz, Praeger, New York, 1976, chapter 14; also, see ref. 41 of chapter 3.
26. This rough estimate can be figured by combining the information on Fig. 11.4 of H. H. Lamb's *Climate, Past, Present and Future,* Methuen, London, 1972, p. 505; and p. 70 of A. N. Strahler's and A. H. Strahler's *Environmental Geoscience,* Hamilton Publishing, Santa Barbara, Calif., 1973.
27. Furthermore, this argument is applicable to the problem of ozone depletion, for some have argued foolishly that a few percent reduction in ozone is harmlessly within natural variability.
28. *The Denver Post* (AP), "Cloud-Seeding Hopes Told," October 17, 1974, p. 23.
29. *Bulletin of the American Meteorological Society,* **55,** 807, 1974.
30. Vern W. Butler, Chairman, North American Interstate Weather Modification Council, Resolution No. 75–3, January 17, 1975, presented this resolution to the council, and it was accepted.
31. Haas, J. E., "Sociological Aspects of Weather Modification," in *Weather and Climate Modification,* edited by Wilmot Hess, John Wiley and Sons, Inc., New York, 1974, p. 788.
32. *Ibid.,* p. 793.
33. *Ibid.,* p. 794.
34. Weisman, J. D., *The Washington Post,* "'72 Flood Laid to Cloud-Seeding," May or June, 1975.
35. Farhar, B. C., *Bulletin of the American Meteorological Society,* **55,** 759, 1974.
36. Phillip C. Jessup, who was a judge at the International Court of Justice from 1961 to 1970, comments on this question in a letter to *Science,* **188,** April 18, 1975.
37. Dotto, L., "Oil Spills Locked in Ice Pose Recovery Problem, Arctic Study Finds," *Toronto Globe and Mail,* November, 1974.

38. Campbell, W. J., and Martin, S., *Science,* **181,** 56, 1973.
39. Ayers, R. C., Jr., Johns, H. O., Glaeser, J. L., *Science,* **186,** 843–844, 1974.
40. *Ibid.*
41. Martin, S., and Campbell, W. J., *Science,* **186,** 845–846, 1974.
42. Dotto, L., "Oil Spills."
43. See the discussion in chapter 6 and ref. 80 of that chapter.
44. Kellogg, W. W., and Schneider, S. H., "Climate Stabilization: For Better or Worse?," *Science,* **188,** 1163–1173, 1974.
45. Lansford, H., *Bulletin of the American Meteorological Society,* **54,** 658, 1973.
46. Margaret Mead, the renowned anthropologist, has already recognized the opportunity to use the problems of the atmosphere as an example of global interdependence. See Margaret Mead's "The Kalinga Prize," *Journal of World History,* **13,** 765–771, 1971. Motivated by the need to show this interdependence, she organized a conference to consider next steps; a number of suggestions can be found in the conference report: *Proceedings of the International Conference on the Atmosphere: Endangered and Endangering,* National Institute of Environmental Health Sciences, Research Triangle Park, North Carolina, October, 1975.

Chapter 8

1. Robbins, W., *The New York Times,* August 25, 1974, p. 1.
2. Mullen, M., *Chicago Tribune,* October 15, 1974, p. 1.
3. Mullen, M., *Chicago Tribune,* October 13, 1974, p. 1.
4. *Ibid.*
5. Weinraub, B., *The New York Times,* December 7, 1974, p. 10.
6. Robbins, W., *The New York Times,* November 12, 1974, p. 3.
7. Farnsworth, C. H., *The New York Times,* November 17, 1974, Sect. 4.
8. Mayer, J., *Boulder Daily Camera,* October 10, 1974, p. 10.
9. *Denver Post* (AP), September 22, 1974, p. 7.
10. *Ibid.*
11. Robbins, W., *The New York Times,* November 17, 1974, p. 1.
12. Robbins, W., *The New York Times,* November 16, 1974, p. 1.
13. *Ibid.*
14. Robbins, W., *The New York Times,* September 4, 1974, p. 1.
15. *Denver Post* (UPI), September 22, 1974, p. 17. In this connection, fertilizer price increases "far exceeded cost increases," according to Ann Crittenden, *The New York Times,* October 20, 1975, p. 1, and "fertilizer-producing companies enjoyed unprecedented earnings."
16. Gelb, L., *The New York Times,* September 21, 1974, p. 4.
17. Robbins, W., *The New York Times,* November 7, 1974, p. 3.
18. Trager, J., *The Great Grain Robbery,* Ballantine, New York, 1975.

19. Barbour, J., *Boulder Daily Camera* (AP), November 3, 1974, p. 4.
20. *Ibid.*
21. Prouty, D., *Denver Post,* October 16, 1974, p. 49.
22. Maidenberg, H. J., *The New York Times,* October 20, 1974, p. 1.
23. *Chicago Tribune,* November 28, 1974, by Don Bacon, Newhouse News Service.
24. *The New York Times* (AP), October 28, 1974.
25. Robbins, W., *The New York Times,* December 10, 1974, p. 65.
26. Robbins, W., *The New York Times,* November 9, 1974, p. 1.
27. Revelle, R., *Scientific American,* **231,** September, 1974, p. 160.
28. In response to an editorial in the *Boulder Daily Camera,* November 6, 1974, p. 4, I wrote the following letter:

"It is a truism that in trying to make a point we often cite experts whose 'wisdom' agrees with our own cherished preconceptions. This is, perhaps, just what the *Camera* has done in its November 6 editorial, 'Optimistic Food Outlook.' The *Camera* quotes Bernard Yo, a U.S. Air Force nutritionist (and optimist), to the effect that hunger and starvation in the world can readily be averted and that existing technologies could feed up to three times the present world population. Says Yo, 'One day I can see the Amazon as the largest food-producing region in the world.' (I wish Mr. Yo a long life; he will need it to see that 'day.') The catch 22, of course, is how far away is that day, and what happens in the interim.

"In fact, Mr. Yo may not be sufficiently optimistic. In a recent *Scientific American* article Harvard's Roger Revelle argues that the earth and modern technology could probably sustain a population ten times present, BUT, he quickly acknowledges that modernization of agriculture in under-developed countries (which the *Camera* and Mr. Yo also cite as an essential step) must proceed rapidly, for, Revelle says, 'if it does not take place quickly it may not be possible at all.' What the *Camera* and Mr. Yo forgot to mention was the capital costs necessary for the rapid improvements in agriculture needed to provide adequate nutrition to today's nearly 4 billion earth people, let alone the costs of providing for a world population expanding at the present rate of some 75 million people per year.

"Let us assume, to pick up on Yo's example, that the Amazon jungles were to be cleared and farmed to keep up with this population growth. Present world figures show that one acre of land feeds about one person (of course, this could be improved severalfold by expensive technologies, but let's use present realities for now). Just to keep up with present population growth, then, each year we need to clear some 75 million acres of remote jungles, to put in agricultural settlements, transport systems, and to train and house the farm workers. The rough cost: about $1,000 per acre, according to some U.N. figures. This very rough estimate shows the staggering monetary realities that lie behind excessive optimism. That is, merely to keep up with present population growth means a total cost of some 75 billion dollars per year—an expenditure roughly equivalent to the present U.S. defense budget! Who will put up the money?

"Other techniques such as use of fertilizers and pesticides can be used to improve food production in poor countries but these have severe ecological costs, not to mention the economic—some $10 billion per year added costs for these inflation-gripped commodities.

"The *Camera* is, of course, right that reducing wastes in underdeveloped countries is essential, and they could add to that list better agricultural and food distribution management as well as reduced local corruption. But, most poor countries are far more pinched by world inflation than we are, and cannot hope to solve their near-term food problems without our help. If the *Camera* really believes 'that over the long haul, the natural law of "survival of the fittest" governs human life,' and that 'the first requisite of survival is to feed ourself,' then they must also be prepared to accept the 'natural law' that desperate countries faced with starvation will not hesitate to resort to resource embargoes, terrorism, or even nuclear blackmail against the 'fitter' rich countries, particularly those that have refused to help them out of their morass. It is obvious that a 'generation of peace' will be hard to achieve on empty stomachs, a point recognized by Secretary of State Kissinger in his sponsorship of the current world food conference in Rome, despite the opposition of prominent Department of Agriculture officials.

"The inevitable consequence of our continued overconsumption coupled with an attitude of 'let them be damned' surely will be an increase in that same defense spending which drains valuable capital resources away from more productive enterprises, like growing food. This, I feel, will be the great challenge to our newly elected leaders: to somehow convince us it is in our interests, both as humans and for our own long-term security, to accept some reductions in the growth of our affluence now so we can help to end the sharp contrast in quality of life between the poor majority and the rich minority of the world's inhabitants. To do otherwise will almost certainly guarantee the specter of Malthusian doom; but now in an age of nuclear proliferation.

"The *Camera* is justified in chiding those 'shortsighted officials and intellectuals' who indulge in overzealous pronouncement of doom, but they should not counterattack with lullabies. 'Efforts to insure an adequate diet for all mankind can no longer concentrate almost wholly on expanding the supply of food,' agriculturalist Lester Brown warned in his recent book, *By Bread Alone*. 'Morally,' Willy Brandt, Nobel Peace Prize winning ex-chancellor of West Germany, told the U.N. last year, 'it makes no difference whether a man is killed in a war or is condemned to starve to death by the indifference of others.' Our new political leaders have, indeed, little time to accomplish a great deal."

29. *The World Food Situation and Prospects to 1985*, Economic Research Service, U.S. Dept. of Agriculture, Foreign Agricultural Economic Report No. 98, 1974, pp. 32–39; see also the discussion of this report in chapter 4.
30. Revelle, in *Scientific American*.
31. *Ibid.*
32. Commoner, B., *Ambio*, **3**, 136–138, 1974.

33. Thompson, L. M., *Science,* **188,** 535–541, 1975.
34. Steinhart, J. and C., *Science,* **184,** 307–316, 1975.
35. Paddock, W. and P., *Famine 1975!,* Little, Brown, and Co., Boston, 1967, p. 230.
36. *Ibid.,* p. 78.
37. McElheny, *New York Times,* November 1, 1974, p. 1.
38. Paddock, W. and P., *Famine 1975!,* p. 79.
39. Brown, L. R., with Eckholm, E. P., *By Bread Alone,* Praeger Publishers, Inc., New York, 1974, pp. 94–96. For an excellent discussion of the world water resources situation, see M. Falkenberg and G. Lindh, in *Ambio,* **3,** 114–122, 1974.
40. Paddock, W. and E., *We Don't Know How,* Iowa State University Press, Ames, 1973, p. 243.
41. Brown, L., with Eckholm, E. P., *By Bread Alone,* pp. 95–96.
42. Weinraub, B., *The New York Times,* March 3, 1975, p. 1.
43. Paddock, W. and E., *We Don't Know How,* p. 250.
44. *Fragile Ecosystems: Evaluation of Research and Applications in the Neotropics,* edited by E. G. Farnworth and F. B. Golley, Springer-Verlag, New York, 1974.
45. *Ibid.,* p. 145.
46. Daniel H. Janzen's "The Unexploited Tropics," *The Bulletin of the Ecological Society of America,* 4–7, September, 1970, is one of them.
47. Farnworth and Golley, *Fragile Ecosystems,* p. 152.
48. *Ibid.*
49. *Ibid.,* pp. 168–169.
50. *Ibid.,* p. 169.
51. Brody, J. E., *The New York Times,* October 28, 1974, p. 1.
52. *Denver Post* (UPI), October 6, 1974, p. 23.
53. Ehrlich, P. R., and Ehrlich, A. H., *Population Resources Environment,* W. H. Freeman, San Francisco, 1972, p. 114.
54. *The New York Times,* June 20, 1974.
55. Myrdal, G., *Scientific American,* September 1974, pp. 172–182.
56. *Ibid.*
57. *Ibid.*
58. Revelle, R., *Scientific American,* September, 1974, pp. 160–170.
59. Schumacher, E. F., *Small Is Beautiful,* Harper and Row, New York, 1973.
60. Revelle, in *Scientific American.*
61. This most fundamental principle of demography is explained in any basic text on population; e.g., Paul R. Ehrlich's and Anne H. Ehrlich's *Population Resources Environment,* W. H. Freeman and Company, San Francisco, 1972, chap. 3.
62. As an example of the debate over the validity of the demographic transition theory, see the exchange of letters between demographers Norman Uphoff and Michael Teitelbaum in *Science,* **190,** 10–11, 1975. The de-

bate was prompted by Teitelbaum's interesting article, "Relevance of Demographic Transition Theory for Developing Countries," *Science,* **188,** 420–425, 1975.
63. Wilson, T. W., *World Population and a Global Emergency,* Aspen Institute for Humanistic Studies, Washington, D. C., 1973, p. 22.
64. Weinraub, B., *The New York Times,* December 7, 1974, p. 10.
65. Schmeck, H. M., Jr., *The New York Times,* October 6, 1974, p. 1.
66. *Denver Post* (AP), September 29, 1974, p. 42, contains the Shankar Rey quote; and Bernard Weinraub, *The New York Times,* September 4, 1974, p. 1, quoted Kulkarni.
67. Weinraub, B., *The New York Times,* December 7, 1974, p. 10.
68. *Time,* June 3, 1974, p. 29.
69. Weinraub, B., *The New York Times,* July 15, 1974.
70. Weinraub, B., *The New York Times,* August 16, 1974.
71. *Ibid.*
72. Weinraub, B., *The New York Times,* September 13, 1974.
73. Weinraub, B., *The New York Times,* December 7, 1974, p. 10.
74. *The New York Times,* November 24, 1974, contains both quotes on family planning.
75. Perhaps the best account of poor Indians' reactions to family planning is in the book by Mahmood Mamdani, *The Myth of Population Control,* Monthly Review Press, New York, 1972.
76. Weinraub, J., *The New York Times,* February 9, 1975, p. 50.
77. Brown, L. R., with Eckholm, E. P., *By Bread Alone,* pp. 188–189.
78. James Howe and the staff of the Overseas Development Council, *The U.S. and World Development: Agenda for Action, 1975,* Praeger Publishers, New York, 1975.
79. Mangala Moonesinghe, M. P., in *The Nation,* Colombo, Sri Lanka, October 25, 1974, p. 3, discusses these issues.
80. Worsnop, R. L., *Boulder Daily Camera,* December 31, 1974, p. 4.
81. Sterling, C., "The Making of the Sub-Saharan Wasteland," *Atlantic Monthly,* May, 1974.
82. *Boulder Daily Camera* (AP), December 27, 1974, p. 1.
83. Weinraub, B., *The New York Times,* December 13, 1974, p. 14.
84. *Ibid.*
85. *Ibid.*
86. *Ibid.*
87. *Ibid.*
88. *Ibid.*
89. Worsnop, R. L., *Boulder Daily Camera.*
90. Lelyveld, J., *The New York Times,* August 28, 1974, p. 33.
91. *Ibid.*
92. Butterfield, F., *The New York Times,* February 9, 1975, p. 17.
93. Robbins, W., *The New York Times,* March 9, 1975, p. 1.
94. *Ibid.*

95. Rensberger, B., *The New York Times,* October 7, 1974, p. 1, Sterling Wortman's visit to China.
96. *Time,* November 11, 1974, p. 80.
97. Kellogg, W. W., Atlas, D., Johnson, D. S., Reed, R. J., and Spengler, K. C., "Visit to the People's Republic of China," *Bulletin of the American Meteorological Society,* **55**, 1291–1330, 1974.
98. Sigurdson, J., *Ambio,* **4**, 112–119, 1975.
99. *Ibid.* This view has been confirmed recently, as reported in an article by Fox Butterfield, *The New York Times,* October 22, 1975, p. 8.
100. *Ibid.*
101. Kandell, J., *The New York Times,* October 30, 1974.
102. *Ibid.*
103. Kandell, J., *The New York Times,* March 10, 1975, p. 2.
104. Rensberger, B., *The New York Times,* November 5, 1974, p. 1.
105. *Ibid.*
106. *Ibid.*
107. Brody, J. E., *The New York Times,* October 28, 1974, p. 1.
108. Rensberger, B., *The New York Times.*

Chapter 9

1. There have been many who have pointed out these problems in relation to the world predicament, but little discussion in detail is available in works prepared for nonspecialists.
2. Heilbroner, R. L., *An Inquiry Into the Human Prospect,* Norton, New York, 1974, p. 52.
3. Oil spills, radioactive contamination, acid rain, or thermal pollution are but a few by-products of energy use that could lead to ecological catastrophies; see Ehrlich, P. R., and Ehrlich, A. H., *Population Resources Environment,* 2nd ed., Freeman, San Francisco, 1972.
4. Heilbroner, R. L., *The Human Prospect,* p. 43.
5. Willrich, M., and Taylor, T., *Nuclear Theft: Risks and Safeguards,* Ballinger Publishing Co., Cambridge, Mass., 1974; and Holdren, J. P., "Hazards of the Nuclear Fuel Cycle," *Bulletin of the Atomic Scientists,* 14–23, October, 1974, are two examples.
6. Mesarovic, M., and Pestel, E., *Mankind at the Turning Point,* E. P. Dutton and Co., New York, 1974, p. 69.
7. *Ibid.,* p. 55.
8. *Ibid.*
9. Boulding, K. E., *The New Republic,* April 29, 1972.
10. Boulding, K. E., *Science,* 187, 1188–1189, 1975.
11. Galbraith, J. K., *The New Industrial State,* Signet Books, New York, 1967, p. 391.
12. American Physical Society Summer Study of Reactor Safety, *Reviews of Modern Physics,* Supplement I, Summer 1975; also Lovins, A. B., *World*

Energy Strategies, Ballinger Publishing Co., Cambridge, Mass., 1975, chapter 6.

13. Excerpts of the debate were taken by Will Schaleben and published in *Free School,* Community Free School, Inc., Boulder, Colorado, January 1975, p. 38.
14. The quote appeared in a copyrighted story from *Newsday* by Stuart Diamond, and was reprinted in the *Boulder Daily Camera,* August 3, 1975, p. 9.
15. Thirty-two scientists, including eleven Nobel prizewinners, signed a pro-nuclear power statement prepared by Hans Bethe, whereas a few months later thousands of scientists, including nine Nobel prizewinners, signed an antinuclear power statement. Both statements appeared in *The Bulletin of the Atomic Scientists* in 1975; see also, the September, 1975, issue of *The Bulletin.*
16. Schumacher, E. F., *Small Is Beautiful,* Harper and Row, 1973.
17. The model of the European commissioners to manage any such World Security Institutions was suggested to me by Harlan Cleveland, who argued that such a model is proving quite effective.
18. For example, the cover story of the January 21, 1974, *Time* magazine read: "Energy Crunch: Real or Phony?"
19. A term dubbed by Paul R. Ehrlich and Anne H. Ehrlich in *The End of Affluence,* Ballantine Books, New York, 1974.
20. Committee on Mineral Resources and the Environment, *Mineral Resources and the Environment,* National Academy of Sciences, Washington, D.C., 1975, p. 2.
21. Schumacher, *Small Is Beautiful.*
22. One private organization that is actively engaged in countering the short-term governmental bias is the Scientists' Institute for Public Information (SIPI), whose objectives are explained in the book by Stuart Blume, *Toward a Political Sociology of Science,* Free Press, New York, 1974.
23. Holdren, "Nuclear Fuel Cycle," contains a number of references to this issue.
24. Galbraith, *New Industrial State,* p. 218.
25. Tugwell, R., *Model for a New Constitution,* James E. Freel and Associates, Berkeley, Calif., 1970.
26. Pirages, D. C., and Ehrlich, P. R., *Ark II,* Freeman, San Francisco, 1974, p. 175.
27. *Ibid.*
28. *Time,* September 2, 1974, p. 42.
29. Harlan Cleveland's introduction to John and Magda Cordell McHale's *Human Requirements, Supply Levels and Outer Bounds: A Framework for Thinking about the Planetary Bargain,* Aspen Institute for Humanistic Studies, Aspen, Colorado, 1975, p. iv.
30. "New Concepts of Development," Resolution of Commission A of the

1st Conference of the International Development Center, Algiers, June 24–27, 1975, Johan Galtung, Rapporteur.
31. Ref. 29, p. iv.
32. The report of the Aspen workshop has been published: Report of an International Workshop, *The Planetary Bargain,* A Policy Paper/Aspen Institute for Humanistic Studies Program in International Affairs, Aspen, Colorado, 1975.
33. Mesarovic and Pestel, p. 63.
34. Holdren, J. P., "Energy and Prosperity," *Bulletin of the Atomic Scientists,* 26–28, January, 1975. The "arithmetic of growth" is clearly explained by Paul R. Ehrlich, Anne H. Ehrlich, and John P. Holdren in *Human Ecology,* Freeman, San Francisco, 1973, p. 214.
35. The situation is described by *Denver Post* staff writer, Todd Phipers, in the *Rocky Mountain News,* December 18, 1974.
36. Reich, C. A., *The Greening of America,* Bantam Books, New York, 1970.
37. Carmody, D., *The New York Times,* February 28, 1975, p. 1.
38. Montgomery, P. L., *The New York Times,* March 1, 1975, p. 10.
39. Carmody, D., *The New York Times,* March 1, 1975, p. 1.
40. Heilbroner, *The Human Prospect,* p. 44.
41. A term attributed to Edward Teller.
42. For example, see Federation of American Scientists' *Public Interest Report, 3,* February, 1975, for a discussion of the "rise and fall of science advice."
43. For a discussion of the role of science advice in the federal government, see Eugene B. Skolnikoff and Harvey Brooks, in "Science Advice in the White House; Continuation of a Debate," *Science,* **187**, 35–41, 1975.
44. Rensberger, B., *The New York Times,* September 3, 1974, p. 1; quotes Norman Borlaug: "If we ever get into the 'publish or perish' game around here we're dead as far as being able to put food into people's bellies is concerned."
45. Timothy Wirth, as quoted in the *Congressional Record,* No. 126-Part II, August 1, 1975.
46. One of the most interesting recent articles on how science can contribute to the solution of today's problems is by Harvey Brooks: "Are Scientists Obsolete?," *Science,* **186,** 501–508, 1974.
47. Everett, B., Food for the World, *American Association of University Women (AAUW) Journal,* **68,** 14–16, November, 1974.
48. Ref. 72 of Chapter 6.
49. Holdren, J. P., and Ehrlich, P. R., "Human Population and the Global Environment," *American Scientist,* **62,** May–June, 1974, p. 291.
50. de Camp, L. S., and de Camp, C. C., *Citadels of Mystery,* Ballantine Books, New York, 1973, p. 261.
51. Huizinga, J., *The Waning of the Middle Ages,* Doubleday Anchor Books, Garden City, New York, 1954, p. 31.
52. *Ibid.,* p. 38.

APPENDICES

1. Much of this appendix is taken from a section I helped to draft for The Institute of Ecology Report *Fragile Ecosystems,* edited by E. A. Farnsworth and F. B. Golley, Springer-Verlag, New York, 1974, Section 6.
2. Newell, R. E., "The Amazon Forest and Atmospheric General Circulation," in *Man's Impact on the Climate,* edited by W. H. Matthews, W. W. Kellogg, and G. D. Robinson, MIT Press, Cambridge, Mass., 1971, pp. 457–459.
3. *Ibid.*
4. Chervin, R. M., Washington, W. W., and Schneider, S. H., *Journal of the Atmospheric Sciences,* 1976.
5. Holdren, J., and Herrera, P., *Energy: A Crisis in Power,* Sierra Club, San Francisco, 1971.
6. Rotty, R. M., and Mitchell, J. M., Jr., *Man's Energy and World Climate,* Institute for Energy Analysis Report, Oak Ridge Associated Universities, 1974.
7. This material is largely excerpted from Schneider, S. H., and Dennett, R. D., "Climatic barriers to long-term energy growth," *Ambio,* **4,** 65–74, 1975.

F

Index

DATE	ISSUED TO